Charles Darwin's *Beagle* Diary

Charles Darwin as a young man, drawn by George Richmond in 1840.

CHARLES DARWIN'S

BEAGLE

DIARY

Edited by Richard Darwin Keynes, CBE, ScD, FRS
Emeritus Professor of Physiology in the University of Cambridge, and
Fellow of Churchill College

CAMBRIDGE
UNIVERSITY PRESS

PUBLISHED BY THE PRESS SYNDICATE OF THE UNIVERSITY OF CAMBRIDGE
The Pitt Building, Trumpington Street, Cambridge, United Kingdom

CAMBRIDGE UNIVERSITY PRESS
The Edinburgh Building, Cambridge CB2 2RU, UK
40 West 20th Street, New York, NY 10011–4211, USA
10 Stamford Road, Oakleigh, VIC 3166, Australia
Ruiz de Alarcón 13, 28014 Madrid, Spain
Dock House, The Waterfront, Cape Town 8001, South Africa

http://www.cambridge.org

First published 1988
First paperback edition 2001
Reprinted 2001

A catalogue record for this book is available from the British Library

Library of Congress Cataloguing in Publication data
Darwin, Charles, 1809–1882.
[Diary of the voyage of H. M. S. Beagle]
Charles Darwin's Beagle diary/edited by Richard Darwin Keynes.
Bibliography
Includes index.
1. Beagle Expedition (1831–1836) 2. Natural history.
3. Geology. I. Keynes, R. D. II. Title. III. Title: Beagle diary.
QH365.A1 1987b
508'.92'4 – dc 19   87–31169

ISBN 0 521 23503 0 hardback
ISBN 0 521 00317 2 paperback

Transferred to digital printing 2004

To my godmother
NORA BARLOW
in the affectionate hope that I have
lived up to her standards

Contents

Contents

Illustrations

The end papers show a track chart of the *Beagle*'s voyage from Captain FitzRoy's *Narrative of the Surveying Voyages of His Majesty's Ships Adventure and Beagle*, Vol. II, 1839

Illustrations

Introduction

The inception of the voyage of the *Beagle*, 1831–1836

In February 1830, Captain Robert FitzRoy RN, newly appointed to command HMS *Beagle*, was engaged on a survey of the western part of Tierra del Fuego in the neighbourhood of the Gilbert Islands. One of the principal obstacles that he encountered in carrying out his allotted task was the incorrigible tendency of the local inhabitants to steal anything on which they could lay their hands, including most exasperatingly the ship's whale-boat. In an attempt to curb their thieving, he captured several of them as hostages, but this move failed to achieve its purpose, because the Fuegians preferred the retention of their booty to the release of their comrades. Thus it was that FitzRoy took on board the *Beagle* the Fuegians who were named Fuegia Basket, Boat Memory and York Minster, adding later the boy Jemmy Button, purchased for the price of a mother-of-pearl button in the Murray Narrow, further to the east. Becoming deeply interested in their welfare, FitzRoy then conceived the notion of taking them back to England to be educated for a while, and later returning them to Tierra del Fuego to pass on the benefits of civilization to their people.

In October 1830 the *Beagle* and *Adventure* arrived back in England, and FitzRoy set out to put his ideas into effect. Boat Memory died soon afterwards of smallpox, but with the aid of the Vicar of Walthamstow the other three began their schooling in English. In the summer of 1831, FitzRoy was summoned to present his Fuegians to the King and Queen, who expressed their gracious approval of his plans for them. However, the Lords of the Admiralty were, perhaps predictably, less than sympathetic with the proposal that the navy should undertake the responsibility of repatriating the three Fuegians, and FitzRoy was obliged to ask for twelve months' leave of absence so that he could charter a boat at his own expense in order to return them to their native land. This leave was duly granted, but in the end did not have to be taken, because FitzRoy's influential relatives were able to prevail on the Admiralty to reappoint him to the command of the *Beagle*, thus enabling him to combine his somewhat eccentric scheme with the more readily acceptable objective of completing the survey of the coast of South America.

In writing his daily journal for 24 January 1830, and concerned at finding substantial compass variations in the neighbourhood of the Fury and Magill Islands in the western part of Tierra del Fuego, FitzRoy had recorded:

> There may be metal in many of the Fuegian mountains, and I much regret that no person in the vessel was skilled in mineralogy, or at all acquainted with geology. It is a pity that so good an opportunity of ascertaining the nature of the rocks and earths of these regions should have been almost lost. I could not avoid often thinking of the talent and experience required for such scientific researches, of which we were wholly destitute; and inwardly resolving that if ever I left England again on a similar expedition, I would endeavour to carry out a person qualified to examine the land; while the officers, and myself would attend to hydrography.[1]

When the *Beagle* was recommissioned in 1831, FitzRoy followed up this resolution by proposing to the Hydrographer of the Navy, Captain Francis Beaufort, 'that some well-educated and scientific person should be sought for who would willingly share such accommodation as I had to offer, in order to profit by the opportunity of visiting distant countries yet little known'.[2]

Captain Beaufort proceeded to consult a friend, George Peacock of Trinity College, Cambridge, for the name of an appropriate candidate, and the post was first offered by Peacock to the Rev. Leonard Jenyns, Vicar of Swaffham Bulbeck. However, Jenyns felt unable to desert his parish, and Peacock therefore turned for further advice to Professor John Stevens Henslow, Professor of Botany and previously of Mineralogy in the University, a leading light in its scientific life who kept open house where undergraduates and senior members could mix. One such undergraduate was Charles Darwin, who through his passionate interest in beetle-collecting had become a close friend of Henslow's, and was known in university circles as 'the man who walks with Henslow'.[3] Darwin had recently been persuaded by Henslow to remedy at the hands of Professor Adam Sedgwick the distaste for geology which he had acquired through the dullness of the lectures in the subject that he had previously attended at Edinburgh University, and was at the time accompanying Sedgwick on a brief geological field trip to North Wales.

So it came about through this somewhat haphazard chain of events that CD found awaiting him at home on 29 August 1831 the fateful letters from Peacock and Henslow suggesting that he should accompany FitzRoy as the *Beagle*'s naturalist and geologist. His immediate reaction was to accept the offer, but his father felt that it would be 'a useless undertaking' most unsuitable to his chosen profession as a clergyman. However, Robert Darwin qualified his opposition by adding, as CD recorded in his *Autobiography*, 'If you can find any man of common sense, who advises you to go, I will give my consent'. The following day, CD rode over to visit his Wedgwood cousins at Maer, and found a strong supporter in his uncle and future father-in-law Josiah Wedgwood II, who at once provided a detailed list of arguments setting Robert's misgivings at rest. Paternal opposition was gracefully withdrawn, and on 1 September 1831 CD wrote to Captain Beaufort 'with my acceptance of the offer of going with Capt. FitzRoy'.[4]

The *Beagle* Diary

Once installed on board the *Beagle*, CD embarked on the manuscript recording his daily activities throughout the voyage of which this volume is a transcript. He generally referred to it as his 'Journal', but it will be entitled here *The Beagle Diary*, in order to avoid confusion with the publications for which it eventually provided the principal source of material. The text has previously been transcribed and edited by Nora Barlow, and appeared as *Charles Darwin's Diary of the Voyage of H.M.S. 'Beagle'*,[5] the first of her many notable contributions to Darwin scholarship. This new version corrects a small number of unimportant errors in the earlier edition, and endeavours to conform with the standards of preserving the original punctuation and spelling that have been set by the editors of *The Correspondence of Charles Darwin*.[6] In order to prepare it, I have mainly used the excellent facsimile of the manuscript that was produced by Genesis Publications in 1979,[7] with reference to the original at Down House in order to check up on certain points of detail.

The manuscript was written throughout in ink on gatherings of paper making pages 20 by 25 cm in size,[8] faintly lined and with a red marginal line. An impressive feature is the manner in which the lay-out adopted by CD for the first entry dated 24 October 1831 was retained almost unchanged through 751 written pages to the final one of 7 November 1836. Each entry opens with the day of the month, and for the first fresh entry on a new page the month itself is given as well. The year and place appear at the top of the page as running headings. A similar lay-out has been retained in the transcript, except that the month is shown only when it changes, and appears instead in the running heading. For the first few months, Mondays are often indicated, and later some but not all Sundays. The pages are numbered top left, and this original pagination is indicated throughout the transcript between vertical lines. Between pp. 15 and 109 of the manuscript, CD has written only on the odd-numbered pages, while on what should be p. 554 he has mistakenly put 534 and continued without correcting his mistake. As a result, the total number of pages of writing is 751 rather than the figure of 779 shown on the final page. His handwriting is almost invariably clear and legible, and despite his complaints (see p. xvii) about the difficulty of composing such a narrative, there are astonishingly few revisions to the sentences that would appear to have been made at the actual time of writing, and thus with no change in ink or pen. The majority of the corrections to the manuscript were evidently made at a later stage, when sections of it were being prepared for publication.

The diary was normally written up only during periods spent on board the *Beagle*, or in a house on shore, and did not accompany CD on most of his excursions inland. Hence although the text always reads as though it was written within a short while of the events described, there are several instances on the occasions of his long journeys on horseback in Patagonia and Chile where many weeks elapsed before the diary could be brought up to date. This is proved by references at several points to occurrences that actually took place long after the date of the entry. Thus in the entry for 4/7 September 1833 he refers to the finding of a horse's tooth at St Fé Bajada, which he did not in fact visit until 10 October, five weeks later. Between the entries for 30 December 1835 and 12 January 1836 there is a note dated February in brackets, suggesting that once again he was catching up with his writing after a gap of a month or more. And there is an even longer gap of some three and a half months between the occasion on 26 March 1835 where he describes being bitten by the Benchuca bug at a village near Mendoza and then mentions 'one which I caught at Iquiqui', and the *Beagle*'s actual arrival at Iquique on 13 July 1835.

While CD was travelling on shore, he kept brief pencil notes in a series of pocket books of which eighteen concerned mainly with details of his travels together with general field notes on geology and natural history are preserved at Down House, except for one that has been lost but was microfilmed in the 1970s.[9] Extracts from some of them were published by Nora Barlow in *Charles Darwin and the Voyage of the Beagle*.[10] For some of his early excursions, the pocket book entries are full of detail, and as in the example quoted in footnote 2 for 8 April 1832, did not require much elaboration when written up for the diary. During the following two months when he lived at Botafogo, he wrote very little in his pocket book, and presumably recorded his activities immediately in the diary. Later pocket book entries vary in the fullness with which place names and other minutiae were recorded, and particularly for the last and longest of his South American expeditions from Valparaiso to Copiapó between 27 April and 4 July 1835, CD appears to have relied on his memory to a considerable extent when producing the account for each successive day in the diary.

The writing of the diary did not at first come easily. On 2 April 1832 CD wrote to Caroline Darwin:

> I am looking forward with great interest for letters, but with very little pleasure to answering them.— It is very odd, what a difficult job I find this same writing letters to be. — I suppose it is partly owing to my writing everything in my journal: but chiefly to the number of subjects; which is so bewildering that I am generally at a loss either how to begin or end a sentence. And this all hands must allow to be an objection. —[11]

Three weeks later, he told her:

> I send in a packet, my commonplace Journal. — I have taken a fit of disgust with it & want to get it out of my sight, any of you that like may read it. —a great deal is absolutely childish: Remember however this, that it is written solely to make me remember this voyage, & that it is not a record of facts but of my thoughts. —& in excuse recollect how tired I generally am when writing it. —
> . . . Be sure you mention the receiving of my journal, as anyhow to me it will [be] of considerable future interest as it [is] an exact record of all my first impressions, & such a set of vivid ones they have been, must make this period of my life always one of interest to myself.— If you speak quite sincerely, —I should be glad to have your criticisms. Only recollect the above mentioned apologies. —[12]

The first response came from Catherine Darwin on 25 July 1832, when she wrote encouragingly:

> I cannot tell you how interesting and entertaining we find your letters and Journal, and what great joy it gives all the house when we have such happy accounts of you in every way . . . If you wish to have *my* Criticisms, I must say I think your descrip-

tions most excellent, and gave me most lively pleasure in reading them . . . Susan read the Journal aloud to Papa, who was interested, and liked it very much. They want to see it at Maer, but we do not know whether you would choose that, and must wait till we hear from you, whether we may or not. It shall be kept most carefully for you. —[13]

On 5 July 1832 CD wrote to Catherine Darwin: 'My journal is going on better, but I find it inconvenient having sent the first first part home on account of dates—'[14] On 3 November 1832 he reported to Caroline Darwin:

Although my letters do not tell much of my proceedings I continue steadily writing the journal; in proof of which the number on the page now is 250. — . . . I am glad the journal arrived safe; as for showing it, I leave that entirely in your hands. — I suspect the first part is abominaly childish, if so do not send it to Maer. — Also, do not send it by the Coach, (it may appear **ridiculous** to you) but I would as soon loose a piece of my memory as it. — I feel it is of such consequence to my preserving a just recollection of the different places we visit. — When I get another opportunity I will send some more. —[15]

Writing again to Catherine from Maldonado on 14 July 1833, CD notified her that he had sent home another instalment (apparently the second) of his diary, but added: 'The journal latterly has not been flourishing, for there is nothing to write about in these well-known-uninteresting countries. —'[16] Caroline Darwin responded on 28 October 1833 with the criticism for which he had asked, when she wrote:

I am very doubtful whether it is not *pert* in me to criticize, using merely my own judgement, for no one else of the family have yet read this last part—but I *will* say just what I think—I mean as to your style. I thought in the first part (of this last journal) that you had, probably from reading so much of Humboldt, got his phraseology & occasionly made use of the kind of flowery french expressions which he uses, instead of your own simple straight forward & far more agreeable style. I have no doubt you have without perceiving it got to embody your ideas in his poetical language & from his being a foreigner it does not sound unnatural in him— Remember, this criticism only applies to parts of your journal, the greatest part I liked exceedingly & could find no fault, & all of it I had the greatest pleasure in reading. —[17]

The other comments from his sisters were uniformly complimentary, except that Susan found fault with his spelling.[18]

Writing to Catherine Darwin in July 1834, CD replied:

I am much pleased to hear my Father likes my Journal: as is easy to be seen I have taken too little pains with it. — My geological notes & descriptions of animals I treat with far more attention: from knowing so little of Natural History, when I left England, I am constantly in doubt whether these will have any value. — . . .

Thank Granny for her purse & tell her I plead guilty to some of her [spelling corrections], but the others are certainly only accidental errors. — Moreover I am much obliged for Carolines criticisms (see how good I am becoming!) they are perfectly just, I even felt aware of the faults she points out, when writing my journal. — [19]

Setting aside his misgivings, however, CD continued to keep his diary up to date with meticulous care and attention to detail. There were few significant references to it in his correspondence with his sisters until on 29 April 1836, when the number on the page had reached 725, he wrote to Caroline Darwin from Mauritius:

Whilst we are at sea, & the weather is fine, my time passes smoothly, because I am very busy. My occupation consists in rearranging old geological notes: the rearrangement generally consists in totally rewriting them. I am just now beginning to discover the difficulty of expressing one's ideas on paper. As long as it consists solely of description it is pretty easy; but where reasoning comes into play, to make a proper connection, a clearness & a moderate fluency, is to me, as I have said, a difficulty of which I had no idea. — I am in high spirits about my geology. — & even aspire to the hope that my observations will be considered of some utility by real geologists . . . The Captain is daily becoming a happier man, he now looks forward with cheerfulness to the work which is before him. He, like myself, is busy all day in writing, but instead of geology, it is the account of the Voyage. I sometimes fear his 'Book' will be rather diffuse, but in most other respects it certainly will be good: his style is very simple & excellent. He has proposed to me, to join him in publishing the account, that is, for him to have the disposal & arranging of my journal & to mingle it with his own. Of course I have said I am perfectly willing, if he wants materials; or thinks the chit-chat details of my journal are any ways worth publishing. He has read over the part I have on board, & likes it. — I shall be anxious to hear your opinions, for it is a most dangerous task, in these days, to publish accounts of parts of the world which have so frequently been visited. It is a rare piece of good fortune for me, that of the many errant (in ships) Naturalists, there have been few or rather no geologists. I shall enter the field unopposed. — I assure you I look forward with no little anxiety to the time when Henslow, putting on a grave face, shall decide on the merits of my notes. If he shakes his head in a disapproving manner: I shall then know that I had better at once give up science, for science will have given up me. — For I have worked with every grain of energy I possess. —[20]

The publication of the *Journal of Researches*

Once back in England, CD's immediate concern was to place his collections of specimens in expert hands for classification, but this done he turned to the question of the publication of his diary. On 7 December 1836 he wrote to Caroline Darwin from London:

My plans have, since being here, become more perplexed, with respect to the Journal part. I am becoming rather inclined to the plan of mixing up long passages with Capt FitzRoy. D[r] Holland looked over a few pages, and evidently thought that it would not be worth while to publish it alone, as it would be partly going over the same ground with the Captain. The little D[r] talked much good sense, and, what was far more surprising much sincerity. I shall go on with the geology and let the journal take care of itself.[21]

However, members of CD's family were unhappy both with Dr Holland's judgement and with FitzRoy's suggestion of joint authorship. On 17 December 1836 Emma Wedgwood, CD's future wife, wrote to Fanny Wedgwood, her sister-in-law:

Catherine tells me they are very anxious to have your and Hensleigh's real opinion of Charles's journal. I am convinced Dr Holland is mistaken if he thinks it not worth publishing. I don't believe he is any judge as to what is amusing or interesting. Cath. does not approve of its being mixed up with Capt. FitzRoy's, and wants it to be put altogether by itself in an Appendix.[22]

This view was reinforced when Hensleigh Wedgwood wrote to CD three days later:

> In short there is more variety and a greater number of interesting portions than in 99.100^{ths} of the travels that are published. I should not have the least doubt of it's success & I think the less it is mixed up with the Captains the better. If Dr Holland read your journey from the Rio Negro & thought it would not do for publication it only affects my opinion of his taste & not the least in the world the merits of the thing itself. I liked your account of Keeling island, but your theory of the lagoon islands seemed to us not quite clearly enough explained.[23]

The matter was quickly settled when FitzRoy himself wrote to CD on 30 December 1836: 'While in London a few days since I consulted Mr Broderip about Capt King's Journal. He recommended a joint publication such as is sketched in the accompanying paper. One volume might be for King — another for you — and a third for me. The *profits* if *any*, to be divided into three equal portions. — What think you of such a plan? — '[24]

CD at once set to work on the preparation of his volume. In August 1837 he was able to send the first proofs to Henslow for criticism, and on 4 November he wrote to his mentor:

> If I live till I am eighty years old I shall not cease to marvel at finding myself an author: in the summer, before I started, if anyone had told me I should have been an angel by this time, I should have thought it an equal improbability. This marvellous transformation is all owing to you. — . . . I sat the other evening gazing in silent admiration at the first page of my own volume, when I received it from the printers![25]

In November 1837 difficulties arose when FitzRoy took sudden and quite unjustified offence at what he regarded as CD's failure to make proper acknowledgement of the help of his shipmates,[26] but this was soon resolved, and by the end of February 1838 both CD's and King's volumes had been printed, while FitzRoy proceeded more slowly with his own.[27] On 1 April 1838 CD was able to report to Susan Darwin:

> The Captain is going on very well, — that is for a man who has the most consummate skill in looking at everything and every body in a perverted manner. — He is working very hard at his book, which I suppose will really be out in June. I looked over a few pages of Captain King's Journal: I was absolutely forced against all love of truth to tell the Captain that I supposed it was very

good, but in honest reality no pudding for little school-boys ever was so heavy. — It abounds with Natural History of a very trashy nature. — I trust the Captain's own volume will be better. —[28]

August 1839 finally saw the publication under the imprint of Henry Colburn of the *Narrative of the Surveying Voyages of His Majesty's Ships Adventure and Beagle between the Years 1826 and 1836, describing their Examination of the southern Shores of South America, and the Beagle's Circumnavigation of the Globe.* Volume I was by Captain King, and Volume II and an Appendix by Captain FitzRoy, while Volume III was sub-titled '*Journal and Remarks. 1832–1836.* By Charles Darwin, Esq., M.A.' In order to provide illustrations for Vols. I and II, FitzRoy obtained from Augustus Earle and Conrad Martens a series of watercolours of various places visited by the *Beagle*, several of which have been reproduced in *The Beagle Record.* These were then engraved by T. Landseer, S. Bull and others, and some are reproduced here. Volume III was not illustrated. The demand for CD's volume immediately exceeded that for the other two, and before the end of the year Colburn brought out a separate second impression with the title changed to *Journal of Researches into the Geology and Natural History of the Countries visited during the Voyage of H.M.S. Beagle round the World, under the Command of Capt. FitzRoy, R.N.* A third issue appeared in 1840. In 1845 CD made extensive revisions to the text, and the order of the wording of the title was changed to become *Journal of Researches into the Natural History and Geology . . .* The copyright was sold for £150 to John Murray, and it appeared with a dozen illustrations as Vol. XII of his Colonial and Home Library. This is the familiar edition of the *Journal of Researches*, which has since been reprinted many times without further alteration, and has been translated into many languages. Of it, CD wrote: 'The success of this my first literary child always tickles my vanity more than that of any of my other works.'[29]

In order to arrive at a text suitable for publication, CD chose to adhere to a geographical rather than a strictly chronological unity, and confined to single chapters his accounts of places such as Tierra del Fuego and the Falkland Islands that were visited by the *Beagle* twice at an interval of a year or more. Although many of the entries in the original diary appeared in print as they stood, about one third were omitted altogether from the published version, and others were somewhat abridged. A considerable amount of scientific material was then added, drawn by CD from the extensive notes on geology, zoology, ornithology and botany, again based on jottings in the little pocket books, that he had entered up separately. In the end, only about half of the 182,000 words in the manuscript diary were incorporated in the *Journal of Researches*, the final length of which was 223,000 words. The picture given in the present volume of CD's share in the voyage of the *Beagle* hence preserves the continuity that he sacrificed to some extent in his better known work, and constitutes an account of his daily activities that is matchless in its immediacy and vivid descriptiveness.

Darwin and FitzRoy

In conclusion, something needs to be said here about FitzRoy's motivation for taking a scientist with him to South America, and about his relations during the voyage with the scientist whom he chose as his companion, since on both questions unfortunate misconceptions have tended sometimes to arise.

It would surely have been one of the major ironies of scientific history had it really been the case, as was claimed by de Beer,[30] Moorehead,[31] Mellersh[32] and others, that FitzRoy's basic purpose in including a naturalist in the complement of the *Beagle* was to establish the literal truth of the account of the Creation given in the first book of Genesis, for the ultimate outcome was precisely the reverse. However, as has already been seen (see p. xii), FitzRoy's original objective was the strictly practical one of having a trained geologist with him, and despite CD's somewhat limited experience in this field beforehand, he must have filled this role to FitzRoy's satisfaction. Moreover, although there is no doubt that FitzRoy was deeply distressed when, long afterwards, his geologist was revealed as the champion of evolutionary biology, he was on his own showing far from being a confirmed believer in the Bible at the actual time of the voyage. In the final chapter of Volume II of the *Narrative*, he wrote:

> While led away by sceptical ideas, *and knowing extremely little of the Bible* [my italics], one of my remarks to a friend, on crossing vast plains composed of rolled stones bedded in diluvial detritus some hundred feet in depth, was 'this could never have been effected by a forty days' flood,'—an expression plainly indicative of the turn of mind, and ignorance of Scripture. I was quite willing to disbelieve what I thought to be the Mosaic account, upon the evidence of a hasty glance, though knowing next to nothing of the record I doubted:—and I mention this particularly, because I have conversed with persons fond of geology, yet knowing no more of the Bible than I knew at that time.[33]

The friend in question was evidently CD, and the occasion was their expedition up the Rio Santa Cruz.

Elsewhere in the chapter, FitzRoy nevertheless attempted vigorously and at inordinate length to reconcile the biblical account of the Flood with the geological evidence. To some extent, he did try to face up to the facts, as for example when he wrote:

> In crossing the Cordillera of the Andes Mr Darwin found pet-rified trees, embedded in sandstone, six or seven thousand feet above the level of the sea: and at twelve or thirteen thousand feet above sea-level he found fossil sea-shells, limestone, sandstone, and a conglomerate in which were pebbles of 'the rock with shells.' Above the sandstone in which the petrified trees were found, is 'a great bed, apparently about one thousand feet thick, of black augitic lava; and over this there are at least five grand alternations of such rocks, and aqueous sedimentary deposits, amounting in thickness to several thousand feet.'[34] These won-derful alternations of the consequences of fire and flood, are, to me, indubitable proofs of that tremendous catastrophe which alone could have caused them;— of that awful combination of water and volcanic agency which is shadowed forth to our minds by the expression 'the fountains of the great deep were broken up, and the windows of heaven were opened.'[35]

But many of his other arguments were palpably absurd. These passages were, however, written early in 1839, by which time, following his marriage soon after the return of the *Beagle* to England, FitzRoy had undergone a religious conversion, and had become a convinced fun-damentalist. There can be no doubt whatever that he did not hold such extreme views when in 1831 he invited CD to sail with him. It was CD rather than FitzRoy who afterwards recalled that: 'Whilst on board the *Beagle* I was quite orthodox, and I remember being heartily laughed at by several of the officers (though themselves orthodox) for quoting the Bible as an unanswerable authority on some point of morality.'[36]

Having disposed of the myth that the voyage was punctuated by quarrels on this particular issue between CD and his Captain, what of their true relations with one another? They can never have been entirely smooth, for it is abundantly clear not only that FitzRoy had a somewhat violent temper, but also that he had manic-depressive tendencies which in the end led to his suicide; and for the two men to have shared extremely cramped quarters on board ship for nearly five years in all weathers must have imposed exceptional strains. Concluding a dispassionate account of

FitzRoy's personal qualities, CD wrote many years later: 'His character was in several respects one of the most noble which I have ever known, though tarnished by grave blemishes.'[37] It would be quite wrong, however, to paint too black a picture of FitzRoy's faults. There is no indication at all of the occurrence of sustained disagreements with him in any of CD's letters home nor in this diary, and one cannot read the letters exchanged between them during the voyage,[38] without receiving the impression that for the bulk of the time they remained on the most cordial terms, with real quarrels that were remarkably few and far between. The final word should remain with CD, when on 20 February 1840 he wrote to FitzRoy:

> However others may look back to the Beagles voyage, now that the small disagreeable parts are well nigh forgotten, I think it far the *most fortunate circumstance in my life* that the chance afforded by your offer of taking a naturalist fell on me. — I often have the most vivid and delightful pictures of what I saw on board pass before my eyes. — These recollections & what I learnt in Natural History I would not exchange for twice ten thousand a year.[39]

Endnotes to Introduction

1 See *Narrative* 1: 385.
2 See *Narrative* 2: 18.
3 See *Autobiography* p. 64.
4 See *Correspondence* 1: 135–6.
5 *Charles Darwin's Diary of the Voyage of H.M.S. 'Beagle'.* Edited by Nora Barlow. Cambridge University Press, 1933.
6 *The Correspondence of Charles Darwin*, Volume 1, *1821–1836*. Cambridge University Press, 1985.
7 *The Journal of a Voyage in H.M.S. Beagle by Charles Darwin.* Genesis Publications, Guildford, 1979.
8 There are 51 gatherings altogether, of which 44 consist of 8 leaves (16 pages), while the seven others consist of 1 leaf (pp. 1–2), 14 (pp. 3–30), 6 (pp. 141–52), 7 (pp. 393–406), 3 (pp. 407–12), 16 (pp. 584–615) and 4 (pp. 777–84) leaves.
9 For a complete list of CD's *Beagle* records see Appendix II in *Correspondence* 1: 545–8.
10 *Charles Darwin and the Voyage of the Beagle.* Edited by Nora Barlow. Pilot Press, London, 1945.
11 See *Correspondence* 1: 219.
12 See *Correspondence* 1: 226–7.
13 See *Correspondence* 1: 253.
14 See *Correspondence* 1: 246–7.
15 See *Correspondence* 1: 276–9.
16 See *Correspondence* 1: 314.
17 See *Correspondence* 1: 345.
18 Susan wrote: 'there is one part of your Journal as your Granny [CD's name for her] I

shall take in hand namely several little errors in orthography of which I shall send you a list that you may profit by my lectures tho' the world is between us. — so here goes. —

wrong	right according to sense
loose. lanscape. higest	lose. landscape. highest
profil. cannabal	profile. cannibal. peaceable
peacible. quarrell	quarrel. — I daresay these errors are the effect of haste, but as your Granny it is my duty to point them out. —'

See *Correspondence* 1: 366.

19 The passage from 'tell her . . .' has been deleted in the manuscript with black ink. See *Correspondence* 1: 392.

20 See *Correspondence* 1: 495–6.

21 See *Correspondence* 1: 524.

22 See *Correspondence* 1: 526.

23 See *Correspondence* 1: 530.

24 See *Correspondence* 1: 535.

25 See *Correspondence* 2: 53–4.

26 See *Correspondence* 2: 57–9.

27 See *Correspondence* 2: 75–6.

28 See *Correspondence* 2: 80–1.

29 See *Autobiography* p. 116.

30 See pp. 22–3 in: Gavin de Beer. *Charles Darwin. Evolution by Natural Selection.* Nelson, London, 1963.

31 See p. 37 in: Alan Moorehead. *Darwin and the Beagle.* Hamish Hamilton, London, 1969.

32 See pp. 73–4 and 175–83 in: H. E. L. Mellersh. *FitzRoy of the Beagle.* Hart-Davis, London, 1968.

33 See *Narrative* 2: 658–9.

34 Quoted from p. 28 of *Extracts from Letters addressed to Professor Henslow by C. Darwin, Esq., read at a Meeting of the Society on the 16th of November 1835.* Printed for distribution among the Members of the Cambridge Philosophical Society, 1 December 1835.

35 See *Narrative* 2: 667–8.

36 See *Autobiography* p. 85.

37 See *Autobiography* pp. 72–6.

38 See *Correspondence* 1: 326, 334–6, 406–7.

39 See *Correspondence* 2: 254–6.

Acknowledgements

I am most grateful to Philip Titheradge, Curator of the Darwin Museum at Down House, Downe, Kent, for access to the manuscript of the *Beagle Diary*, and to Darwin's field notebooks. The cost of obtaining copyflow prints from microfilms of the notebooks was met by a grant from the Darwin Fund of the Royal Society.

My greatest indebtedness is to the editors of the *Correspondence*, on whom I have drawn heavily for their listing in their Appendix II to Volume 1 of Darwin's *Beagle* records and in their Appendix IV to Volume 1 of the books on board the *Beagle*, for much helpful material in their footnotes, and in preparation of the Biographical Register. My thanks are also due to the following for help and advice in different ways on a variety of points: Dr Arturo Amos, Silvia Aramayo, Professor Duccio Bonavia, Professor Carlos Chagas, Dr Gordon Chancellor, Paul Downham, Dr Ian Forster, Dr Patricio Garrahan, Peter Gautrey, Natalie Goodall, Randal Keynes, Dr Simon Keynes, Dr David Kohn, Edgar Krebs, Professor Aristides Leão, Professor Carlos Monge, Dr Luis Rinaldini, David Scrase, David Stanbury, Dr David Stoddart, Professor Rudolf Trümpy, Mrs Margaret Twinn. Not least, I owe a great deal to my wife for her forbearance, help and encouragement at all times.

Note on editorial policy

❖

As far as possible, my aim has been to adopt the practices laid down and explained in full by the editors of *The Correspondence of Charles Darwin*, though I am very conscious that I have not always achieved the same high standards. Thus CD's own spelling, idiosyncratic though it may be for certain words – *broard, throughily, untill, neighbourhead, yatch, monotomous*, are some of them – has been retained throughout, even where the mistake is a clear slip of the pen. Where there is doubt, and there is no difficulty in deciding what CD's exact intention should have been, for example in the case of adding the final *s* to the plural of a noun, I have generally given him the benefit of it. Similarly, where it is hard to decide whether a word starts with a lower case or capital letter, I have used a capital in the cases of proper names and places. His abbreviations appear as nearly as possible as they are written, with '&' almost invariably used in place of 'and' during the first years of the voyage, but less so in its last few months. In other instances where, as Sulloway has shown (see *Journal of the History of Biology* **16**: 361–90, 1983), useful information for dating purposes may be derived from a systematic study of CD's spelling habits, it should be noted that some of his mistakes were evidently put right only after the end of the voyage. In such cases I have shown the final corrected spelling, and reference has to be made to the original manuscript or the facsimile to see what was actually written.

CD's somewhat erratic punctuation has in the main been respected, and I have restored the numerous dashes that were cut out in Nora Barlow's edition. However, an awkward problem is raised in this respect by the many dots that might either be regarded merely as 'pen rests', or taken as commas or full stops. I have tried to deal with the frequent uncertainties that thus arise either by omitting these dots, or by showing them as appropriate punctuation marks, in accordance with modern usage and the sense of the passage concerned. In doing this, my decisions have often been rather arbitrary, although in general erring on the side of helping the reader. The extracts quoted from the field notes preserved at Down House have been treated in the same fashion.

In order to avoid confusion with editorial footnotes, some of the marginal notes added by CD himself have been incorporated in brackets in the text, and identified as *Note in margin*. Marginal notes in another hand, any of CD's notes that seem to require an explanation, and those which do not fit in at an obvious point, have been dealt with in footnotes, as have all revisions to the text that materially affect its meaning. It is important in this connexion to appreciate that the manuscript Diary is not the exact document that was forwarded by CD to his family in instalments from South America, or brought back with him in 1836. It provided CD with the backbone of the account of the voyage that he gave in his *Journal of Researches*, and the majority of the revisions to the text and the marginal notes were probably made in 1837 while he was editing it for publication. Although in theory it might be possible to discriminate between corrections made at the time of writing without a change in ink or pen, and those made later, it would not be at all easy. The transcript presented here represents the final text embodying all CD's corrections, whenever they were made. In the many places where the order of the wording was revised within a sentence with no significant alteration in the information conveyed, the final version appears without comment. Wherever whole sentences or parts of them were deleted, this is recorded in a footnote. All the changes have been listed in 'Alteration Notes' similar to those appearing in *The Correspondence of Charles Darwin*, which have been deposited in the Cambridge University Library for use by anyone wishing to consult them, but in view of the public availability of the facsimile for detailed study, are not included here.

The basic layout of the manuscript diary has been slightly modified in that for the daily entries, the month is given as well as the day of the month only when it changes, but is shown with the year as a running heading at the top of the page along with the place. CD's original paragraphing has been respected throughout, except that wherever a date in the margin or body of the text indicates the start of a different day, a fresh entry is begun. The pagination of the manuscript is shown by the numbers between vertical lines, thus |000|.

Words underlined once by CD are shown in italics, and those underlined twice in bold type. His own round brackets are retained, while editorial interpolations of obviously missing words or letters are shown within square brackets. The rather few words that are illegible are so marked.

Principal sources of references

Narrative 1

Narrative of the surveying voyages of His Majesty's Ships Adventure and Beagle, between the years 1826 and 1836, describing their examination of the southern shores of South America, and the Beagle's circumnavigation of the globe. Volume I. Proceedings of the first expedition, 1826–1830, under the command of Captain P. Parker King, R.N., F.R.S. Henry Colburn, London, 1839.

Narrative 2

Narrative of the surveying voyages . . . of the globe. Volume II. Proceedings of the second expedition, 1831–1836, under the command of Captain Robert Fitz-Roy, R.N. Henry Colburn, London, 1839.

Journal of Researches

1st edition: Narrative of the surveying voyages . . . of the globe. Volume III. Journal and remarks. 1832–1836. By Charles Darwin Esq., M.A. Sec. Geol. Soc. Henry Colburn, London, 1839.

2nd edition: Journal of Researches into the Natural History and Geology of the Countries visited during the Voyage of H.M.S. Beagle round the World under the Command of Capt. FitzRoy, R.N. By Charles Darwin, M.A., F.R.S. John Murray, London, 1845.

Zoology 1

The zoology of the voyage of H.M.S. Beagle, under the command of Captain FitzRoy, R.N., during the years 1832 to 1836. Edited and superintended by Charles Darwin, Esq. M.A. F.R.S. Sec.G.S. Naturalist to the expedition. Part I. Fossil mammalia: by Richard Owen, Esq. F.R.S. Smith, Elder and Co., London, 1840.

Zoology 2

The zoology of the voyage of H.M.S. Beagle . . . Part III. Birds, by John Gould, Esq. F.L.S. Smith, Elder and Co., London, 1841.

Autobiography
The Autobiography of Charles Darwin 1809–1882. Edited by Nora Barlow. Collins, London, 1958.

CD and the Voyage
Charles Darwin and the Voyage of the Beagle. Edited by Nora Barlow. Pilot Press, London, 1945.

Beagle Record
The Beagle Record. Selections from the original pictorial records and written accounts of the voyage of H.M.S. Beagle. Edited by R. D. Keynes. Cambridge University Press, 1979.

Correspondence **1**
The correspondence of Charles Darwin. Volume 1. 1821–1836. Cambridge University Press, 1985.

Correspondence **2**
The correspondence of Charles Darwin. Volume 2. 1837–1843. Cambridge University Press, 1986.

Charles Darwin's *Beagle* Diary

1831–1836

I had been wandering about North Wales on a geological tour with Professor Sedgwick[1] when I arrived home on Monday 29th August. My sisters first informed me of the letters from Prof. Henslow[2] & M[r] Peacock[3] offering to me the place in the Beagle which I now fill. — I immediately said I would go; but the next morning finding my Father so much averse to the whole plan, I wrote to M[r] Peacock to refuse his offer. — On the last day of August I went to Maer,[4] where everything soon bore a different appearance. — I found every member of the family so strongly on my side that I determined to make another effort. — In the evening I drew up a list of my Fathers objections, to which Uncle Jos wrote his opinion & answer.[5] This we sent off to Shrewsbury early the next morning & I went out shooting. — About 10 oclock Uncle Jos sent me a messuage, to say he intended going to Shrewsbury & offering to take me with him. — When we arrived there, all things were settled, & my Father most kindly gave his consent. —

I shall never forget what very anxious & uncomfortable days these two were. — My heart appeared to sink within me, independently of the doubts raised by my Fathers dislike to the scheme. I could scarcely make up my mind to leave England even for the time which I then thought the voyage would last. Lucky indeed it was for me that the first picture of the expedition was such an highly coloured one. — |2|

In the evening I wrote to M[r] Peacock & Cap[t] Beaufort[6] & went to bed very much exhausted. On the 2nd I got up at 3 oclock & went by the Wonder coach as far as Brickhill, I then proceeded by postchaises to Cambridge. I there staid two days consulting with Prof. Henslow. At this point I had nearly given up all hopes, owing to a letter from Cap. FitzRoy to M[r] Wood,[7] which threw on every thing a very discouraging appearance. On Monday 5th I went to London & that same day saw Caps. Beaufort & FitzRoy. The latter soon smoothed away all difficulties & from that time to the present has taken the kindest interest in all my affairs. — On Sunday 11th sailed by Steamer to Plymouth in order to see the Beagle. I returned to London on 18th.[8] On Monday the 19th by mail to Cambridge, where after taking leave of Henslow on Wednesday night I got to St Albans & so by the Wonder to Shrewsbury on Thursday 22nd. — I left home on October 2nd for London, where I remained after many & unexpected delays till the 24th on which day I arrived at Devonport & this journal begins. — |3|

(16th December)

[1] Almost CD's entire field experience as a geologist before the voyage consisted in a tour of North Wales with Adam Sedgwick from 5 to 20 August 1831 (see *Autobiography* pp. 68–71, and letter from Sedgwick of 4 September 1831, *Correspondence* 1: 137–9). After staying with friends at Barmouth he returned to Shrewsbury on 29 August.
[2] See *Correspondence* 1: 128–9.
[3] See *Correspondence* 1: 129–30.
[4] Home of his uncle Josiah Wedgwood II.
[5] See *Correspondence* 1: 132–5.
[6] See *Correspondence* 1: 135–6.
[7] Alexander Charles Wood, Robert FitzRoy's cousin.
[8] This date should be the 17th (see letters to Susan Darwin and Henslow, *Correspondence* 1: 155–7).

Monday, October 24th   Arrived here in the evening after a pleasant drive from London.

25th   Went on board the Beagle, found her moored to the Active hulk & in a state of bustle & confusion. — The men were chiefly employed in painting the fore part & fitting up the Cabins. — The last time I saw her on the 12th of Sept[r] she was in the Dock yard & without her masts or bulkheads & looked more like a wreck than a vessel commissioned to go round the world.

26th   Wet cold day, went on board, found the Carpenters busy fitting up the drawers in the Poop Cabin.[1] My own private corner looks so small that I cannot help fearing that many of my things must be left behind. —

[1] See plan on p. 63 and sketch labelled in CD's hand (CUL DAR 44), *Beagle Record* p. 103.

27th   Went on board.

28th   A fine day. — M[r] Earl[1] arrived from London after having had a most stormy passage. — It blew a SW gale for the whole week, & the Steam Packet during this whole time was pitching about. I think if I had gone by it, this journal book would have been as useless to me as so much waste paper.

[1] Augustus Earle was the *Beagle*'s first official artist, serving in this capacity until ill health forced his resignation at the end of 1832.

29th   A beautiful day, dined at 5 oclock with Gun-room officers. — They amused themselves with giving most terrific accounts of what Neptune would do with me on crossing the Equator. — Mr Earl mentioned, that some years ago when after having crossed the Line, they fell in with a ship, all her sails set. — Not a man could they see on deck, but on boarding her & going below, they found every body, even the Captain & his wife, so very drunk|4| that they could not move. — They had been making merry after Neptune's revels. —

30^(th)   Dined at one oclock with the Mids—after that had a sail, & landed at Millbrook. — Stokes, Musters & myself then took a long scrambling walk. —

Monday 31^(st)   Went with Mr Stokes to Plymouth & staid with him whilst he prepared the astronomical house belonging to the Beagle for observations on the dipping needle. The gardens belonging to the Athæneum were fixed upon as being a place well known & easily described.[1]

[1]The Plymouth Athenaeum had been founded as the local Literary and Philosophical Society in 1812, and construction of its Athenaeum building had been begun in 1818. For an account of its history see M. A. Wilson "A hundred and fifty years of the Plymouth Athenaeum", *Proceedings of the Plymouth Athenaeum* 1: 17–30, 1967.

November 1^(st)   A very wet day, staid in the house in consequence. Captain King & his son[1] arrived in the evening & dined with us. — The latter is going out in Beagle as Midshipman.

[1]Captain Philip Parker King, RN, FRS, had been Commander of the *Adventure* and *Beagle* on the first surveying expedition to South America 1826–30. He was the author of Vol. 1 of the *Narrative*. He settled in Australia, where CD met him on 26 January 1836 (see p. 405). His eldest son Philip Gidley King, then aged 14, was a midshipman on the *Beagle* from 1831 to 1836, after which he too lived in Australia.

2^(nd)   Went on board.

3^(rd)   Walked to Plymouth with Cap^s King & FitzRoy.

4^(th)   Cap FitzRoy took me in the Commissioners boat to the breakwater, where we staid for more than an hour. Cap. FitzRoy was employed in taking angles, so as to connect a particular stone, from which Cap King commenced for the last voyage his longitudes, to the quay at Clarence Baths, where the true time is now taken. — Sir J. Rennie, the architect, was on the Breakwater, & gave some interesting accounts of the effects of various severe gales. — In 1826 several blocks of stone weighing 10 tuns each, were considerably displaced. — It now offers a much better resistance to a heavy sea than it formerly did. — It is now constructed of the shape of a roof of a house placed on the ground; before this alteration, it was that of a roof on a low wall, so that the sea acted on a perpendicular|5| surface. — Every.body agrees in the Breakwater being as useful as it is a most stupendous work of art. — In the evening dined with Mr Harris, (the author of several papers on Electricity) and met there several very pleasant people. — Colonel Hamilton Smith, who is writing on fishes with Cuvier. —Cap^s King & Lockier. The former mentioned an anecdote showing how completely civilization & dram-drinking were synonymous things in New S. Wales. — A native asked him one day for some rum; which being refused & wine offered, he seemed discontented. Upon Cap King remonstrating with & asking him what he did before the English came there; he answered Oh! we were not civilized then. —

5th Wretched, miserable day, remained reading in the house.

6th Went with Musters to the Chapel in the Dock-yard. — It rained torrents all the evening. — It does not require a rain gauge to show how much more rain falls in the Western than in the Central & Eastward parts of England. —

Monday 7th Staid at home.

8th In the morning, marked the time whilst Stokes took the altitude of the sun. — Went on board the Beagle; she now begins for the first time to look clean & well arranged. — Was introduced to Cap FitzRoys two brothers, who have come down from London to wish him farewell. —

9th Walked to Plymouth with Cap^s Fitz & Videl & called on Mr Harris.

10th Assisted Cap. FitzRoy at the Athæneum in reading the various angles of the dipping needle, after that |6| heard the Russian horn band. And in the evening dined at the Admirals, Sir Manley Dixon: every body there except myself was a naval officer & of course the conversation was almost exclusively nautical. — This made the evening very pleasant to me, but I could not help thinking how very different it would have been under different circumstances.

11th Breakfasted with Mr Harris & went again to the Athæneum & spent the whole day at the dipping needle. — The end, which it is attempted to obtain, is a knowledge of the exact point in the globe to which the needle points. The means of obtaining it is to take, under all different circumstances, a great number of observations, & from them to find out the mean point. — The operation is a very long & delicate one. —

12th Breakfasted with Col. Hamilton Smith & spent some pleasant hours in talking on various branches of Natural history. Took a walk to some very large Limestone quarries, returned home & then went on board the Beagle. — The men had just finished painting her & of course the decks were clear & things stowed away. — For the first time I felt a fine naval fervour; nobody could look at her without admiration; & as for the Poop Cabin it would [be] superfluous to wish for anything more spacious & comfortable. — The day has been an excellent one for the paint drying, so calm & so truly Autumnal that it gives one hopes that the Westerly gales have tired themselves with blowing. — It is a|7| great consolation to know, that even if we had sailed at the beginning of October, it is probable we should have scarcely reached Madeira. —

13th   Walked to Saltram & rode with Lord Borrington to Exmoor to see the Granite formation, the road passed through very extensive oak woods situated on the side of hills at the bottom [of] which were running very clear & broard brooks. — Exmoor geographically is the same as Dartmoor & extends to Exeter. It has a desolate appearance, the tops of the hills only showing the mossy forms of the Granite. — In the evening the Fuegians[1] arrived by Steam Packet together with their school master Mr Jenkins. Their names are York Minster, Jemmy Button & Fuegia. — Matthews the missionary arrived also at the same time. —

[1] Four natives from Tierra del Fuego named York Minster, Boat Memory, James Button and Fuegia Basket, had been brought back by Robert FitzRoy on the *Beagle* in October 1830 (see *Narrative* 1: 391–444 and 2: 1–16; and *Beagle Record* pp. 4–5). Boat Memory died of smallpox, but one of the objectives of the *Beagle's* second voyage to South America was to repatriate the other three.

Monday 14th   Cap. FitzRoy removed the Chronometers on board & placed the books in the Poop Cabin. — Went on board, the paint is not yet fixed, so that nothing can be done. — In the evening the Instructions from the Admiralty arrived. — They are in every respect most perfectly satisfactory, indeed exactly what Cap Fitz himself wished. — The orders merely contain a rough outline. — There could not be a greater compliment paid to Cap FitzRoy than in so entirely leaving the plans to his own discretion. — |8|

15th   Went with Cap FitzRoy to Plymouth & were unpleasantly employed in finding out the inaccuracies of Gambeys new dipping needle.

16th   Went on Board & spent the whole day in idly but very agreeably wandering up & down the streets with Cap FitzRoy.

17th   A very quiet day.

18th   Cap FitzRoy has been busy for these last two days with the Lords of the Admiralty. —

19th   I have now a regular employment every morning taking & comparing the differences in the Barometers. In the evening drank tea with Capt Vidal. He has seen a great deal of the same sort of service that we are going to be employed on; he was eight years surveying the African coast. — during this time he buried 30 young officers; a boat never was sent up a river, without its causing the death of some of the party. —

20th   Went to Church & heard a very stupid sermon, & afterwards took a long walk in a very picturque country, between Mount Edgecombe & Mill Brook.

Monday 21st   Carried all my books & instruments on board the Beagle. — In the evening went to the Athæneum & heard a popular lecture from Mr Harris on his lightning conductors. By means of making an Electric machine, a thunder cloud—a tub of water the sea, & a toy for a line of battle ship he showed the whole process of it being struck by lightning & most satisfactorily proved how completely his plan|9| protects the vessel from any bad consequences. This plan consists in having plates of Copper folding over each other, let in in the masts & yards & so connected to the water beneath. — The principle, from which these advantages are derived, owes its utility, to the fact that the Electric fluid is weakened by being transmitted over a large surface to such an extent that no effects are perceived, even when the mast is struck by the lightning. — The Beagle is fitted with conductors on this plan; it is very probable, we shall be the means of trying & I hope proving the utility of its effects. —

About six oclock, a Marine, being drunk & whilst crossing from the Hulk to another vessel slipped overboard & was not seen again. His body has not been found. —

22nd   Went on board & returned in a panic on the old subject want of room. returned to the vessel with Cap FitzRoy, who is such an effectual & goodnatured contriver that the very drawers enlarge on his appearance & all difficulties smooth away. — In the evening dined & spent a very pleasant afternoon with Capt Vidal. —

23rd   This has been a very important day in the annals of the Beagle; at one oclock she was loosed from the moorings & sailed about a mile to Barnett pool. Here she will remain till the day|10| of sailing arrives. This little sail was to me very interesting, everything so new & different to what one has ever seen, the Coxswains piping, the manning the yards, the men working at the hawsers to the sound of a fife, but nothing is so striking as the rapidity & decision of the orders & the alertness with which they are obeyed. — There remains very little to be done to make all ready for sailing. All the stores are completed & yesterday between 5 & 6 thousands canisters of preserved meat were stowed away. — Not one inch of room is lost, the hold would contain scarcely another bag of bread. My notions of the inside of a ship were about as indefinite as those of some men on the inside of a man, viz a large cavity containing air, water & food mingled in hopeless confusion. —

24th A very fine day & an excellent one for obtaining sights. — Every body hailed the sun with joy, for untill the time is well taken, we cannot leave harbour. — I went on board several times in the course of the day; but did not succeed in doing any good, as they were changing the place of anchorage & that is not the time for a Landsman to give trouble about his own lumber. —

25th Very busily employed on board in stowing away my clothes & after that in arranging the books, did not leave the vessel till it was dark. — |11|

26th Again employed all day long in arranging the books; we (Stokes & myself) succeded in leaving the Poop Cabin in very neat order. After having finished this & bringing on board some things of my own, King & I walked on the sea shore & returned home through a part of Lord Mount Edgcombe's park. — The day has been a very fine one & the view of Plymouth was exceedingly striking. The country is so indented with arms of the sea that there is a very new & different scene from every point of view. —

27th An idle day, had a pleasant sail in Captain FitzRoy boat & then called on several people.

Monday 28th Cap. FitzRoy gave a very magnificent luncheon to about forty persons: it was a sort of ships warming; & every thing went off very well, in the evening a Waltz was raised which lasted till every body went away. —

29th To day the Captain has had another large party, but not being very well, I have not gone to it. — In the evening dined with Sir Manley Dixon.

30th Cap King was here the whole morning & I had with him some very interesting conversation on Meteorology, he paid great attention to this subject during the last voyage. — Afterwards I took a very pleasant walk to Corsan, all my thoughts are now centered in the future & it is with great difficulty that I can talk or think on any other subject; |12| When I first had the offer of the voyage I was in the same state & a very uncomfortable one it is; but this present time has the great & decided advantage of everything being fixed & settled. —

December 1st Breakfasted with Cap King. — The Commissioner took Lord Graves party to see the Caledonia & offered me a place in the Yatch. — The Caledonia is generally considered one of the finest vessels in the world, she carries 120 32 pounders. — So large a vessel is an astonishing sight, one wonders by what contrivance everything is governed with such regularity & how amongst such numbers such order prevails. On coming near her the hum is like that of [a] town heard at some distance in the evening. —

2nd Worked all day long in arranging & packing my goods in the drawers. — Erasmus[1] arrived in the afternoon & I spent with him a very pleasant evening. —

[1] Erasmus Alvey Darwin, CD's elder brother.

3rd Incessantly busy in ordering, paying for, packing all my numberless things; how I long for Monday even sea-sickness must be better than this state of wearisome anxiety. — Erasmus being here is a great pleasure, but I do not see much of him.

4th I am writing this for the first time on board, it is now about one oclock & I intend sleeping in my hammock. — I did so last night & experienced a most ludicrous difficulty in getting into it; my great fault|13| of jockeyship was in trying to put my legs in first. The hammock being suspended, I thus only succeded in pushing [it] away without making any progress in inserting my own body. —the correct method is to sit accurately in centre of bed, then give yourself a dexterous twist & your head & feet come into their respective places. — After a little time I daresay I shall, like others, find it very comfortable. — I have spent the day partly on board & partly with my brother: in the evening, Cap King & son, Stokes, my brother & myself dined with Cap FitzRoy. —

In the morning the ship rolled a good deal, but I did not feel uncomfortable; this gives me great hopes of escaping sea sickness. — I find others trust in the same weak support. — May we not be confounded. — It is very pleasant talking with officer on Watch at night — every thing is so quiet & still, nothing interrupts the silence but the half hour bells. — I will now go and wish Stuart (officer on duty) good night & then for practising my skill in vaulting into my hammock. —

Monday 5th It was a tolerably clear morning & sights were obtained, so now we are ready for our long delayed moment of starting. —it has however blown a heavy gale from the South ever since midday, & perhaps we shall|14| not be able to leave the Harbour. The vessel had a good deal of motion & I was as nearly as possible made sick. I returned home very disconsolate, but mean to treat myself with sleeping, for the last time, on a firm flat steady bed. — In the evening dined with Erasmus. I shall not often have such quiet snug dinners. — I take the opportunity of mentioning a very curious circumstance which the watermen here have observed. — When building the walls of the Victualling office in 6 fathom water, the men made signals by tapping on the inside [of] the diving bell.— This the Watermen used every where to hear, even at Torpoint, a distance of two miles.— it sounded like a person hitting the boat with a small hammer, & for a long time it quite puzzled the men, so much so that they hauled the boat up, thinking it was a crab or some animal.—

6th Again sailing has been deferred. In the morning the wind was SW, but light; afterwards it increased into a gale from the South. Stokes & myself arranged the Poop Cabin, after which I was forced to beat a hasty retreat on shore. I could not even for a short time have stood the motion, had I not been hard at work. Dined in the evening with Erasmus.—[1]|15|

[1] From here up to p. 110 of the manuscript CD writes only on alternate pages.

7th It is daily becoming more wearisome remaining so long in harbour; at least I have nothing more to do. Every thing is on board & we only wait for the present wind to cease & we shall then sail.— This morning it blew a very heavy gale from that unlucky point SW.— The Beagle struck her Top Gallant masts & veered her yards to the wind.—

8th I am writing this & the two last days journal in my own corner.— The cabin begins now to look comfortable, but yet very much crowded.— It is a miserable wet day & no hopes of the wind changing; my first question every morning how is the wind? Oh for the lucky day, when the answer is NE.—

9th Finally arranged the Poop Cabin.— Erasmus & myself then took a long & very pleasant walk on Mount Edgcombe. the view from it is of a most striking & uncommon kind, a birds eye view of three large towns, Devonport Stonehouse & Plymouth, situated on arms of the sea, seen from a most beautiful & picturesque hill. In the evening, dined for the last time with my brother.—

10th Early in the morning torrents of rain; the sky then became very clear, with a light wind from SW. We all thought we should have settled weather.— The Captain said last night, that if|17| it was possible he would sail to day; accordingly at 9 oclock we weighed our anchors, & a little after 10 sailed.— Erasmus was on board & we had a pleasant sail till we doubled the Breakwater; where he left us & where my misery began. I was soon made rather sick, & remained in that state till evening, when, after having received notice from the Barometer, a heavy gale came on from SW. The sea run very high & the vessel pitched bows under.— I suffered most dreadfully; such a night I never passed, on every side nothing but misery; such a whistling of the wind

& roar of the sea, the hoarse screams of the officers & shouts of the men, made a concert that I shall not soon forget. —

11th It lasted till the Sunday morning, when it was determined to put back to Plymouth & there remain for a more fortunate wind. — We got to our anchorage at Barnett Pool at 12 oclock, & are now lying quiet & snug. — Some short time afterward, Musters, a fellow companion in misery, & myself took a good walk, which considerably revived us — but even yet my head is giddy & uncomfortable. — I was surprised to find, that leaving England as I then thought for four years, made little or no impression on my feelings. — I did expect, to have felt some of the same heart-sinking sensations which I experienced when I first had the offer of the voyage. I left harbour as placidly|19| as if I was merely going a trip to France: I suppose I have so often & so throughily considered the subject, that no new & fresh ideas connected with it can arise in my mind; & it is their newness which gives intensity to ones feelings. After having had so much time to make up my mind, I am decided I did right to accept the offer; but I yet think it is doubtful how far it will add to the happiness of ones life. — If I keep my health & return, & then have strength of mind quietly to settle down in life, my present & future share of vexation & want of comfort will be amply repaid. — I find it necessary to forget the many little comforts which one enjoys on shore almost without perceiving them. Nothing can be done without so much extra trouble, even a book cannot be taken from the shelves or a piece of soap from the washing stand, without making it doubtful whether in the one case it is worth while to wash ones hands, or in the other to read any passage. —

Monday 12th Boisterous weather, the ship rolled a good deal; & I actually felt rather uncomfortable: I look forward to sea-sickness with utter dismay, not so much as regards the misery of a fortnight or three weeks, as the being incapacitated for a much longer time from any active employment. — In middle of day walked to Corsan bay &|21| there enjoyed the sight of the sea lashing itself & foaming on the rocks. — There is no pleasure equal to that which fine scenery & exercise creates, it is to this I look forward to with more enthusiasm than any other part of our voyage. — Dined with Sir Manley Dixon, a pleasant quiet party, or rather to speak more truly, I suspect very dull to every body but myself, for the Beagle was the chief subject of conversation, & it is now the only one that at all interests me. — It is no easy matter at any time, but now a most painful one to make conversation at a regular party. —

We have had a long & rough pull to the vessel, but I am now seated in my own corner, snug & quiet & am listening to the wind roaring through the rigging with same sort of feeling that I often have when sitting round a Christmas fire. — Eight bells have struck, or it is 12 oclock, so I will turn into my hammock. —

13th An idle day; dined for the first time in Captains cabin & felt quite at home. — Of all the luxuries the Captain has given me, none will be so essential as that of having my meals with him. — I am often afraid I shall be quite overwhelmed with the numbers of subjects which I ought to take|23| into hand. It is difficult to mark out any plan & without method on ship-board I am sure little will be done. — The principal objects are 1st, collecting observing & reading in all branches of Natural history that I possibly can manage. Observations in Meteorology. — French & Spanish, Mathematics, & a little Classics, perhaps not more than Greek Testament on Sundays. I hope generally to have some one English book in hand for my amusement, exclusive of the above mentioned branches. — If I have not energy enough to make myself steadily industrious during the voyage, how great & uncommon an opportunity of improving myself shall I throw away. — May this never for one moment escape my mind, & then perhaps I may have the same opportunity of drilling my mind that I threw away whilst at Cambridge. —

14th A beautiful day giving great hopes of a fair wind. Took my usual & delightful walk in the beautiful country around Mount Edgcombe. — Everything connected with dressing & sleeping have hitherto been my greatest drawbacks to comfort. — But even these difficulties are wearing away. My hammock after endless alterations has been made flat & I have trained myself to a regular method in dressing & undressing. —
    Orders are issued for sailing tomorrow morning. — |25|

15th The wind continues in the old point SW, which independently of detaining us appears invariably to bring bad [weather] with it. — The ship is full of grumblers & growlers, & I with sea-sickness staring me in the face am as bad as the worst. — The time however passes away very pleasantly, but instead of working, the whole day is lost between arranging all my nick-nackiries & reading a little of Basil Halls fragments. —

16th This day is come to its close much in same way as yesterday. I am now sitting in my own corner feeling most comfortably at home. — This is the first time that I have not left the vessel during the whole day. —

The wind with torrents of rain is sweeping down upon us in heavy gusts.

17th   Walked with Sullivan & King to the coast near the Ramhead & there saw a wild stormy sea breaking on the rocks. We passed through a village of the name of Corsan,[1] one of the most curiously built places I ever saw. — None of the streets are for thirty yards in the same straight line, & all so narrow that a cart certainly could not pass up them. — It is situated in a very pretty little bay, which shelters numerous fishing & smugling boats from the sea. — Our old enemy the SW Gale is whistling throug the rigging: today it drove back a Brig which left Plymouth three weeks ago, so that we ought to be instead|27| of discontented, most thankful for remaining in our present snug anchorage. — The novelty of finding myself at home on board a ship is not as yet worn away, nor have I ceased to wonder at my extraordinary good fortune in obtaining what in the wildest Castles in the air I never had even imagined. If it is desirable to see the world, what a rare & excellent opportunity this is. — It is necessary to have gone through the preparations for sea to be throughily aware what an arduous undertaking it is. It has fully explained to me the reasons so few people leave the beaten path of travellers. —

[1] Correctly spelt Cawsand.

18th   Dined at 12 oclock with the Midshipmen, & then with Bynoe & Stokes walked to Whitson bay: the sea here presented a most glorious & sublime appearance. — For nearly quarter of a mile it was a confused mass of breakers & from the white covering of foam looked like so much snow. Each wave as it dashed against the rocks threw its spray high on the hill & wetted our faces. — To perfect the scene a single man was watched from a rock to spy out any chance wreck. —

Monday 19th   A fine calm day with a gentle breeze from the North. — There is every probability of sailing tomorrow morning. — The weighing of our anchor will be hailed with universal joy.|29|

20th   The rain fell in torrents & the South W wind blew all the morning; but now the moon is shining bright on the sea, which looks so calm, that one would think it never would again be troubled by a storm. — Nothing can be more beautiful than the view from our present anchorage, on such a clear night as this is; the Sound looks like a lake. — May these not turn out false signs, that our disappointment to be the more bitter. — The sailors declare there is somebody on shore keeping a black cat under a tub, which it stands to reason must keep us in harbour. —

21st The morning was very calm & the sun shone red through the mist: every thing gave us hopes of a steady NE wind, —& a prosperous voyage. — But here we are yet to remain alternately praying to & abusing the SW gales. — From weighing to again letting down our anchor everything was unfortunate. — We started at 11 oclock with a light breeze from NW & whilst tacking round Drakes Island, our ill luck first commenced. It was spring tide & at the time lowest ebb; this was forgotten, & we steered right upon a rock that lies off the corner. — There was very little wind or swell on the sea so that, although the vessel stuck fast for about half an hour, she was not injured. Every mæneuvre was tried to get her off; the one that succeeded best was making every person on board run|31| to different parts of the deck, by this means giving to the vessel a swinging motion. — At last we got clear & sailed out of harbour not a jot the worse from our little accident. When we were on the open sea I soon became sick: at 4 oclock I went down to the Captains cabin & there slept till 8 oclock, after that I retreated to my hammock & enjoyed a most comfortable sleep till morning. — As soon as it was light Stokes & myself looked at a pocket compass, which we agreed was bewitched, for it pointed to NE instead of to where we were sailing W by S—Our doubts were cleared up by Wickham putting his head in & telling us we should be in Plymouth Sound in the course of an hour. — During the middle watch the wind began to change its direction & at 4 oclock, when we were only 11 miles from the Lizard, it blew a gale from SW. — Upon this the Captain wared the ship & we returned to our old home at the rate of eleven knots an hour. —

22nd I have not felt at all comfortable all this day; took a long walk with Stokes & Bynoe, during the whole time torrents of rain were pouring down. — By some mischance in dropping the anchor it got twisted with the chain: they were hard at work for eight hours in getting all clear. — In the evening double|33| allowance was served out to the men. — Several vessels which sailed with us have all been likewise forced to put back. —

23rd In the morning Sullivan Bynoes & myself shot matches with the rifle for sundry bottles of wine to be paid for & drunk at the Madeira islands. — in the evening went with Stokes to a bad concert. — Although I am continually lamenting in the bitterness of my heart against all the long delays & vexations that we have endured, I really believe they have been much to my advantage. — for I have thus become broken in to sea habits without having at the same time to combat with the miseries of sickness. —

24th   A blank & idle day.

25th   Christmas day, in morning went to Church & found preaching there an old Cambridge friend Hoare. — Dined at 4 oclock with Gunroom officers, it does me good occasionally dining there, for it makes me properly grateful for my good luck in living with the Captain. — The officers are all good friends yet there is a want of intimacy, owing I suppose to gradation of rank, which much destroys all pleasure in their society. — The probability of quarrelling & the misery on ship board consequent on it produces an effect contrary to what one would suppose. —instead of each one endeavouring to encourage habits of friendship, it seems a generally received maxim|35| that the best friends soon turn out the greatest enemies. — It is a wonder to me that this independence one from another, which is so essential a part of a sailors character, does not produce extreme selfishness. — I do not think it has this effect, & very likely answers their end in lessening the number of quarrels which always must necessarily arise in men so closely united. — Let the cause be what it may, it is quite surprising that the conversation of active intelligent men who have seen so much & whose characters are so early & decidedly brought out should be so entirely devoid of interest. —

Christmas day is one of great importance to the men: the whole of it has been given up to revelry, at present there is not a sober man in the ship: King is obliged to perform duty of sentry, the last sentinel came staggering below declaring he would no longer stand on duty, whereupon he is now in irons getting sober as fast as he can. — Wherever they may be, they claim Christmas day for themselves, & this they exclusively give up to drunkedness—that sole & never failing pleasure to which a sailor always looks forward to. —

Monday 26th   A beautiful day, & an excellent one for sailing, —the opportunity has been lost owing to the drunkedness & absence of nearly the whole crew. —the ship has been all day|37| in state of anarchy. One days holiday has caused all this mischief; such a scene proves how absolutely necessary strict discipline is amongst such thoughtless beings as Sailors. — Several have paid the penalty for insolence, by sitting for eight or nine hours in heavy chains. — Whilst in this state, their conduct was like children, abusing every body & thing but themselves, & the next moment nearly crying. — It is an unfortunate beginning, being obliged so early to punish so many of our best men there was however no choice left as to the necessity of doing it. Dined in gun-room & had a pleasant evening.

27th   I am now on the 5th of Jany writing the memoranda of my misery for the last week. A beautiful day, accompanied by the long wished for E wind. — Weighed anchor at 11 oclock & with difficulty tacked out. — The Commissioner Capt Ross sailed with us in his Yatch. — The Capt Sullivan & myself took a farewell luncheon on mutton chops & champagne, which may I hope excuse the total absence of sentiment which I experienced on leaving England. — We joined the Beagle about 2 oclock outside the Breakwater, — & immediately with every sail filled by a light breeze we scudded away at the rate of 7 or 8 knots an hour. — I was not sick that evening but went to bed early. —[1]|39|

[1] CD wrote many years later: 'These two months at Plymouth were the most miserable which I ever spent, though I exerted myself in various ways. I was out of spirits at the thought of leaving all my family and friends for so long a time, and the weather seemed to me inexpressibly gloomy. I was also troubled with palpitations and pain about the heart, and like many a young ignorant man, especially one with a smattering of medical knowledge, was convinced that I had heart-disease. I did not consult any doctor, as I fully expected to hear the verdict that I was not fit for the voyage, and I was resolved to go at all hazards.' See *Autobiography* pp. 79–80.

28th   Waked in the morning with an eight knot per hour wind, & soon became sick & remained so during the whole day. — My thoughts most unpleasantly occupied with the flogging of several men for offences brought on by the indulgence granted them on Christmas day. — I am doubtful whether this makes their crime [of] drunkedness & consequent insolence more or less excusable.

29th   At noon we were 380 miles from Plymouth the remaining distance to Madeira being 800 miles. — We are in the Bay of Biscay & there is a good deal of swell on the sea. — I have felt a good deal [of] nausea several times in the day. — There is one great difference between my former sea sickness & the present; absence of giddiness: using my eyes is not unpleasant: indeed it is rather amusing whilst lying in my hammock to watch the moon or stars performing their small revolutions in their new apparent orbits. — I will now give all the dear bought experience I have gained about sea-sickness. — In first place the misery is excessive & far exceeds what a person would suppose who had never been at sea more than a few days. — I found the only relief to be in a horizontal position: but that it must never be forgotten the more you combat with the enemy the sooner will he yield. I found in the|41| only thing my stomach would bear was biscuit & raisins: but of this as I became more exhausted I soon grew tired & then the sovereign remedy is Sago, with wine & spice & made very hot. — But the only sure thing is lying down, & if in a hammock so much the better. —

The evenings already are perceptibly longer & weather much milder.

30th At noon Lat. 43, South of Cape Finisterre & across the famous Bay of Biscay: wretchedly out of spirits & very sick. — I often said before starting, that I had no doubt I should frequently repent of the whole undertaking, little did I think with what fervour I should do so. — I can scarcely conceive any more miserable state, than when such dark & gloomy thoughts are haunting the mind as have to day pursued me. —

I staggered for a few minutes on deck & was much struck by the appearance of the sea. — The deep water differs as much from that near shore, as an inland lake does from a little pool. — It is not only the darkness of the blue, but the brilliancy of its tint when contrasted with the white curling tip that gives such a novel beauty to the scene. — I have seen paintings that give a faithful idea of it. — |43|

31st In the morning very uncomfortable; got up about noon & enjoyed some few moments of comparative ease. — A shoal of porpoises dashing round the vessel & a stormy petrel skimming over the waves were the first objects of interest I have seen. — I spent a very pleasant afternoon lying on the sofa, either talking to the Captain or reading Humboldts glowing·accounts of tropical scenery. — Nothing could be better adapted for cheering the heart of a sea-sick man.

January 1st The new year to my jaundiced senses bore a most gloomy appearance. — In the morning almost a calm, but a long swell on the sea. —in the evening it blew a stiff breeze against us. — This & three following days were ones of great & unceasing suffering. —

Monday 2nd Heavy weather. — I very nearly fainted from exhaustion.

3rd We looked for the eight stones & passed over the spot where they are laid down in the charts. — Perhaps their origin might have been Volcanic & have since disappeared.

4th We heaved to during the night & at day break saw Porto Santo, in few hours we passed Madeira, leaving it on our West. — As the anchorage there is bad & the landing difficult, it was not thought worth while to beat dead to Windward in order to reach it. — |45|accordingly we steered for Teneriffe. — I was so sick that I could not get up even to see Madeira when within 12 miles—in the evening a little better but much exhausted. —

5th Passed this morning within a few miles of the Piton rock: the most Southern of the Salvages: it is a wild abrupt rock & uninhabited. — At noon we were 100 miles from Teneriffe. — The day has been beautiful & I am so much better that I am able to enjoy it; the air is very mild & warm: something like a spring day in England, but here the sky is much brighter & atmosphere far more clear. — There was a very long gradual swell on the sea, like what is seen on the Pacific: The ocean lost its flat appearance & looked more like an undulating plain. —

6th After heaving to during the night we came in sight of Teneriffe at day break, bearing SW about 12 miles off. — We are now a few miles tacking with a light wind to Santa Cruz. — Which at this distance looks a small town, built of white houses & lying very flat. — Point Naga, which we are doubling, is a rugged uninhabited Mass of lofty rock|47| with a most remarkably bold & varied outline. — In drawing it you could not make a line straight. — Every thing has a beautiful appearance: the colours are so rich & soft. — The peak or sugar loaf has just shown itself above the clouds. — It towers in the sky twice as high as I should have dreamed of looking for it. — A dense bank of clouds entirely separates the snowy top from its rugged base. — It is now about 11 oclock and I must have another gaze at this long wished for object of my ambition. — Oh misery, misery—we were just preparing to drop our anchor within ½ a mile of Santa Cruz when a boat came alongside bringing with it our death-warrant. — The consul declared we must perform a rigorous quarantine of twelve days. — Those who have never experienced it can scarcely conceive what a gloom it cast on every one: Matters were soon decided by the Captain ordering all sail to be set & make a course for the Cape Verd Islands. —[1] And we have left perhaps one of the most interesting places in the world, just at the moment when we were near enough for every object to create, without satisfying, our utmost curiosity. — The abrupt vallies which divided|49| in parallel rows the brown & desolate hills were spotted with patches of a light green vegetation & gave the scenery to me a very novel appearance. — I suppose however that Volcanic islands under the same zone have much the same character. — On deck to day the view was compared as very like to other places, especially to Trinidad in West Indies. — Santa Cruz is generally accused of being ugly & uninteresting, it struck me as much the contrary. The gaudy coloured houses of white yellow & red; the oriental-looking Churches & the low dark batteries, with the bright Spanish flag waving over them were all most picturesque. — The small trading vessels with their raking masts & the magnificent back ground of Volcanic rock would together have made a most beautiful picture. — But it is past & tomorrow morning we shall probably only see the grey outline of the surrounding hills. — We are however as yet only a few miles from the town. —it is now about 10 oclock & we have been becalmed for several hours. — The night does its best to smooth our sorrow—the air is still & deliciously warm—the only sounds are the waves rippling on the stern & the sails idly flapping round the masts. — |51|Already can I understand Humboldts enthusiasm about the tropical nights, the sky is so clear & lofty, & stars innumerable shine so bright that like little moons they cast their glitter on the waves.

[1] FitzRoy wrote: 'This was a great disappointment to Mr Darwin, who had cherished a hope of visiting the Peak. To see it—to anchor and be on the point of landing, yet be obliged to turn away without the slightest prospect of beholding Teneriffe again—was indeed to him a real calamity.' See *Narrative* 2: 49.

7*th*   We were beating about during the night with a light baffling wind & in the morning a most glorious view broke upon us. — The sun was rising behind the grand Canary & defined with the clearest outline its rugged form. — Teneriffe, grey as yet from the morning mist, lay to the West: some clouds having floated past, the snowy peak was soon in all its grandeur. As the sun rose it illumined this massive pyramid, parts of which either stood relieved against the blue sky or were veiled by the white fleecy clouds: all rendered the scene most beautiful & varied. — Such moments can & do repay the tedious suffering of sickness. — We stood on a tack in direction of Santa Cruz; but were soon becalmed before reaching it. — The day has been one of great interest to me: every body in the ship was in activity, some shooting, others fishing, all amused. — No one could withstand such delightful weather, |53| nothing reminded one that there are such extremes as hot or cold. — During the day we frequently saw the Cone, but the rest of the mountain even to the waters edge was hidden. —it is then that its extreme height most strikes one. — Some old paintings, where you see Jupiter & other gods quietly conversing on a rock amongst the clouds do not give a very exaggerated idea of the Peak of Teneriffe. —

A fine breeze is now blowing us from its coast: one has read so many accounts of this island, that it is like parting from a friend; a different feeling from what I shall experience when viewing the Andes. —

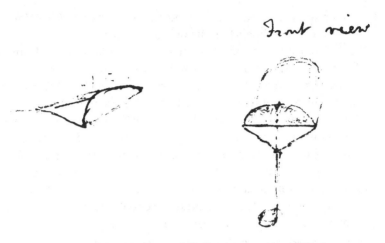

CD's sketch of his plankton net (MS diary, opposite p. 55).

8th & Monday 9th   These two days have passed quietly reading. — there was nothing to remind you that you were not sailing in the English Channel. —

10th   We crossed the Tropic this morning, if our route did not extend further, Neptune would here celebrate the aweful ceremonies of the Equator. — The weather is beautiful, & very little hotter than the middle of our summer: we have all put on our light clothes; what a contrast one fortnight has brought about as compared to the miserable wet weather of Plymouth. — |55| There was a glorious sunset this evening & is now followed by an equally fine moonlight night. — I do not think I ever before saw the sun set in a clear horizon. I certainly never remarked the marvellous rapidity with which the disk after having touched the ocean dips behind it. — I proved to day the utility of a contrivance which will afford me many hours of amusement & work. — it is a bag four feet deep, made of bunting, & attached to [a] semicircular bow this by lines is kept upright, & dragged behind the vessel. — this evening it brought up a mass of small animals, & tomorrow I look forward to a greater harvest. — [1]

[1] This appears to be only the second recorded use of a plankton net, the first being that of J. V. Thompson a few years earlier, of which CD may have learnt from Professor Grant in Edinburgh. It is evidently based on the oyster-trawl recommended to him in a letter from John Coldstream dated 13 September 1831 (see *Correspondence* 1: 151–3), which, however, was designed with a bar to scrape along the bottom rather than collecting at the surface in open water.

11th   I am quite tired having worked all day at the produce of my net. — The number of animals that the net collects is very great & fully explains the manner so many animals of a large size live so far from land. — Many of these creatures so low in the scale of nature are most exquisite in their forms & rich colours. — It creates a feeling of wonder that so much beauty should be apparently created for such little purpose. — The weather is beautiful & the blueness of the sky when contrasted with white clouds is certainly striking. —|57| Again did I admire the rapid course of the setting sun. — It did not at first occur to me that it was owing to the change of Latitude: I forgot that the same vertical motion of the sun which causes the short twilight at the Equator, must necessarily hasten its disappearance beneath the horizon. — The mean Temp from 12 observations for the 10th was 73½. — [1]

[1] In CD's zoological notes (CUL DAR 30.1), a description of a *Medusa* caught this day is followed by: 'Caught a Portugeese Man of War, Physalia. — Getting some of the slime on my finger from the filaments it gave considerable pain, & by accident pulling my finger into my mouth I experienced the sensation that biting the root of the Arum produces. —'

12th, 13th   These have been two quiet uninteresting days: my time since the making of the net has been fully occupied with collecting & observing the numerous small animals in the sea. — I find sea-life so far from unpleasant, that I am become quite indifferent whether we arrive a week sooner or later at any port. — I cannot help much regretting we were unable to stay at Teneriffe: St Jago is so miserable a place that my first landing in a Tropical country will not make that lasting impression of beauty which so many have described. —

14th, 15th   These like the last two days have rapidly glided past with nothing to mark their transit. — The weather has been light & to sailors very annoying: all the 15th we were tacking about the NW end of St Jago. —making so little way from the effects of a strong current, that after some hours we scarcely got on a mile. — Some few birds have been hovering about the vessel|59| & a large gay coloured cricket found an insecure resting place within the reach of my fly-nippers. — He must at the least have flown 370 miles from the coast of Africa.

Monday 16th   At about 11 oclock we neared the Western coast of St Jago[1] & by about three we anchored in the bay of Porto Praya. — St Jago viewed from the sea is even much more desolate than the land about Santa Cruz. — The Volcanic fire of past ages & the scorching heat of a tropical sun have in most places rendered the soil sterile & unfit for vegetation. — The country rises in successive steps of table land, interspersed by some truncate conical hills, & the horizon is bounded by an irregular chain of more lofty & bolder hills. — The scene when viewed through the peculiar atmosphere of the tropics was one of great interest: if indeed a person fresh from sea & walking for the first time in a grove of Cocoa-nut trees, can be a judge of anything but his own happiness. — At three oclock I went with a party to announce our arrival to the "Governador". — After having found out the house, which certainly is not suited to the grandeur of his title we were ushered into a room where the great man most courteously received us. — After having made out our story in a very|61| ludicrous mixture of Portugeese, English & French, we retreated under a shower of bows. — We then called on the American Consul who likewise acts for the English. — The Portugeese might with great advantage have instilled a little of his well-bred politesse into this quarter. — I was surprised at the houses: the rooms are large & airy, but with uncommonly little furniture, & that little in vile taste. — We then strolled about the town, & feasted upon oranges: which I believe are now selling a hundred per shilling. I likewise tasted a Banana: but did not like it, being maukish & sweet with little flavor. — The town is a miserable place, consisting of a square & some broard streets, if indeed they deserve so respectable a name. — In the middle of these "Ruas" are lying together goats, pigs & black & brown children: some of whom boast of a shirt, but quite as many not: these latter look less like human being than I could have fancied any degradation could have produced. — There are a good many black soldiers, it would be difficult I should think to pick out a less efficient body of men. — Many of them only possess for arms a wooden staff. — Before returning to our boat, we walked across the town & came to a deep valley. — Here I first saw the glory of tropical|63| vegetation. Tamarinds, Bananas & Palms were flourishing at my feet. — I expected a good deal, for I had read Humboldts descriptions[2] & I was afraid of disappointments: how utterly vain such fear is, none can tell but those who have experienced what I to day have. — It is not only the gracefulness of their forms or the novel richness of their colours, it is the numberless & confusing associations that rush together on the mind, & produce the effect. — I returned to the shore, treading on Volcanic rocks, hearing the notes of unknown birds, & seeing new insects fluttering about still newer flowers. — It has been for me a glorious day, like giving to a blind man eyes. — he is overwhelmed with what he sees & cannot justly comprehend it. — Such are my feelings, & such may they remain. —

[1] The *Journal of Researches* pp. 1–7 opens with an account of St Jago based on the entries for
16 January to 6 February.
[2] In his *Autobiography* pp. 67–8, CD wrote: 'During my last year at Cambridge I read with
care and profound interest Humboldt's *Personal Narrative*. This work and Sir J. Herschel's
Introduction to the Study of Natural Philosophy stirred up in me a burning zeal to add even
the most humble contribution to the noble structure of Natural Science. No one or a dozen
other books influenced me nearly so much as these two.' CD had with him on the *Beagle*
a copy of the 1822 English translation of Vols. 1 and 2 of the *Personal Narrative*, 3rd edition,
combined in one volume, inscribed 'J. S. Henslow to his friend C. Darwin on his
departure from England upon a voyage around the World 21 Septr 1831'.

17th   Immediately after breakfast I went with the Captain to Quail
Island. — This is a miserable desolate spot, less than a mile in circumfer-
ence. It is intended to fix here the observatory & tents; & will of course
be a sort of head quarters to us. — Uninviting as its first appearance was,
I do not think the impression this day has made will ever leave me. —
The first examining of Volcanic rocks must to a Geologist be a memorable epoch, & little less so to the naturalist is the first burst of admiration
at seeing Corals growing on their|65| native rock. — Often whilst at
Edinburgh, have I gazed at the little pools of water left by the tide: &
from the minute corals of our own shore pictured to myself those of
larger growth: little did I think how exquisite their beauty is & still less
did I expect my hopes of seeing them would ever be realized. — And in
what a manner has it come to pass, never in the wildest castles in the air
did I imagine so good a plan; it was beyond the bounds of the little
reason that such day-dreams require. — After having selected a series
of geolog specimens & collected numerous animals from the sea, — I sat
myself down to a luncheon of ripe tamarinds & biscuit; the day was hot,
but not much more so than the summers of England & the sun tried to
make cheerful the dark rocks of St Jago. — The atmosphere was a
curious mixture of haziness & clearness. — distant objects were blended
together: but every angle & streak of colour was brightly visible on the
nearer rocks.[1]

Let those who have seen the Andes be discontented with the scenery
of St Jago. I think its unusually sterile character gives it a grandeur
which more vegetation might have spoiled. — I suppose the view is
truly African, especially to our left, where some round sandy hills were
only broken|67| by a few stunted Palms. — I returned to the ship heavily
laden with my rich harvest, & have all evening been busily employed in
examining its produce. —

[1] In his *Autobiography* p. 81, CD wrote: 'The geology of St Jago is very striking yet simple:
a stream of lava formerly flowed over the bed of the sea, formed of triturated recent shells

and corals, which it has baked into a hard white rock. Since then the whole island has been upheaved. But the line of white rock revealed to me a new and important fact, namely that there had been afterwards subsidence around the craters, which had since been in action, and had poured forth lava. It then first dawned on me that I might perhaps write a book on the geology of the various countries visited, and this made me thrill with delight. This was a memorable hour to me, and how distinctly I can call to mind the low cliff of lava beneath which I rested, with the sun glaring hot, a few strange desert plants growing near, and with living corals in the tidal pools at my feet. Later in the voyage Fitz-Roy asked to read some of my Journal, and declared it would be worth publishing; so here was a second book in prospect!'

The absence of a reference in the Diary to the line of white rock may be because at this stage CD had far from satisfied himself as to its geological origin. In one of the first entries in his notebooks for 1832 (Down House Notebook 1.4; and see *CD and the Voyage* pp. 155–7), the white layer is described as 'White sand made up of shells upon which rests carious rock, then prismatic feldspar. Between these & former ones, hard white rock with yellow spots. Heap of white balls beneath white sand. — . . . Same process now going on shore: living Iron found in it . . . I should think this coast one of short duration. — Sand white from decomposition of Feldspar(?). In places every rock is covered with pisiform concretions. In places impossible to tell whether it is Breccia of modern or older days. — Going few hundred yards more South, lower beds of white sand become filled with large boulders of lower rocks: beneath this comes a line of another stratum: more soily & contains large and more numerous shells. A regular bed of oysters remains attached to the rocks on which they grew . . . The lower carious part of Feldspar perhaps owing to the expansive power of steam. — Although in parts this old sea coast is 30 or 40 feet above present level of ocean, yet in others the present breccia again covers it, owing to its having sunk again most likely, as it agrees with that which has been raised. — What confusion for geologists. —' A few days later he reminds himself to perform an experiment on 'Shells action under Blowpipe??', as a test of the suggested action of the burning lava on the subaqueous layers of shells.

The watercolour of St Jago painted by Conrad Martens on 9 June 1833 (CM No. 30, *Beagle Record*) shows clearly the line of white rock.

18th   I have been excessively busy all day & have hardly time to write my days log: the little time I was out of my cabin, I spent geologising on Quail Island. — The day has been very hot: & I have feasted on Tamarinds & a profusion of oranges. — for dinner I had Barrow Cooter for fish & sweet potatoes for vegetables: quite tropical and correct. —

19th   I took a walk with Musters. I went to the West along the coast, & then returned by a more inland path. — My imagination never pictured so utterly barren a place as this is. — it is not the absence of vegetation solely that produces this effect: every thing adds to the idea of solitude: nothing meets the eye but plains strewed over with black & burnt rocks rising one above the other: And yet there was a grandeur in such scenery & to me the unspeakable pleasure of walking under a tropical sun on a wild & desert island. — It is quite glorious the way my collections are increasing. I am even already troubled with the vain fear that there will be nobody in England who will have the courage to

examine some of the less known branches. – I have been|69| so
incessantly engaged with objects full of new & vivid interest: that the
three days appear of an indefinite length. – I look back to the 16th as a
period long gone by. –

20th I took a long walk with Maccormick into the interior. – Although in
such a country the objects of interest are few, yet perhaps from this very
reason, each individual one strikes the imagination the more. – We
followed one of the broard water courses, which serves as a road for the
country people, by the greatest good luck it lead us to the celebrated
Baobob trees. – I had forgotten its existence, but the sight immediately
recalled a description of it which I had formerly read. – This enormous
tree measured 36^f.2 inches at the height of 2..8 from ground. Its altitude
in no way corresponds with its great thickness. – I should not suppose
it was 30 feet high. – This tree is supposed to be one of longest lived that
exists. – Adanson supposed that some reached to the age of 6000
years. – This one bears on its bark the signs of its notoriety – it is as
completely covered with initials & dates as any one in Kensington
Gardens. – We passed on with nothing except the novelty of the scene
that could give us any enjoyment: the glowing sun above our heads was
the only thing that reminded us we were in the tropics. – Nature is
here|71| sterile, nothing breaks the absolute stillness, nothing is seen to
move: we may indeed except a gay coloured kingfisher & its prey, the
less gaudy grasshopper. At midday, we seated ourselves under the
shade of a Tamarind & measured out our small portion of water. – The
bluish green tint of its colour & the extreme lightness of its pinnate
foliage gives to this Acacia a most pleasing appearance. – We then left
the valley & crossed over to Red hill, which is 1300 [ft.] high &
composed of more recent Volcanic rocks. – On [the] road, two black
men brought us some goats milk, to pay them we put some copper
money on our open hands: they took a farthing, & when we gave them
a penny, we hardly could prevent them pouring down a quart of milk
into our very throats. – These merry simple hearted men left us in roars
of laughter. – I never saw anything more intelligent than the Negros,
especially the Negro or Mulatto children. – they all *immediately* perceived & are astonished at the percussion guns. – they examine every
thing with the liveliest attention, & if you let them the children
chattering away, will pull everything out of your pockets to examine
it. – My silver pencil case was pulled out & much speculated upon. –
When catching a stinging ichneumon, the children pinched themsel-
ves|73| in order to show that the insect would pain me. – We scaled the

top of the Red hill & from it had a good view of the most desolate countrys in the world. — Our road home, near to Praya, lay through a more fertile valley & few will imagine how refreshing is the sight of the dark green of the Palm. — We returned to the vessel very thirsty & covered with dust, but not much fatigued, neither did I suffer much from the heat of the sun. —

21st All day I have been working at yesterdays produce. — Geology is at present my chief pursuit[1] & this island gives full scope for its enjoyment. — There is something in the comparative nearness of time, which is very satisfactory whilst viewing Volcanic rocks. —

There have been two bright meteors passing from East to West.

[1] In his *Autobiography* p. 77, CD wrote: 'The investigation of the geology of all the places visited was far more important [than its natural history], as reasoning here comes into play. On first examining a new district nothing can appear more hopeless than the chaos of rocks; but by recording the stratification and nature of the rocks and fossils at many points, always reasoning and predicting what will be found elsewhere, light soon begins to dawn on the district, and the structure of the whole becomes more or less intelligible. I had brought with me the first volume of Lyell's *Principles of Geology*, which I studied attentively; and this book was of the highest service to me in many ways. The very first place which I examined, namely St. Jago in the Cape Verde islands, showed me clearly the wonderful superiority of Lyell's manner of treating geology, compared with that of any other author, whose works I had with me or ever afterwards read.' Some pages later (*Autobiography* p. 101) he says: 'The science of Geology is enormously indebted to Lyell — more so, as I believe, than to any other man who ever lived. When I was starting on the voyage of the *Beagle*, the sagacious Henslow, who, like all other geologists believed at that time in successive cataclysms, advised me to get and study the first volume of the *Principles*, which had then just been published, but on no account to accept the views therein advocated. How differently would any one now speak of the *Principles*!'

Writing to W. D. Fox in August 1835 he said: 'I am glad to hear you have some thoughts of beginning geology. — I hope you will, there is so much larger a field for thought, than in the other branches of Nat: History. — I am become a zealous disciple of Mr Lyells views, as known in his admirable book. — Geologizing in S. America, I am tempted to carry parts to a greater extent, even than he does. Geology is a capital science to begin, as it requires nothing but a little reading, thinking & hammering. —' See *Correspondence* 1: 460.

The volumes of Lyell's *Principles* that CD had on the *Beagle* were inscribed as follows: Vol. 1 (published 1830) 'Given me by Capt. FR C. Darwin'; Vol. 2 (1832) 'Charles Darwin M: Video. Novem^r 1832'; Vol. 3 (1833) 'C. Darwin'. The third volume had reached him some time before he wrote to Henslow from Valparaiso on 24 July 1834. See *Correspondence* 1: 397–403.

22nd This day has passed (& it is a subject for wonder) very much like any other Sunday out of the Magic line of the Tropics. — In the evening I strolled about Quail Island & caught myself thinking of England & its politicks. — it is my belief that the word reform has not passed the lips of any man on board since we saw Madeira. — So absorbing is the interest of a new country. — |75|

Monday 23rd Walked with Maccormick to Flag Staff Hill. — We passed
over an extended plain of table land. — There was scarcely one green
leaf on the whole tract, yet large flocks of goats, together with some
cattle, contrive to live. — It rains but very seldom in this country & when
it does a mass of vegetation springs up; this soon drys up & withers: &
upon such miserable sort of hay the animals exist: at present it has not
rained for a year, & I suppose will not till the proper time next year, viz.
November & October. — At these periods the island is very unhealthy:
one ship some years past lost six of its junior officers. — A little to the
North of the hill, we found a very curious ravine, not much above 30
yards across. —about 200 feet high. We with some difficulty found one
single path at the very end, where we descended. — In this wild dell we
found the building places of many birds. — Hawks & Ravens & the
beautiful Tropic bird were soaring about us: a large wild cat bounded
across & reached its den before Maccormick could shoot it. — The place
seemed formed for wild animals: large blocks of rocks, entwined with
succulent creepers & the ground strewed over with bleached bones of
Goats would have been a fine habitation for a Tiger.|77|

24th After our one oclock dinner, Wickham, the Captain & myself
walked to the famous Baobob tree & measured it more accurately. —[1]
Cap FitzRoy first took an angle by a pocket sextant & afterward climbed
the tree & let down a string, both ways gave the same result, viz 45 feet
in height. — Its circumference measured 2 feet from the grounds (there
being no projecting roots) gave 35. — Its form is oval, & its greatest
visible diameter was 13 feet. — So that in an accurate drawing its height
would be 3·4 of its breadth. Cap FitzRoy made a sketch which gave a
good idea of its proportion, yet in this the height was only about 2·4 of
[its] breadth. Proving, what one so often observes, that a faithful
delineation of Nature does not give an accurate idea of it. — We
returned home, after our merry & pleasant walk, just as it was dark. —

A very pretty schooner came in this morning: it is strongly suspected
that she is a slaver in disguise, she says she is a general trader to the
coast of Africa. — The Captain means to overhaul her in the morning &
make out what she is. — I suppose every thing is well concealed, else
she would not have come into a harbour where a pennant was fly-
ing.|79|

[1] FitzRoy wrote: 'In a valley near the town is a very remarkable tree of the Baobab kind,
supposed to be more than a thousand years old; but I am not aware of the grounds upon
which this assertion is made.' See *Narrative* 2: 54.

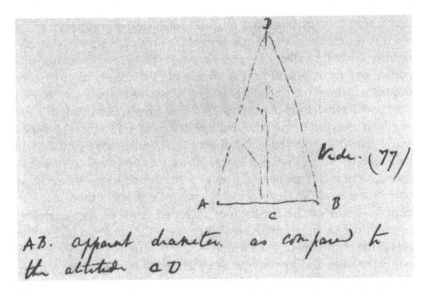

FitzRoy's sketch of the baobob tree (MS diary, opposite p. 69).

25th Collected some marine animals at Quail Island & spent most part of the day in examining them. —

26th Rowlett, Bynoe & myself started early in the morning on a riding expedition to Ribera Grande. — We went to Praya to get our horses & there had our breakfast: The greatest shopkeeper in the place was our host: He is an American & has married a Spanish woman & seems one of the most influential people in the place. — After we had finished our Coffee in his large & airy rooms, we mounted our ponys. — The road to Ribera for the first six miles is totally uninteresting & till we arrived at the valley of St Martin the country presented its usual dull brown appearance: here our eyes were refreshed by the varied & beautiful forms of the tropical trees. The valley owes its fertility to a small stream & following its course Papaw trees, Bananas & Sugar cane flourished. — I here got a rich harvest of flowers, & still richer one of fresh water shells. — After having watered our active & sure footed little horses, we again commenced climbing. — In the course of an hour, we arrived at Ribera & were astonished at the sight of a large ruined fort & a Cathedral: — Ribera Grande lies|81| 9 miles to the West of Praya & was till within later years the principal place in the island. — The filling up of its harbour has been the cause of the overthrow of its grandeur. — It now presents a melancholy, but very picturesque appearance. The town is situated at the foot of a high black precipice, through which a narrow & abrupt valley has cut its way. — The vegetation in this little corner was most beautiful; it is impossible sufficiently to admire the exquisite form of the Cocoa-nut tree, & when, as in this case, they are seen waving their lofty heads above the dark green of an Orange Grove, one feels convinced that all the praise bestowed on tropical scenery is just. — Having procured a black padre for a guide, & a Spaniard who had served in the Peninsular war for our interpreter, we visited a collection of buildings of which an antient Church forms the principal part. — It is here the Governors & Captain Generals of the Islands are buried, — some of the tombstones recorded dates of the fourteenth century; the heraldic ornaments were the only things in this retired place that reminded one of Europe. — This Church or Chapel, formed one of the sides of a Quadrangle, in which Bananas were|83| growing. — On two of the others were the buildings in which the people connected with the institution lived. — On the fourth was a hospital, containing about a dozen of miserable looking inmates. — In one of the rooms, to our surprise, we were shown a collection of tolerable paintings. — the colouring & drawing of the drapery was excellent. — We then, accompanied as before, returned to the "Venda" & eat our dinner. — to see which operation a concourse of black men, women & children had collected. We luckily had brought some cold meat, as the only things the men helped us to were wine & crumbs made from Indian corn. — Certainly the whole scene was most amusing, our companions the blacks were extremely merry, every thing we said or did was followed by their hearty laughter. Our Spanish interpreter now left us, before mounting his donkey, he loaded a formidable pistol with slugs. — quietly remarking "this very good for black man". — Before leaving Ribera, we visited the Cathedral. — It is a building of some size, but does not appear from the absence of plate to be so rich|85| as the smaller Church. — It boasts however of a small organ, which sent forth most singularly inharmonious notes. — We presented our friend the black priest (which the Spaniard with much candour said he thought made no difference) with a few shillings & wishing him good morning returned as fast as the Ponys would carry us to Porto Praya.

27th Employed in working at yesterdays produce.

28th Collected a great number of curious & beautiful animals from the little pools left by the tide. The colours of the sponges & corallines are extremely vivid & it is curious how all animated nature becomes more gaudy as it approaches the hotter countrys. — Birds, fishes, plants, shells are familiar to every one. —but the colours in these marine animals will rival in brilliancy those of the higher classes.[1]

[1] In *Journal of Researches* pp. 6–7, CD wrote: 'I was much interested, on several occasions, by watching the habits of an Octopus or cuttle-fish. Although common in the pools of water left by the retiring tide, these animals were not easily caught. By means of their long arms and suckers, they could drag their bodies into very narrow crevices; and when thus fixed, it required great force to remove them. At other times they darted tail first, with the rapidity of an arrow, from one side of the pool to the other, at the same time discolouring the water with a dark chestnut-brown ink. These animals also escape detection by a very extraordinary, chameleon-like, power of changing their colour. They appear to vary the tints, according to the nature of the ground over which they pass: when in deep water, their general shade was brownish purple, but when placed on the land, or in shallow water, this dark tint changed into one of a yellowish green. The colour, examined more carefully, was a French gray, with numerous minute spots of bright yellow: the former of these varied in intensity; the latter entirely disappeared and appeared again by turns. These changes were effected in such a manner, that clouds, varying in tint between a hyacinth red and a chestnut brown, were continually passing over the body.'

29th Divine service was performed on Board. —it is the first time I have seen it: it is a striking scene & the extreme attention of the men renders it much more imposing than I had expected. Every thing on board on Sunday is most delightfully clean —the lower decks would put to shame many gentlemens houses. — |87|

30th Walked to the coast West of Quail Island with King, & collected numerous marine animals, —all of extreme interest. — I am frequently in the position of the ass between two bundles of hay. —so many beautiful animals do I generally bring home with me. — In the morning a few drops of rain fell.

31st This morning the view was very fine. —the air being singularly clear. —& the mountains were projected against dark blue or black clouds. — Judging from their appearances I should have thought the air was saturated with moisture. — The Hygrometer proved the contrary, the diff: between Temp & Dew point being 29·6: this is nearly double what it has been any other morning: on the 20th & 21st it was 15·5. — The dew formed at 42·2 & atmosphere was 71·8. — On the previous morning the diff was only 8·8: & dew point 64·4. — This uncommon dryness of the air was accompanied by continued flashes of lightning —consequent I suppose on the great change from unusual dampness to such extreme dryness. —

The whole of this day I have been working very hard with microscope at yesterdays produce.

February 1ˢᵗ   Busy with my usual employment, viz marine animals. |89|

2ⁿᵈ   We (*Note opposite:* Rowlett, Bynoe & myself) started by day-break on a riding excursion to St Domingo. — For the first 5 miles the road passed over one of the numerous plains of table-land: The country here has not quite so sterile an appearance owing to the stunted Acacia trees which are sparing[ly] scattered over its faces. — These trees are curiously bent by the prevailing wind & I should think formed an excellent average wind vane for the Island. Their direction is exactly NE & SW (Magnetic), & by its force their tops are often bent into an exact right angle. — At the foot of a pyramidal hill of scoriæ I tied up my pony to examine the rocks. — The road makes so little impression on the barren soil, that we here missed our track & took that to Fuentes. — This we did not find out till we arrived there, & we were afterwards very glad of our mistake. — Fuentes is a pretty village with a small stream & everything appears to prosper well. — Excepting indeed that which ought to do so most—its inhabitants. — The black children, perfectly naked & looking very wretched, were carrying bundles of fire wood half as big as their own bodies. — The men & women badly clothed looked much over-worked. |91|

We gladly left Fuentes & passed along a wild narrow road to St. Domingo, which lay about a league to the East. — Before we arrived at Fuentes, we saw a large flock of the wild Guinea fowl: they were extremely wary & would not allow us to approach near. — Their manner of avoiding us was like that of Partridges on a rainy day in September, no sooner do they alight than with their heads cocked up they run away & then if approached fly again. — On approaching St Domingo a turn in the road first showed us the background of wild peaked rocks. —their forms are most fantastic; one part looks like a castle wall, others like towers & pyramids. — Every thing betrays marks of extreme violence: & which is better shown by the rocks being in horizontal beds. — As the road approaches the sides of the hill or precipice, the town & valley of Sᵗ Domingo are seen. — I can imagine no contrast more striking than that of its bright vegetation against the black precipices that surround it. — A clear brook gives a luxuriance to the spot that no other part of the Island would lead you to expect. — Nothing has surprised me so much as the very dark green of the oranges; |93| some tropical forms can easily be imagined either from hot-house specimens or from drawings, —such as Bananas. —but I do not think any adequate idea of the beauty of Oranges or Cocoa Nut trees can be formed without actually seeing them on their own proper soil —.

We had an introduction to a most hospitable Portugeese, who treated us most kindly & feasted us with a most substantial dinner of meat cooked with various sorts of herbs & spices, & Orange Tart. — This man is a principal owner of the plantation & apparently lives in great comfort: his house is simple, but he has perhaps the Utopian felicity of growing every thing he wants on his own ground. — We were told there was a lake about 2 miles from St. Domingo: after dinner we started to see, & followed a path by the side of a brook. — On each side were flourishing Bananas, Sugar Cane, Coffee, Guavas, Cocoa Nuts, & numberless wild flowers. — None can conceive such delight but those who fond of Natural history have seen such scenes. — We at last arrived at the lake: one certainly on the smallest scale, for it was not 20 feet across. —by such great names in this dry country do they|95| designate a small puddle of fresh water. — After again & again admiring this beautiful & retired valley, we returned to our ponys, & wishing our most hospitable entertainer buenas dias, we took the direct road for Praya. — The day was a grand feast day & the village very full of people—a little distance out of it we overtook about 20 young black girls. —dressed in most excellent taste. —their black skins & snow white linen were adorned with a gay coloured turbans & large shawls. — When we approached them they suddenly all turned round & covered the path with their shawls. —they sung with great energy a wild song: beating time with their hands upon the legs. — We threw them some Vintem, which were received with screams of laughter, & we left them redoubling the noise of their song. — We arrived after it was dark at Praya & with our tired ponys had some difficulty in picking out our way. —

3rd A blowing day: I observe it feels quite cool when thermometer is under 75 if at the time there is a fresh breeze. — Walked along Eastern coast & found some beautiful corals. —|97|

4th Walked with Musters to a high hill N by E of Praya. — On the road saw a large flock of guinea fowl, & their usual companion & destroyer the wild cat. — These animals appear to be very common in the island, so many have been seen since we were here. —

5th This day or rather the 6th was originally fixed for sailing but the Captain is so much engaged with experiments on Magnetism, that the time is put off till tomorrow. — I was engaged with my usual occupation of collecting marine animals in the middle of the day & examining them in the evening. — Daily do I feel myself very hardly used, when on returning to the ship I find it growing dark soon after six oclock. — The days are exactly the same as in a dry hot summer in England, but it is very surprising the sun choosing to set before its accustomed time about 8 oclock. —

6ᵗʰ Went in a boat dredging for Corals; but did not succeed in obtaining any. Tomorrow we certainly sail. And I am glad of it, for I am becoming rather impatient to see tropical Vegetation in greater luxuriance than it can be seen here. — Upon the whole the time has been for me of a proper length & has flown away very pleasantly. — It is|99| now three weeks, & what may appear very absurd it seems to me of less duration than one of its parts. — During the first week every object was new & full of uncommon interest & as Humboldt remarks the vividness of an impression gives it the effect of duration. — in consequence of this, those few days appeared to me a much longer interval than the whole three weeks does now. —

8ᵗʰ The dates for the few last days are wrong, for we certainly sailed on the 8th after noon. — Again I admired the varied outline of the hills round Praya; the memory of which will never be effaced from my mind.

9ᵗʰ Beautiful & calm day, but I could not enjoy it, as to my great indignation I felt squeamish & uncomfortable.

10ᵗʰ In the morning a vessel was in sight.[1] We chased her all day & have just come up to her this evening. — She is a Packet bound for Rio & in the morning I intend sending a letter to England[2] via Rio de Janeiro, as possibly it may sooner arrive there by this than any other conveyance. I have felt a little sea-sickness to day: which is too bad, as objects of interest are continually occurring. — There were plenty of flying fish round the vessel but no large ones. Everybody is much pleased with the Beagles sailing, it certainly is something|101| extraordinary so very easily to beat a packet, which is built as a man of war & without her guns. — It is rather unaccountable the extreme interest that is universally felt at speaking a ship in "blue water". — We expected no news & we received none yet I believe a great disappointment to every person in the ship if we had not boarded her. — To our shame be it spoken, we entirely forgot the Cholera Morbus, & although ourselves having smarted from the quarantine at Teneriffe, yet we made no enquiries about our friends in England. —

[1] FitzRoy wrote: 'On the 10th we spoke the Lyra packet, going from England to Rio de Janeiro, and received a box from her containing six of Massey's sounding-leads, those excellent contrivances which we frequently found so useful.' See *Narrative* 2: 55.

[2] CD wrote to his father this day: 'I have a long letter, all ready written, but the conveyance by which I send this is so uncertain. — that I will not hazard it, but rather wait for the chance of meeting a homeward bound vessel. — Indeed I only take this opportunity as perhaps you might be anxious, not having sooner heard from me . . . Natural History goes on excellently & I am incessantly occupied by new & most interesting animals. — . . . I think, if I can so soon judge. — I shall be able to do some original work in Natural History. — I find there is so little known about many of the Tropical animals.' The longer letter was eventually posted on 5 March 1832 from Bahia. See *Correspondence* 1: 201–5.

11th We are rapidly gaining on our voyage to the Equator.

12th There has been a little swell on the sea to day, & I have been very uncomfortable: this has tried & quite overcome the small stock of patience that the early parts of the voyage left me. — Here I have spent three days in painful indolence, whilst animals are staring me in the face, without labels & scientific epitaphs. —

13th This has been the first day that the heat has annoyed us, & in proportion all have enjoyed the delicious coolness of the moonlight evenings: but when in bed, it is I am sure just like what one would feel if stewed in very warm melted butter. — This morning a glorious fresh trade wind is driving us along; I call it glorious because others|103| do; it is however bitter cruelty to call anything glorious that gives my stomach so much uneasiness. — Oh a ship is a true pandemonium, & the cawkers who are hammering away above my head veritable devils. —

14th To day at noon we were 150 miles from the Equator, & have experienced the weather which is so frequent in these regions. — The wind has been light & variable accompanied by small squalls & much rain. — The thermometer is night & day between 75 & 80. —air is very damp & oppressive. — The appearance of the sky is in these parts generally striking: the scene' after sunset was particularly so. Every class & form of clouds was present, & by their shadows gave to the sea a dead black colour. The sails were flapping against the mast & a long swell quietly rolled the ship. The place where the sun had set was marked by a long red streak on the horizon & higher above it by a clear yellow space, which cast a glare on that part of the ocean. — It is in such moments that one fully recollects the many miles that separates our ship from any land. —

Every body is alive with the anticipation about Neptunes appearance, & I hear of nothing but razors sharpened with a file & a lather made of paint & tar, to be used by the gentlest valet de chambre.|105|

15th, 16th Saw the rocks of St Pauls right ahead: heaved to during the night, & this morning we were a few miles distant from them. — When within 3 miles, two boats were lowered, one with Mr Stokes for surveying the island, the other with Mr Wickham & myself for geologizing & shooting. — St Pauls may be considered as the top of a submarine mountain. — It is not above 40 feet above the sea, & about ½ a mile in circumference. — Bottom could not be found within a mile of the Island, & if the depth of the Atlantic is as great as it is usually supposed, what an enormous pyramid this must be. —

We had some difficulty in landing as the long swell of the open sea broke with violence on the rocky coast. — We had seen from a distance large flocks of sea-birds soaring about, & when we were on the Island a most extraordinary scene was presented. — We were surrounded on every side by birds,[1] so unaccustomed to men that they would not move. — We knocked down with stones & my hammer the active and swift tern. — Shooting was out of the question, so we got two of the boats crew & the work of slaughter commenced. They soon collected a pile of birds, & hats full of eggs.|107|

Whilst we were so active on shore, the men in the boat were not less so. — They caught a great number of fine large fish & would have succeeded much better had not the sharks broken so many of their hooks & lines: they contrived to land three of these latter fish, & during our absence 2 large ones were caught from the ship. — We returned in great triumph with our prey, but were a good deal fatigued. — The island is only 50 miles from the Equator, & the rocks being white from the birds dung, reflected a glaring heat. — The birds were only of two sorts, Booby and Noddys, & these with a few insects were the only organized beings that inhabited this desolate spot.[2] In the evening the ceremonies for crossing the line commenced: The officer on watch reported a boat ahead. — The Captain turned "hands up, shorten sail", and we heaved to in order to converse with Mr Neptune. The Captain held a conversation with him through a speaking trumpet, the result of which was that he would in the morning pay us a visit. —

[1] FitzRoy wrote: 'When our party had effected a landing through the surf, and had a moment's leisure to look about them, they were astonished at the multitudes of birds which covered the rocks, and absolutely darkened the sky. Mr Darwin afterwards said, that till then he had never believed the stories of men knocking down birds with sticks; but there they might be kicked, before they would move out of the way. The first impulse of our invaders of this bird-covered rock, was to lay about them like schoolboys; even the geological hammer at last became a missile. "Lend me the hammer?" asked one. "No, no," replied the owner, "you'll break the handle;" but hardly had he said so, when, overcome by the novelty of the scene, and the example of those around him, away went the hammer, with all the force of his own right-arm.' See *Narrative* 2: 56.

[2]CD wrote: 'Not a single plant, not even a lichen, grows on this island; yet it is inhabited by several insects and spiders. The following list completes, I believe, the terrestrial fauna: a species of Feronia and an acarus, which must have come here as parasites on the birds; a small brown moth, belonging to a genus that feeds on feathers; a staphylinus (*Quedius*) and a woodlouse from beneath the dung; and lastly, numerous spiders, which I suppose prey on these small attendants on, and scavengers of the waterfowl. The often-repeated description of the first colonists of the coral islets in the South Sea, is not, probably, quite correct: I fear it destroys the poetry of the story to find, that these little vile insects should thus take possession before the cocoa-nut tree and other noble plants have appeared.' See *Journal of Researches* p. 10.

17th   We have crossed the Equator, & I have undergone the disagreeable operation of being shaved. About 9 oclock this morning we poor "griffins", |109| two & thirty in number, were put altogether on the lower deck. — The hatchways were battened down, so we were in the dark & very hot. — Presently four of Neptunes constables came to us, & one by one led us up on deck. — I was the first & escaped easily: I nevertheless found this watery ordeal sufficiently disagreeable. — Before coming up, the constable blindfolded me & thus lead along, buckets of water were thundered all around; I was then placed on a plank, which could be easily tilted up into a large bath of water. — They then lathered my face & mouth with pitch and paint, & scraped some of it off with a piece of roughened iron hoop. — a signal being given I was tilted head over heels into the water, where two men received me & ducked me. — at last, glad enough, I escaped. — most of the others were treated much worse, dirty mixtures being put in their mouths & rubbed on their faces. — The whole ship was a shower bath: & water was flying about in every direction: of course not one person, even the Captain, got clear of being wet through.[1] |110|

[1]FitzRoy wrote: 'The disagreeable practice alluded to has been permitted in most ships, because sanctioned by time; and though many condemn it as an absurd and dangerous piece of folly, it has also many advocates. Perhaps it is one of those amusements, of which the omission might be regretted. Its effects on the minds of those engaged in preparing for its mummeries, who enjoy it at the time, and talk of it long afterwards, cannot easily be judged of without being an eye-witness.'

18th   At last I certainly am in the Southern hemisphere, & whilst enjoying the cool air of the evening, I can gaze at the Southern Cross, Magellans cloud & the great crown of the South. — In August quietly wandering about Wales, in February in a different hemisphere; nothing ever in this life ought to surprise me. — I find I had formed a very exaggerated idea of the heat in these zones during their cooler months. — I have often grumbled at a hot summers day in England in much more earnest than I do at present. —

Crossing the line, by T. Landseer after A. Earle (*Narrative* **2**: 57).

19th This morning a vessel was in sight, but would not show her colours. — An hour before sunset Fernando was clearly visible — it appears an extraordinary place. — there is one lofty mountain that at a distance looks as if it was overhanging. — We are at present lying off & on, & as soon as the moon gets up we shall anchor in the harbor. — Just before it was dark Sullivan harpooned a large porpoise. The instrument was hurled with such force that it passed through the entire body. — In a few minutes a fine animal about five feet long was lying|111| on the deck & in a still less time a dozen knives were skinning him for supper. — The view of the group of Islands was very grand by the clear moonlight, & I felt rather disappointed when I found at day-break

20th that the hills are by no means lofty. — I have written one account of the Island in my geology and it is much too hard work to copy anything when the sun is only a few degrees from the Zenith. — I spent a most delightful day in wandering about the woods. — The whole island is one forest, & this is so thickly intertwined that it requires great exertion to crawl along. — The scenery was very beautiful, & large Magnolias & Laurels & trees covered with delicate flowers ought to have satisfied me. — But I am sure all the grandeur of the Tropics has not yet been seen by me. — We had no gaudy birds, No humming birds. No large flowers — I am glad that I have seen these islands, I shall enjoy the greater wonders all the more from having a guess what to look for. — All the trees either bearing some fruit or large flowers is perhaps one of the most striking things that meet one whilst wandering in a|112| wood in these glorious regions. — I joined the Captain in the evening & was informed that we should sail that very evening. — What decided his plans is the great difficulty in landing in the surf. —

21st We sailed at night, but have not made much way this morning. — latterly it has been a dead calm, the ships head standing the wrong way. — As long as one was motionless the extreme heat is rather enjoyable — but after any bodily or mental exertion a most helpless degree of languor comes over every faculty. During the night it is like sleeping in a warm bath. I am forced to get out & lie on the table, the hardness of which is delightful after the round soft hammock. —

22nd The wind has continued so variable that this morning we were yet in sight of Fernando Noronha. — The day has been uncomfortably hot & the evening deliciously cool. — The most serious discomfort which affects me, is the difficulty of sleeping: before going to bed it is next to impossible to keep the head from falling on the book, but the instant one is in the hammock all sleep deserts you. — |113|

San Salvador, Bahia, by T. A. Prior after A. Earle (*Narrative* **2**: 62).

23rd, *24th* & *25th* These three days have passed by quietly & without note. — On the 23rd we had scarcely got out of the "variables" which are so common in the Equatorial regions, but for the two last days we have been driving with a steady Trade wind for the continent of S America. —

Since leaving Teneriffe the sea has been so calm that it is hard to believe it the same element which tossed us about in the Bay of Biscay. This stillness is of great moment to the quantity of comfort which is attainable on ship-board, hitherto I have been surprised how enjoyable life is in this floating prison. — But the greatest & most constant drawback to this is the very long period which separates us from our return. — Excepting when in the midst of tropical scenery, my greatest share of pleasure is in anticipating a future time when I shall be able to look back on past events; & the consciousness that this prospect is so distant never fails to be painful. — To enjoy the soft & delicious evenings of the Tropic; to gaze at the bright band of Stars which stretches from Orion to the Southern Cross, & to enjoy such pleasures in quiet solitude, leaves an impression which a few years will not destroy. — |114|

26th For the first time in my life I saw the sun at noon to the North: yesterday it was very near over our heads & therefore of course we are a little to the South of it. — I am constantly surprised at not finding the heat more intense than it is; when at sea & with a gentle breeze blowing one does not even wish for colder weather. — I am sure I have frequently been more oppressed by a hot summers day in England.

27th Quietly sailing. tomorrow we shall reach Bahia.

28th About 9 oclock we were near to the coast of Brazil; we saw a considerable extent of it, the whole line is rather low & irregular, & from the profusion of wood & verdure of a bright green colour. — About 11 oclock we entered the bay of All Saints, on the Northern Side of which is situated the town of Bahia or St Salvador. It would be difficult [to] imagine, before seeing the view, anything so magnificent. — It requires, however, the reality of nature to make it so. — if faithfully represented in a picture, a feeling of distrust would be raised in the mind, as I think is the case in some of Martins[1] views. — The town is fairly embosomed in a luxuriant wood & situated on a steep bank overlooks the calm waters of the great bay of All Saints. |115| The houses are white & lofty & from the windows being narrow & long have a very light & elegant appearance. Convents, Porticos & public buildings vary the uniformity of the houses: the bay is scattered over with large ships; in short the view is one of the finest in the Brazils.— But their beauties are as nothing compared to the Vegetation; I believe from what I have seen Humboldts glorious descriptions are & will for ever be unparalleled: but even he with his dark blue skies & the rare union of poetry with science which he so strongly displays when writing on tropical scenery, with all this falls far short of the truth. The delight one experiences in such times bewilders the mind. —if the eye attempts to follow the flight of a gaudy butter-fly, it is arrested by some strange tree or fruit; if watching an insect one forgets it in the stranger flower it is crawling over. —if turning to admire the splendour of the scenery, the individual character of the foreground fixes the attention. The mind is a chaos of delight, out of which a world of future & more quiet pleasure will arise.— I am at present fit only to read Humboldt; he like another Sun illumines everything I behold.—

[1] Presumably John Martin, historical and landscape painter.

29th   The day has passed delightfully: delight is however a weak term for such transports of pleasure: I have been wandering by|116| myself in a Brazilian forest: amongst the multitude it is hard to say what set of objects is most striking; the general luxuriance of the vegetation bears the victory, the elegance of the grasses, the novelty of the parasitical plants, the beauty of the flowers.—the glossy green of the foliage, all tend to this end.— A most paradoxical mixture of sound & silence pervades the shady parts of the wood. —the noise from the insects is so loud that in the evening it can be heard even in a vessel anchored several hundred yards from the shore.— Yet within the recesses of the forest when in the midst of it a universal stillness appears to reign.— To a person fond of Natural history such a day as this brings with it pleasure more acute than he ever may again experience.— After wandering about for some hours, I returned to the landing place.— Before reaching it I was overtaken by a Tropical storm.— I tried to find shelter under a tree so thick that it would never have been penetrated by common English rain, yet here in a couple of minutes, a little torrent flowed down the trunk. It is to this violence we must attribute the verdure in the bottom of the wood. —if the showers were like those of a colder clime, the moisture would be absorbed or evaporated before reaching the ground.|117|

March 1st   I can only add raptures to the former raptures. I walked with the two Mids a few miles into the interior. The country is composed of small hills & each new valley is more beautiful than the last.— I

collected a great number of brilliantly coloured flowers, enough to
make a florist go wild. — Brazilian scenery is nothing more nor less than
a view in the Arabian Nights, with the advantage of reality. — The air is
deliciously cool & soft; full of enjoyment one fervently desires to live in
retirement in this new & grander world. —

2nd & 3rd   I am quite ashamed at the very little I have done during these
two days; a few insects & plants make up the sum total. — My only
excuse is the torrents of rain, but I am afraid idleness is the true
reason. — Yesterday Cap Paget dined with us & made himself very
amusing by detailing some of the absurdities of naval etiquette. — To
day Rowlett & myself went to the city & he performed the part of
Cicerone to me. —in the lower part near to the wharfs, the streets are
very narrow & the houses even more lofty than in the old town of
Edinburgh. the smell is very strong & disagreeable, which is not to be
wondered at, since I observe they have the same need of crying "gardez
l'eau" as in|118| Auld Reekie. — All the labor is done by the black men,
who stand collected in great numbers round the merchants
warehouses. — The discussions which arise about the amount of hire
are very animated; the negroes at all times use much gesticulation &
clamor & when staggering under their heavy burthens, beat time &
cheer themselves by a rude song. — I only saw one wheel carriage; but
the horses are by no means scarce; they are generally small & well
shaped & are chiefly used for the merchants to ride. — We paid a visit to
one of the principal churches, we here found for a guide, a little Irish
boy about 13 years old. — His father was buried there two months ago,
& was one of the unfortunate people whom Don Pedro enticed into the
country under the pretence of settling them. — This little fellow cont-
rives to support his mother & sister by the few Vintems which in the
course of the day he earns by messages. — Mr Gond, one of the
principal merchants in the place, offered to lend us horses, if we would
walk to his country house. — We gladly accepted his offer & enjoyed a
most delightful ride; one beautiful view after another opening upon us
in endless succession.|119|

4th   This day is the first of the Carnival, but Wickham, Sullivan & myself
nothing undaunted were determined to face its dangers. — These
dangers consist in being unmercifully pelted by wax balls full of water
& being wet through by large tin squirts. — We found it very difficult to
maintain our dignity whilst walking through the streets. — Charles the
V has said that he was a brave man who could snuff a candle with his

fingers without flinching; I say it is he who can walk at a steady pace, when buckets of water on each side are ready to be dashed over him. After an hours walking the gauntlet, we at length reached the country & there we were well determined to remain till it was dark. — We did so, & had some difficulty in finding the road back again, as we took care to coast along the outside of the town. — To complete our ludicrous miseries a heavy shower wet us to the skins, & at last gladly we reached the Beagle. — It was the first time Wickham had been on shore, & he vowed if he was here for six months it should be [the] only one. —

5th King & myself started at 9 oclock for a long naturalizing walk. — Some of the valleys were even more beautiful than any I have yet seen. — There is a wild luxuriance in these spots that is|120| quite enchanting. — One of the great superiorities that Tropical scenery has over European is the wildness even of the cultivated ground. Cocoa Nuts, Bananas, Plantain, Oranges, Papaws are mingled as if by Nature, & between them are patches of the herbaceous plants such as Indian corn, Yams & Cassada: & in this class of views, the knowledge that all conduces to the subsistence of Mankind, adds much to the pleasure of beholding them. We returned to the ship about ½ after 5 oclock & during these eight hours we scarcely rested one. — The sky was cloudless & the day very hot, yet we did not suffer much: It appears to me that the heat merely brings on indolence, & if there is any motive sufficient to overcome this it is very easy to undergo a good deal of fatigue. — During the walk I was chiefly employed in collecting numberless small beetles & in geologising. — King shot some pretty birds & I a most beautiful large lizard. — It is a new & pleasant thing for me to be conscious that naturalizing is doing my duty, & that if I neglected that duty I should at same time neglect what has for some years given me so much pleasure. — |121|

6th I pricked my knee some days since, & it is now so much swolen that I am unable to walk. — The greater part of the day has been spent in idly lying on deck. — I am not surprised that people are so indolent in a hot country; neither mind or body require any exercise; watching the sky is sufficient occupation for the former & the latter seems well contented with lying still. —

12th Since the 6th I have been for the greater part of the time in my hammock; my knee continued to swell & was exceedingly painful. — To day is the first I have been able to sit up for many hours together. — It has been mortifying to see the clear blue sky above my head & not be able to enjoy it. — I have heard of interesting geological facts & am disabled from examining them; but instead of grumbling I must think myself lucky in having at all seen the glorious city of Bahia. — We have had some festivities on board; the day before yesterday there was a grand dinner on the quarter deck. — Cap Paget has paid us numberless visits & is always very amusing: he has mentioned in the presence of those who would if they could have contradicted him, facts about slavery so revolting, that|122| if I had read them in England, I should have placed them to the credulous zeal of well-meaning people: The extent to which the trade is carried on; the ferocity with which it is defended; the respectable (!) people who are concerned in it are far from being exaggerated at home. — I have no doubt the actual state of by far the greater part of the slave population is far happier than one would be previously inclined to believe. Interest & any good feelings the proprietor may possess would tend to this. — But it is utterly false (as Cap Paget satisfactorily proved) that any, even the very best treated, do not wish to return to their countries. — "If I could but see my father & my two sisters once again, I should be happy. I never can forget them." Such was the expression of one of these people, who are ranked by the polished savages in England as hardly their brethren, even in Gods eyes. — From instances I have seen of people so blindly & obstinately prejudiced, who in other points I would credit, on this one I shall never again scruple utterly to disbelieve: As far as my testimony goes, every individual|123| who has the glory of having exerted himself on the subject of slavery, may rely on it his labours are exerted against miseries perhaps even greater than he imagines. —

13th Unable as yet to leave the ship.

14th Hired a boat & went some miles up the harbour. — I found some interesting geological appearances & spent some pleasant hours in wandering on the beach.

15th The Beagle weighed anchor this morning & proceeded to sound the bank, which runs out at the head of the bay. — As it was intended to come in again I was landed on shore & was very glad to have one other opportunity of admiring the beautiful country round Bahia. I procured an Irish boy as an interpreter & again started to revisit the same place as I did yesterday. — After walking for some time in the heat of the sun, we entered a Venda & drank some most excellent Sangaro. — As is generally the case we were soon surrounded by black men, women & children. I do not know whether they afforded me or I them the most

amusement; their astonishment was great at the Fly net, small pistol &
compass: as one thing came out after another from my most capacious
pockets, they cried "full, full of sins". — Doubtless|124| thinking all my
instruments were related "al Diabolo". — Every body is delighted with
the excellent manners of the Negros. — I gave my friends at the Venda
some wine & when I parted with them it is my firm belief, no Dutchess
with three tails could have given such courtlike & dignified bows as the
black women saluted me with. — In the evening I went to the Hotel d
Universe, where by the help of the three words "comer" to eat, "cama"
a bed & "pagar" my host & myself contrived to agree very well.

16*th*   The next morning I took a long walk & collected a great number of
plants & insects; it was a fine glowing day; but it is quite delightful to
find, so contrary to what I had expected, that the heat by no means
incapacitates one for exercise. — In the middle of the day went on board
the Samarang & dined there. The difference between a surveying vessel
& one in real fighting order is very striking. In the Samarang at any time
under five minutes they could fire an effective broadside. I spent most
part of the evening with the Mids; & such a set of young unhanged|125|
rogues the young gentlemen" are, is sufficient to astonish a shore-going
fellow. — About 9 oclock the Beagle came in & anchored & instead of
sleeping on board the Samarang I went to my own hammock. — It was
a piece of high good luck that I remained on shore during the two days:
the ship rolled & pitched so much, that the greater part of the junior
officers were sick. — People in general are not at all aware what a lasting
misery sea-sickness is. Continually one meets men who having been at
sea during their whole life yet are uncomfortable in every breeze.

17*th*   Took a farewell stroll with King: the evening was bright & exceedingly clear; not a breath of air moved the leaves; every thing was quiet;
nothing could be better adapted for fixing in the mind the last &
glorious remembrances of Bahia. — If to what Nature has granted the
Brazils, man added his just & proper efforts, of what a country might
the inhabitants boast. But where the greater parts are in a state of
slavery, & where this system is maintained by an entire stop to
education, the mainspring of human actions, what can be expected; but
that the whole would be polluted by its part. —|126|

18*th*   We got under weigh early in morning & cruized about the harbor
untill the charts were finished. — Against a strong tide we slowly stood
out of the bay of All Saints & took a lasting farewell of Bahia: if I have
already seen enough of the Tropics to be allowed to judge, my report

would be most favourable; nothing can be more delightful than the
climate, & in beauty the sky & landscape are unparalleled in a colder
zone. —

19th The next morning from the light winds & strong current we were yet
in sight of the coast of Brazil:

20th & this morning to the astonishment of every body the opening into
Bahia was distinctly visible. — In the forenoon a water-spout took place
at a few miles distance & was to me a very interesting phenomenon. —
From a stratus or black bank of clouds, a small dark cylinder (shaped
like a cows tail) depended & joined it self to a funnel shaped mass which
rested on the sea. — It lasted some moments & then the whole appearance vanished into an exceedingly heavy rain storm. — When they
approach near to a vessel, it is usual to fire a big gun in order to break
them.

   A large shark followed the ship, & was first struck by a harpoon; after
this he was hooked by a bait & again being struck broke the hook &
escaped. — Such an adventure creates great interest all over the whole
ship. — |127|

21st The greatest event of the day has been catching a fine young shark
with my own hooks: It certainly does not require much skill to catch
them, yet this no way diminishes the interest. — In this case the hook
was bigger than the palm of the hand & the bait only a bit of salted pork
just sufficient to cover the point. Sharks when they seize their prey turn
on their backs; no sooner was the hook astern, than we saw the silvery
belly of the fish & in a few moments we hauled him on deck. —

22nd & 23rd The wind yet continues very light & contrary; there is
however to my cost a little swell, enough to make me all day long rather
uncomfortable: Occupation is the best cure, & I always have, when
leaving a port, the pleasant one of arranging the collections. —

24th, 25th & Monday 26th These three days, like the weather, have passed
away with quietness & enjoyment. — We are nearly 4 degrees from the
coast of Brazil & about 2 from the Albrolhos, from which islands a long
shoal extends itself. — The Lead has been regularly cast at every two
hours. — to day after finding no bottom at 230 fathoms we suddenly
came on the bank with between 30 and 40. We are now steering for the
islands. —

   |128| I find living on board a most excellent time for all sorts of study;
& I cannot imagine why anybody who is not sick should make objec-

tions on that score. — There is little to interrupt one, for instance since leaving Bahia the only living things that we have seen were a few sharks & Mother Carys chickens. — At night in these fine regions of the Tropics there is one certain & never failing source of enjoyment, it is admiring the constellations in the heaven. — Many of those who have seen both hemispheres give the victory to the stars of the North. — It is however to me an inexpressible pleasure to behold those constellations, the first sight of which Humboldt describes with such enthusiasm. — I experience a kindred feeling when I look at the Cross of the South, the phosphorescent clouds of Magellan & the great Southern Crown. —

27ᵗʰ, 28ᵗʰ During these two days the labours of the expedition have commenced. — We have laid down the soundings on parts of the Abrolhos, which were left undone by Baron Roussin. — The depth varied to an unusual extent: at one cast of the lead there would be 20 fathoms & in a few minutes only 5. — The scene being quite new to me was very interesting. — Everything in such a state of preparation; Sails all shortened & snug: anchor ready to let |129|fall: no voice or noise to be heard, excepting the alternate cry of the leadsmen in the chains. — We anchored for the night

29ᵗʰ & next morning we altered our place to within 2 miles of the groupe of Islands. — The Abrolhos consisted of 5 small rocky islands, which although uninhabited are not unfrequently visited by fishermen. — Two parties landed directly after breakfast.[1] I commenced an attack on the rocks & insects & plants. — the rest began a more bloody one on the birds. — Of these an enormous number were slaughtered by sticks, stones & guns; indeed there were more killed than the boats could hold. — We all returned for dinner & after that a boat was given to the midshipmen in order that they might see the islands. — I took the opportunity & had another ramble on this solitary spot. — Whilst pulling back to the ship we saw a turtle; it immediately went down, nothing certainly could be imagined worse for surprising an animal than a boat full of midshipmen.

[1] FitzRoy wrote: 'We anchored near the islets, at dusk, on the 28th, after being in frequent anxiety, owing to sudden changes in the depth of water; and next morning, moved to a better berth at the west side, very near them. They are rather low, but covered with grass, and there is a little scattered brushwood. The highest point rises to about a hundred feet above the sea. Their geological formation, Mr Darwin told me, is of gneiss and sandstone, in horizontal strata. When our boats landed, immense flights of birds rose simultaneously, and darkened the air. It was the breeding and moulting season; nests full of eggs, or young unfledged birds, absolutely covered the ground, and in a very short time our boats were laden with their contents. See *Narrative* 2: 64–5.

30th All to day we have been cruizing in sight of the Islands & have been employed in|130| sounding & taking angles. — I have been most pleasantly employed in working at yesterdays produce. — We are now (at night) sailing with a fine breeze abaft the beam for Rio. —

31st A fine rattling breeze.

April 1st All hands employed in making April fools. — at midnight nearly all the watch below was called up in their shirts; Carpenters for a leak: quarter masters that a mast was sprung. — midshipmen to reef top-sails; All turned in to their hammocks again, some growling some laughing. — The hook was much too easily baited for me not to be caught: Sullivan cried out, "Darwin, did you ever see a Grampus: Bear a hand then". I accordingly rushed out in a transport of Enthusiasm, & was received by a roar of laughter from the whole watch. —

2nd A rainy, squally morning, very unusual at this time of year in these Latitudes; being now about 130 miles East of Rio. A large flock of Mother Carys chicken are hovering about the stern in same manner as swallows do on a calm summer evening over a lake. — A flying fish fell on the deck this morning; it struck the mast high up near the main yard: sticking to the fish was a crab, the pain of which caused perhaps this unusual degree of action. —|131|

3rd This morning Cape Frio was in sight: it is a memorable spot to many in the Beagle, as being the scene of the disgraceful wreck of the Thetis.[1] All day we ran along the coast & in the evening drew near to the harbour of Rio. — The whole line is irregularly mountainous, & interspersed with hills of singular forms. — The opening of the port is recognised by one of these, the well known Sugar-loaf. — As it would be impossible to get a good anchorage or enjoy the view so late in the evening, the Captain has put the ships head to the wind & we shall, to my great joy, cruize about for the night. — We have seen great quantities of shipping; & what is quite as interesting, Porpoises, Sharks & Turtles; altogether, it has been the most idle day I have spent since I left England. — Everybody is full of anxiety about letters & news papers, tomorrow morning our fates will be decided.

[1] A full account of the loss of the frigate HMS *Thetis* with 25 of her crew in a squall off Cape Frio on 5 December 1830 is given by FitzRoy in *Narrative* 2: 67–72. He concludes: 'Those who never run any risk; who sail only when the wind is fair; who heave to when approaching land, though perhaps a day's sail distant; and who even delay the performance of urgent duties until they can be done easily and quite safely; are, doubtless, extremely prudent persons — but rather unlike those officers whose names will never be forgotten while England has a navy.'

Mole, palace and cathedral at Rio de Janeiro, by T. Hain after A. Earle (*Narrative* **1**: 106).

4th The winds being very light we did not pass under the Sugar loaf till
after dinner: our slow cruize was enlivened by the changing prospect of
the mountains; sometimes enveloped by white clouds, sometimes
brightened by the sun, the wild & stony|132| peaks presented new
scenes. — When within the harbor the light was not good, but like to a
good picture this evenings view prepared the mind for the morrows
enjoyment. — In most glorious style did the little Beagle enter the port
& lower her sails alongside the Flag ship. We were hailed that from
some trifling disturbances we must anchor in a particular spot. Whilst
the Captain was away with the commanding officer, we tacked about
the harbor & gained great credit from the manner in which the Beagle
was manned & directed. — Then came the ectacies of opening letters,
largely exciting the best & pleasantest feelings of the mind; I wanted not
the floating remembrance of ambition now gratified, I wanted not the
real magnificence of the view to cause my heart to revel with intense
joy; but united with these, few could imagine & still fewer forget the
lasting & impressive effect. —

5th In the morning I landed with Earl at the Palace steps; we then
wandered through the streets, admiring their gay & crowded appearance. — The plan of the town is very regular, the lines, like those in
Edinburgh, running parallel, & others crossing them at right angles. —
The principal streets leading from the squares are straight|133| &
broard; from the gay colours of the houses, ornamented by balconys,
from the numerous Churches & Convents & from the numbers hurry-
ing along the streets, the city has an appearance which bespeaks the
commercial capital of Southern America. — The morning has been for
me very fertile in plans: most probably I shall make an expedition of
some miles into the interior, — & at Botofogo Earl & myself found a most
delightful house which will afford us most excellent lodgings. —[1]

I look forward with the greatest pleasure to spending a few weeks in
this most quiet & most beautiful spot. — What can be imagined more
delightful than to watch Nature in its grandest form in the regions of the
Tropics? — We returned to Rio in great spirits & dined at a Table d Hote,
where we met several English officers serving under the Brazilian
colours. — Earl makes an excellent guide, as he formerly lived some
years in the neighbourhead: it is calamitous how short & uncertain life
is in these countries: to Earls enquiries about the number of young men
whom he left in health & prosperity, the most frequent answer is he is
dead & gone. — The deaths are generally to be attributed to drinking:

few seem able to resist the temptation, when|134| exhausted by business in this hot climate, of strongly exciting themselves by drinking spirits. —

[1] See drawing and watercolour by Conrad Martens, CM Nos. 42 and 43, *Beagle Record* p. 52. 'Botofogo' is more correctly spelt 'Botafogo'.

6th   The day has been frittered away in obtaining the passports for my expedition into the interior. — It is never very pleasant to submit to the insolence of men in office; but to the Brazilians who are as contemptible in their minds as their persons are miserable it is nearly intolerable. — But the prospect of wild forests tenanted by beautiful birds, Monkeys & Sloths, & Lakes by Cavies & Alligators, will make any naturalist lick the dust even from the foot of a Brazilian. —

7th   I finally made the few but necessary arrangements for my riding excursion to Rio Macaè:[1] & in the evening moved some of my goods & chatels to Botofogo. Earl & King likewise prepared themselves for residing there.

[1] The accent sometimes inserted by CD is wrong. A modern map shows the name of the town and river as Macaé.

8th   At 9 oclock I joined my party at Praia Grande,[1] a village on the opposite side of the Bay. — We were six in number & consisted of Mr Patrick Lennon, a regular Irishman, who when the Brazils were first opened to the English made a large fortune by selling spectacles, Thermometers &c &c. — About 8 eight years since he purchased a tract of forest country on the Macae & put an English agent over it. — Communication is so difficult|135| that from that time to the present he has been unable to obtain any remittances. — After many delays M[r] Patrick resolved in person to visit his estate. — It was easily arranged that I should be a companion & certainly in many respects it has been an excellent opportunity for seeing the country & its inhabitant. — M[r] Lennon has resided in Rio 20 years & was in consequence well qualified to obtain information—in his disposition very shrewd & intelligent. He was accompanied by his nephew a sharp youngster following the steps of his Uncle & making money. — Thirdly came Mr Lawrie, a well informed clever Scotchman, *selfish unprincipled* man, by trade partly Slave-Merchant partly Swindler. He brought a friend a Mr Gosling an apprentice to a Druggist. Mr Lawries brother married a handsome Brazilian lady, daughter of a large landed proprietor, also on the Macaè, & this person Mr Lawrie was going [to] visit. — A black boy as guide & myself completed the party. — And the wilds of Brazils have seldom seen a more extraordinary & quixotic set of adventurers. —

Our first stage was very interesting,[2] the day was powerfully hot & as we passed through the woods, every thing was still, excepting the large & brilliant|136| butterflies, which lazily fluttered about. — The view seen when crossing the hills behind Praia Grande is most sublime & picturesque. — The colours were intense & the prevailing tint a dark blue, the sky & calm waters of the bay vied with each other in splendor. — After passing through some cultivated country we entered a Forest, which in the grandeur of all its parts could not be exceeded. — As the gleams of sunshine penetrate the entangled mass, I was forcibly reminded of the two French engravings after the drawings of Maurice Rugendas & Le Compte de Clavac. — In these is well represented the infinite numbers of lianas & parasitical plants & the contrast of the flourishing trees with the dead & rotten trunks. I was at an utter loss how sufficiently to admire this scene. — We arrived by mid-day at Ithacaia; this small village is situated on a plain & round the central houses are the huts of the negroes. — These from their regular form & position reminded me of the drawings of the Hottentot habitations in Southern Africa.[3] As the moon would rise early, we determined to start that evening for our sleeping place at the Lagoa Marica.|137|

As it grew dark we passed under one of the massive bare & steep hills of granite which are so common in this country. — This spot is notorious as having been for a long time the residence of some run-away slaves, who by cultivating a little ground near the top contrived to eke out a subsistence.[4] We continued riding for some hours; for the few last miles the road was intricate, it passed through a desert waste of marshes & lagoons. — The scene by the dimmed light of the moon was most desolate; a few fire-flies flitted by us & the solitary snipe as it rose uttered its plaintive cry. — the distant & sullen roar of the sea scarcely broke the stillness of the night. —[5] We arrived at last at the Venda, & were very glad to lie down on the straw mats. —|138|

[1] Now part of Niteroi.

[2] The entry for this Sunday in CD's second field notebook (Down House Notebook 1.10) runs as follows: 'Hills generally rounded, often bare; — flat alluvial valley between them. Village of Itho-caia 12 miles from Rio. — Temp. in white sand 104° in shade. — View at first leaving Rio sublime, picturesque, intense colours, blue prevailing tint — large plantations of sugar rustling & coffee — Mimosa natural veil — Forest like but more glorious than those in the engraving; gleams of sunshine; parasitical plants; bananas; large leaves; sun sultry. — All still, but large & brilliant butterflies; Much water; surprised to see Guinea fowls; our calvacade very Quixotic; the banks most teeming with wood & beautiful flowers; village of Ith regular like the Hottentots; the poor blacks thus perhaps try to persuade themselves that they are in the land of their Fathers. — The rock from which the old woman threw herself. — Temp. of room 80° — Our dinner; eggs and rice; our host saying we could have anything. — About 4 oclock, & arrived at our sleeping place about

9. — Sand & swampy plains & thickets alternating—passed through by a dim moonlight—
the cries of snipes; fire flies & a few noisy frogs & goat suckers.'
[3]The next sentence has been deleted. It runs: 'Thus perhaps do these poor people in the
midst of their slavery call to their minds the home of their fathers.—'
[4]Followed by another deleted passage: 'At length some soldiers were sent & secured
them all, excepting one old woman, who sooner than be again taken, dashed herself to
pieces from the very summit.— I suppose in a Roman matron this would be called noble
patriotism, in a negress it is called brutal obstinacy!—
[5]The words 'scarcely . . . night' are substituted for 'kept concert with our feelings'.

9th   We left our miserable sleeping place before sunrise.— The road
passed through a narrow sandy plain, lying between the sea & the
interior salt lagoons.— The number of beautiful fishing birds such as
Egrets, Cranes &c & the succulent plants assuming such fantastical
forms gave to the scene an interest which it would not otherwise have
possessed.— The few stunted trees were loaded by parasitical plants,
amongst which the beauty & delicious fragrance of some of the
Orchideæ were most to be admired.— As the sun rose, the day became
very hot, & the reflection of the light & heat from the white sand was
very distressing. The thermometer in my pocket stood at 96°.— Dined
at Mandetiba: therm. in shade 84°.— The beautiful view quite refreshed
us; the distant wooded hills were seen over & reflected in the perfectly
calm water of an extensive lagoon.— As the Venda here was a very
good one, & I have the pleasant but rare remembrance of an excellent
dinner, I will be grateful & describe it as the type of its Class. These
houses are often large, & are built of thick upright posts, with boughs
interwoven, which are afterwards plastered, & they seldom have
floors, & never glazed windows,|139| but are generally pretty well
roofed.— Universally the front part is open, forming a kind of veran-
dah; in which are placed tables & benches. On each side are the bed
rooms where the passenger may sleep, as comfortably as he is able, on
wooden platforms, covered by a thin straw mat.[1]

The Venda stands in a court, where the horses are fed.— On first
arrival we unsaddle our horses & give them their Indian corn.— Then
with a low bow ask the Signor to do us the favor to give us something
to eat.— "Anything you choose Sir" is his answer.— For the few first
times vainly I thanked providence for guiding us to so good a man.—
The conversation procceeding, the case usually became deplorable:
"Any fish can you do us the favor of giving?".— "Oh no Sir." "Any
soup." No Sir." Any bread." "Oh no Sir."— Any dried meat. "Oh no
Sir.— If we were lucky, by waiting 2 hours we obtained fowls rice &
farinha.— It not unfrequently happens that the guest is obliged to kill
with stones the poultry, for his own dinner.|140|

When really exhausted with fatigue & hunger, we timorously hinted we should be glad of our meal. — The pompous, &, though true, most unsatisfactory answer was given, "it will be ready when it is ready". — If we had dared to remonstrate any further, we should have been told to proceed on our journey as being too impertinent. — Their charges are, however, exceedingly moderate, but they will, if they are able, cheat. — The hosts are most ungracious & disagreeable in their manners. — their houses & their persons are often filthily dirty. — the want of the common accomodation of forks, knives, spoons is even common. I am quite sure no cottage, no hut in England could be found in a state so utterly destitute of what we considered comforts. — At Campos Novos, we fared sumptuously, having rice & fowls, biscuit & wine & spirits for dinner, coffee in the evening. & with it for breakfast fish. — good food for the horses, & this only cost 2^s 6^d per head. — Yet this same man, being asked if he knew anything of a whip which one of the party lost, gruffly answered, "How should I know? Why did you not take care of it. — I suppose the dogs have eat it". |141|

Leaving Mandetiba, we continued to pass through an intricate wilderness of lakes, in some of which were fresh, in others salt water shells.

We at last entered the forest; the trees were very lofty, & what was always to be remarked in them was the whiteness of the boles, this at a distance adds much to their effect. — I see by my note book, "wonderful, beautiful flowering parasites" invariably this strikes me as the most novel object in a Tropical forest. — On the road we passed through tracks of pasturage, much injured by the enormous conical ants nests, which in height were about 12 feet. — they give to the plain exactly the appearance of the Mud Volcanoes at Jorullo, figured by Humboldt. — We arrived after it was dark at Ingetado: having been 10 hours on horseback. I never ceased to wonder, from the beginning to the end of the journey, at the amount of labor which these horses are capable of enduring: I presume it is from being in a country more congenial to their original nature. — and from the same cause they seem far better than English horses to recover [from] injuries & wounds. — |142|

[1] The words 'where the passenger may sleep . . . thin straw mat' are substituted for 'with either Lathes or platforms: each person has a mat & with the knapsack for a pillow makes as good a bed as he is able'.

10th We all started before it was light in high spirits; but 15 miles of heavy sand before we got our breakfast at Addea de St Pedro[1] nearly destroyed the whole chivalrous party. — After another long ride we arrived at our sleeping place, Campos Novos. — It was a very pleasant cool evening. Thermom. on the turf 74°: I went out collecting & found some fresh water shells. —

[1] Now São Pedro D'aldeia.

11th   Passed through several leagues of a thick wood. — I felt unwell, with a little shivering & sickness: crossed the Barra de St Jaôa in a canoe, swimming alongside our horses: could eat nothing at one oclock, which was the first time I was able to procure anything. — Travelled on till it was dark, felt miserably faint & exhausted; I often thought I should have fallen off my horse. — Slept at the Venda da Matto, 2 miles S of the entrance of the Rio Macaè into the sea. — All night felt very unwell; it did not require much imagination to paint the horrors of illness in a foreign country, without being able to speak one word or obtain any medical aid. —

12th   The next morning, I nearly cured myself by eating cinnamon & drinking port wine; gladly in the evening|143| did I arrive at Socêgo, the house of Signor Figuireda, the elder Mr Lawrie's father in law. —

13th   Felt much better & able throughily to enjoy our days rest here. — In this case the Fazenda consists of a piece of cleared ground cut out of the almost boundless forest. — On this are cultivated the various products of the country: Coffee is the most profitable: the brother of our host has 100,000 trees, producing on an average 2 lb per tree, many however singly will bear 8 lb. or even more. Mandeika (or Cassada) is likewise cultivated in great quantity: every part is useful. — the leaves & stalks are eat by the horses; the roots, ground into pulp, pressed dry, & then baken makes the Farinha; by far the most import[ant] article of subsistence in the Brazils. From this is prepared the Tapioka of commerce. — It may be mentioned as a curious though well known fact that the expressed juice is a most deadly poison; a few years ago at this Fazenda a Cow died from drinking some of it. — Feijôa or beans are much cultivated & form a most excellent vegetable: one bag bringing sometimes 80. — Sugar Cane is also grown. And rice in the swampy parts. Signor Fig. planted three bags & they produced 320.|144|

The house was simple & uncomfortable, & formed like an English barn: it was well floored, & thatched with reeds. — The windows merely had shutters. Interiorly it was divided into rooms by partitions which did not reach the roof. At one end was a sitting room of the whole breadth. — the gilded chairs & sofas were oddly contrasted by the white washed walls. — Beyond this was a longitudinal division, one side of which was the dining room, on the other, 4 bedrooms belonging to the family. Separated from this building only by a few inches was another long shed, the adjoining end formed the kitchen: the other, large storehouses & granaries. — These formed one line on the other side of a cleared space where coffee was drying, were the bedrooms for guests, stables & working shops for the blacks, who had been taught different trades. Surrounding these were the huts of about 110 negroes, whom Signor & one white man as a manager contrive to keep in perfect order. — The house, built on a hill at the foot of which a brook runs, overlooked the cultivated|145| ground, & was bounded by an horizon of green luxuriant forest. — The pasturage abounded in cattle, goats, sheep & horses; near the house, oranges, Bananas flourished almost spontaneously. — The woods are so full of game, that they had hunted & killed a deer on each of the three days previous to our arrival. — This profusion of food shows itself at the dinners, when if the tables do not groan, the guests surely do. — Each person is expected to eat of every dish; one day having, as I thought, nicely calculated so that nothing should go away untasted, to my utter dismay a roast turkey & a pig appeared in all their substantial reality. — During the meals, it was the employment of a man to drive out sundry old hounds & dozens of black children which together at every opportunity crawled in. — As long as the idea of slavery could be banished, there was something exceedingly fascinating in this simple & patriarchal style of living. — It was a such perfect retirement & independence of the rest of the world. — As soon as any stranger is seen arriving, a large bell is set tolling & generally some small cannon are fired; thus it is announced|146| to the rocks & woods & to no one else. — One morning I walked out before daylight to admire the solemn stillness, when it was broken by the morning hymn raised on high by the whole body of the blacks; in this manner do they generally begin their dayly work. — In such Fazendas as these I have no doubt the slaves pass contented & happy lives. — Signor Manoel Joaquem da Figuireda is a man of an intelligent & enterprising character. — Some of the roads through his estate were cut in a European fashion; in a years time he believes he shall [be] able so to shorten the road to Campos (a large city) that instead of two days ride it will be only one: He has likewise fixed a saw-mill, which answers admirably in sawing the rose-wood. — This cut into thick planks is floated down to Macaè. — If many were to imitate the example of this man, what a difference a few years would produce in the Brazils. —

14th   Started at midday for Mr Lennons estate; the road passed through a vast extent of forests; on the road we saw many beautiful birds, Toucans & Bee-eaters. We slept at a Fazenda a league from our journeys end; the agent received us hospitably & was the only Brazilian|147| I have seen with a good expression: the slaves here appeared miserably over-worked & badly clothed. — Long after it was dark they were employed. The common method of maintaining the slave, as at Signor Figuireda, is to give them two days, Saturday & Sunday, the produce of which is sufficient to support them & their families for the ensuing five. —

15th   We were obliged to have a black man to clear the way with a sword; the woods in this neighbourhead contain several forms of vegetation which I had not before seen. —some species of most elegant tree ferns. —a grass like the Papyrus; & the Bamboo, the circumference of the stems were 12 inches. —[1] On arriving at the estate, there was a most violent & disagreeable quarrell between Mr Lennon & his agent, which quite prevented us from wishing to remain there. — This Fazenda is the most interior piece of cleared ground, untill you pass the mountains. —its length is 2 & ½ miles, Mr Lennon is not sure how many broard. —it may be guessed what|148| a state the country must be in when I believe every furlong of this might be cultivated. — In the evening it rained very hard, I suffered from the cold, although the thermometer was 75°. — During Mr Lennons quarrell with his agent, he threatened to sell at the public auction an illegitimate mulatto child to whom Mr Cowper was much attached: also he nearly put into execution taking all the women & children from their husbands & selling them separately at the market at Rio. — Can two more horrible & flagrant instances be imagined? —& yet I will pledge myself that in humanity & good feeling Mr Lennon is above the common run of men. — How strange & inexplicable is the effect of habit & interest!. — Against such facts how weak are the arguments of those who maintain that slavery is a tolerable evil!

[1] Followed by a deleted sentence: 'I was rather disappointed in them & can hardly believe they were good specimens. —'

16th   Started early in the morning to Signor Manuel at Socêgo, whom it was agreed upon should be arbitrator: Again I enjoyed the never failing delight of riding through the forests.

17th & 18th   These two days were spent at Socêgo, & was the most enjoyable part of the whole expedition; the greater part of them was spent in the woods, & I succeded in collecting many insects & reptiles. — |149|The woods are so thick & matted that I found it quite impossible to leave the path. — the greater number of trees, although so lofty, are not more than from 3 to 4 feet in circumference. These are interspersed with others of a much greater size. — Signor Manuel was making a canoe 70 feet long, & on the ground was left 40 feet, so that there were 110 feet of straight solid trunk. — The contrast of the Palms amongst other trees never fails to give the scene a most truly tropical appearance: the forests here are ornamented by one of the most elegant, the Cabbage-Palm; with a stem so narrow, that with the two hands it may be clasped, it waves its most elegant head from 30 to 50 feet above the ground. — The soft part, from which the leaves spring, affords a most excellent vegetable. — The woody creepers, themselves covered by creepers, are of great thickness, varying from 1 to nearly 2 feet in circumference. — Many of the older trees present a most curious spectacle, being covered with tresses of a liana, which much resembles bundles of hay. — If the eye is turned from the world of foliage above, to the ground, it is|150| attracted by the extreme elegance of the leaves of numberless species of Ferns & Mimosas. —[1] Thus it is easy to specify individual objects of admiration; but it is nearly impossible to give an adequate idea of the higher feelings which are excited; wonder, astonishment & sublime devotion fill & elevate the mind. —

[1] The words 'Mimosas' and 'Ferns' are marked to be reversed in order, and CD has written in the margin 'Effect of walking on Mimosa'.

19th   Left Socêgo, crossed the Rio Macaè & slept at the Venda de Matto: in the evening walked on the beach & enjoyed the sight of a high & violent surf.

20th   Returned by the old route to Campos Novos; the ride was very tiresome, passing over a heavy & scorching sand.[1] Whilst swimming our horses over the St Joâo, we had some danger & difficulty. — the animals became exhausted & we had two drunken Mulattos in the boat. —

[1] CD has written in the margin 'Chirping sand', presumably because sand of this type sometimes emits a metallic creaking sound when one walks over it.

21st   Started at day-break & proceeded for some leagues on the former road; we then turned off, being determined to reach the city by the interior line. — Our party was reduced to M[r] Lennon, his nephew & myself. — We arrived in the evening, almost without having rested our horses, at the Rio Combrata: this country was much more cultivated. The Venda was beyond anything miserable, we were obliged to sleep on the Indian corn. — |151|

22nd As usual started sometime before daylight & proceeded to Madre de Dios[1] where we breakfasted, had it not been for the torrents of rain this would have been a very interesting ride; the country is richly cultivated, the Sugar Cane being the chief produce. — The woods contained numbers of beautiful birds; the hedges were decorated by several species of passion flowers. — Madre de Dios, like all the villages is extremely foreign looking & picturesque. — The houses are low & painted with gay colours; the tops of the windows & doors being arched takes away the still effect so universal in an English town. — One or two handsome Churches in the centre of the village completes the picture.

It continued to rain & we started for our sleeping place, Fregueria de Tabarai. —[2] This interior road is the best I have seen, but it is much inferior to the worst turnpike road. — I do not think a gig could travel on it. — Yet this is one of the principal passes in the Brazils. — We met a good many people on horseback. — The only vehicle is a most rude cart with almost solid wheels, it is drawn by eight oxen yoked together: as it moves it makes a most extraordinary creeking noise. — We did not pass over one stone bridge. Where any exist, they are made of logs of wood;|152| they were sometimes in so bad a state that we were obliged to leave the road to avoid them. — The distances are inaccurately known, no two people at all agreeing in their accounts. — Instead of milestones, the roadside is often marked by crosses, to signify where human blood has been spilled. — The evening was so cold that I fairly trembled with it yet the thermometer was $62\frac{1}{2}$.

[1] Now Rio Bonito.
[2] Now Itaborai.

23rd The number of pretty & gay houses showed our approach to the city. — During the day we passed through a wood of Acacias, the finely pinnate foliage makes for the sky a most delicate veil. — And casts on the ground a pleasing kind of shade; from the softness of the leaves, no rustling is heard when a breeze moves them. We arrived in the evening at Praia Grande, where owing to having lost our pass ports, we were plagued to prove that our horses were not stolen. —

24th To my joy I at last gained the Beagle. I found a days rest so delightful that I determined idly to remain on board. — During my absence several political changes have taken place in our little world. — Mr Maccormick has been invalided, & goes to England by the Tyne. Mr Derbyshire by his own request was discharged the service. — In his place Mr Johnstone will be moved into the Beagle from the Warspite. —[1]|153|

[1] Robert McCormick was Surgeon in the *Beagle* from 1831 to 1832, and Alexander Derbishire was Mate. In a letter to his sister Caroline, CD wrote: 'I had sealed up the first letter, all ready to be sent off during my absence: but no good opportunity occurred so it & this will go together. — I take the opportunity of Maccormick returning to England, being invalided, i.e. being disagreeable to the Captain & Wickham. — He is no loss. — Derbyshire is also discharged the service, from his own desire not choosing his conduct which has been bad about money matters to be investigated. —' In his own memoirs published in 1884, McCormick wrote: 'Having found myself in a false position on board a small and very uncomfortable vessel, and very much disappointed in my expectations of carrying out my natural history pursuits, and every obstacle having been placed in the way of my getting on shore and making collections, I got permission from the admiral in command of the station here to be superseded and allowed a passage home in H.M.S. *Tyne*.' See *Correspondence* 1: 225–7.

25th Moved all my things from the Beagle to Botofogo. Whilst landing on the beach I suffered on a small scale, sufficient however to paint some of the horrors of shipwreck. — Two or three heavy seas swamped the boat, & before my affrighted eyes were floating books, instruments & gun cases & everything which was most useful to me. — Nothing was lost & nothing completely spoiled, but most of them injured. —[1]

[1] In a letter of this date to his sister Caroline, CD wrote: 'I send in a packet, my commonplace Journal. — I have taken a fit of disgust with it & want to get it out of my sight, any of you that like may read it. —a great deal is absolutely childish: Remember however this, that it is written solely to make me remember this voyage, & that it is not a record of facts but of my thoughts. — & in excuse recollect how tired I generally am when writing it. —' Later in the same letter he said: 'Be sure you mention the receiving of my journal, as anyhow to me it will [be] of considerable future interest as it [is] an exact record of all my first impressions, & such a set of vivid ones they have been, must make this period of my life always one of interest to myself. — If you will speak quite sincerely, —I should be glad to have your criticisms. Only recollect the above mentioned apologies. —' See *Correspondence* 1: 225–7.

26th Employed all day in restoring the effects of yesterdays disaster.

27th In the morning arranged my collections from the Interior, & after dinner went with the Captain to M[r] Aston, the English minister. — The evening passed away very pleasantly, & from the absence of all form almost resembled a Cambridge party. — The Captain has informed me of the important fact that the Beagle will return to Bahia for a few days. — There has been a long dispute about the longitude of Rio, & everybody thought that when that was settled the whole coast of S America would likewise be so. — To the Captains astonishment he finds there is a difference between Bahia & Rio; that is, one side is right at the former place, |154|the other at the latter. — It is in order to verify this, that the second trip is undertaken. — I have made up my mind quietly to remain here & be picked up on the Beagles return.

28th Breakfasted on board, & in the evening went to a pleasant dinner at the Admirals, Sir Thomas Baker.

29th Delightfully quiet day, employed in writing up my journal during the Macaè excursion.

30th Dined with Mr Aston.

May 1st Worked at a host of fresh water animals with which every ditch abounds.

2nd Walked to Rio: the whole day has been disagreeably frittered away in shopping. —

3rd Went on board the Warspite, a 74 line of battle ship, to see her inspected by the Admiral. — It was one of the grandest sights I ever witnessed. — When the Admiral arrived the yards were manned by about 400 seamen; from the regularity of their movements & from their white dresses, the men really looked more like a flock of wild-fowl than anything else. — When a ship is inspected, everything is done precisely the same as if she was engaged with an enemy; & although on paper it may sound like childs play, in the reality it was most animating. — One almost wished for an enemy, when the aweful|155| words were shouted to the great batteries below — "Clear for Action". — After having mæn-euvered the enormous guns & proved how well & easily it was done. — "Fire in the Cock-pit rung through the decks. — in perfect order, the guns yet working, the pumps were rigged, the fire engines brought into play, & all the firemen with their buckets. — The action became hot-ter. — (nobody knew what was coming). The Admiral sung out "a Raking shot has cut our fore-shrouds". "Captain Talbot wared ship: cut away the mizen mast. — in an instant men with their axes sprung to their places: & then it was truly wonderful how soon the store rooms were opened & vast ropes brought to support the tottering fore mast. — The admiral was determined to puzzle them: during all this bustle he ordered a broardside, & shouted the main shrouds & fore stay arè gone. — In short in a few minutes all our principal ropes were cut through & joined. —

Perhaps however the most glorious thing was when the Bugle gave the signal for the Boarders; the very ship trembled at so dense a|156| body rushing a long with their drawn cutlasses. — The appalling shout, with which the English seamen executes the most dangerous service he is ever called upon to perform, was the only thing that was absent. —

In the evening dined with the Admiral & afterwards enjoyed the calmer pleasure of reading letters from Shropshire.

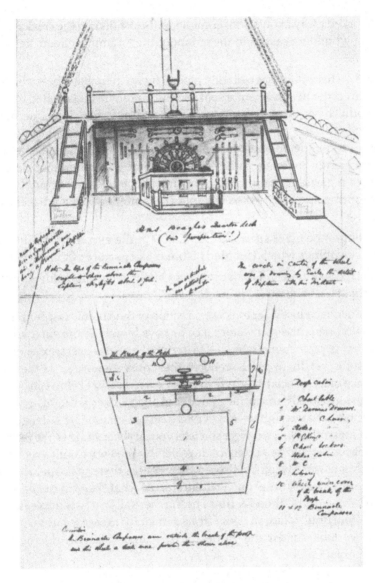

The Beagle's quarter deck, drawn from memory by Philip Gidley King for Mr
Hallam Murray in 1890 (from the archives of John Murray Ltd).

4th   Worked away at my usual employments, & filled up the cracks in the time by building castles in the air about the "pomp & circumstance of war". —

5th & 6th   These days have quietly glided away; there have been torrents of rain, & the fields are quite soaked with water; if I had wished to walk it would have been very disagreeable, but as it is, I find one hours collecting keeps me in full employment for the rest of the day. — The naturalist in England enjoys in his walks a great advantage over others in frequently meeting with something worthy of attention; here he suffers a pleasant nuisance in not being able to walk a hundred yards without being fairly tied to the spot by some new & wondrous creature. —

7th   Went on board & spent the day there, in the evening brought with me a few things which I wanted before the departure of the Beagle. —

8th   Torrents of rain. — I am at present chiefly collecting spiders. In the course of a few hours ·26 rain fell. — |157|

9th   Went out collecting & took the direction of the Botanic Garden; I soon came to one of the salt water lakes or bays by which the surrounding country is often penetrated. — Many of the views were exceedingly beautiful; yet in tropical scenery, the entire newness, & therefore absence of all associations, which in my own case (& I believe in others) are unconsciously much more frequent than I ever thought, requires the mind to be wrought to a high pitch, & then assuredly no delight can be greater; otherwise your reason tells you it is beautiful but the feelings do not correspond. — I often ask myself why can I not calmly enjoy this; I might answer myself by also asking, what is there that can bring the delightful ideas of rural quiet & retirement, what that can call back the recollection of childhood & times past, where all that was unpleasant is forgotten; untill ideas, in their effects similar to them, are raised, in vain may we look amidst the glories of this almost new world for quiet contemplation. —

The Captain called in the evening & says the Beagle sails tomorrow. — We also today heard the bad news that three of the party, who went up in the Cutter to Macucù for snipe shooting, are taken seriously ill with Fevers. — There is reason to fear that others were to day beginning to feel the bad effects of their excursion. — The first case occurred 4 days after the arrival|158| of the party on board on the 2nd. — I very nearly succeeded in joining it; my good star presided over me when I failed. — Four of us belonging to the Beagle are now living here. — Earl, who is unwell & suffers agonies from the Rheumatism. — The serjeant of Marines, who is recovering from a long illness, & Miss Fuegia Basket, who daily increases in every direction except height. —

10ᵗʰ The Beagle sailed for Bahia this evening.

11ᵗʰ, 12ᵗʰ & Sunday 13ᵗʰ These four days I have been almost laid up by an inflammation in my arm. — Any small prick is very apt to become in this country a painful boil. — Earl continues very ill & is in bed. — This is the winter season; a great deal of rain falls, but chiefly by night; in other respects the weather is most delightful & cool. — The temperature in a room generally varies from 70°–75°. —

14ᵗʰ My arm is nearly well. I took the opportunity of paying several calls; that most empty yet burdensome form of civility. —

15ᵗʰ Went out collecting & had a most delightful walk: —It is now full moon. I do not know whether the clear outline of the view seen by night is most admirable, or when lighted up by the gorgeous colours of a Tropical sun. —|159|

16ᵗʰ Examined the rich produce of yesterdays collecting. — Earl is considerably better. —

17ᵗʰ Heavy rain; in the course of the day 1·6 inches fell. —as the storm passed over the Caucovado the sound produced by the drops pattering on the countless number of leaves was very singular. — It might be heard for ¼ of a mile. — I jumped up to see what it was; for it sounded like the rushing of a large body of water. —

18ᵗʰ & 19ᵗʰ These days have glided away very pleasantly, but with nothing particular to mark their passage. — What will not habit do? I find my eye wanders idly from the Orange to the Banana & from it to the Cocoa Nut; whilst I take no more notice than if they were laurel or apple trees. It is very amusing to hear people complaining of the extreme cold. —the depth of winter, however, brings not with it its usual & solitary silence. In the evening various species of frogs make an almost musical concert; this, as the night advances, is taken up in a higher key by a multitude of Cicadas & Crickets. —

Sunday 20ᵗʰ Mʳ Derbyshire, who after leaving the Beagle has remained in the city, paid us a visit. — In the evening Earl (who is nearly well) & we two walked round the Botofogo bay.

21ˢᵗ Took a long scramble through the woods; the bottom is so thickly strewed over with dry sticks & leaves, that in walking one makes as much noise as a large quadruped would.|160| This is very disagreeable, as it puts all birds & animals to flight, & likewise destroys that quietness which is the principal charm of these forests. — This morning has been the fourth attempt to reach the sea by crossing a mere band of wood, each time I followed a track made by the woodmen, but as soon as that ended I was utterly disabled by the thickets from proceeding even five yards further. To night there has been a good deal of lightning, & the air very sultry. Therm. 75°. — As far as I am able to judge, it would seem that in hot countries, the effect produced on the body increases in a greater ratio than the temperature; that is to say, if at present the thermometer was to rise to 85° the debilitating effects would be *more* than double, than if it was at 80°. —

22*nd* This has been my alternate day of rest, whilst working at the yesterdays collecting. — I give up the evenings to reading & writing; in the latter, the number of friends to whom I am in debt keeps me in full employment. I have just finished Ansons voyage,[1] my pleasure in reading such works is at least trebled by expecting to see some of the described places & in knowing a little about the sea. —

[1] George Anson. *A Voyage round the world, in the years MDCCXL, I, II, III, IV by George Anson Esq . . . compiled from papers . . . of . . . Lord Anson . . . by Richard Walter.* London, 1748.

23*rd* Collected numerous animals on the sandy plain, which skirts the sea at the back of the Sugar loaf. — The ground|161| here being cleared of Cactuses & bushes is for many acres planted with Pineapples. They are cultivated in straight rows, & at a considerable distance apart. — Thus does this fruit nursed with so much care in England here occupy land, which for all other purposes is entirely sterile & unproductive. — The number of oranges which the trees in the orchards here bear, is quite astonishing. I saw one to day where I am sure there were lying on the ground sufficient to load several carts, besides which the boughs were almost cracking with the burthen of the remaining fruit. —

24*th* Remained at home.

25*th* Walked to the city to procure some things which I wanted, then joined Earl & Derbyshire & we proceeded together to ascend the Caucovado. — The path for the few first miles is the Aqueduct; the water rises at the base of the hill & is conducted along a sloping ridge to the city. — At every corner alternate & most beautiful views were presented to us. — At length we commenced ascending the steep sides, which are universally to the very summit clothed by a thick forest. — The water-courses were ornamented by that most elegant of all vegetable forms, the tree fern. — they were not of a large size, but in the

vividness of the green lightness of the foliage, & in the beautiful curve
of head, they were most classically admirable. — |162| We soon gained
the peak & beheld that view, which perhaps excepting those in Europe,
is the most celebrated in the world. — If we rank scenery according to
the astonishment it produces, this most assuredly occupies the highest
place, but if, as is more true, according to the picturesque effect, it falls
far short of many in the neighbourhead. — Everybody has remarked
that a landscape seen from an eminence loses much of its beauty, &
although here the two elements are largely present, which perhaps are
least injurious from this cause, viz. an extent of forest land & of open
sea, yet the observation holds good. — The Caucovado is about 2000
feet high, one side of it for nearly 1000 is so precipitous, that it might be
plumbed with a lead. —at the foot there is a large wood; nothing pleased
me so much as the beautiful appearance this presented when seen so
nearly vertically. — It would lead one to suppose that the view from a
Balloon would be exceedingly striking. — Some years ago a poor insane
young woman threw herself from this summit; in few places could a
more horrible lovers leap be found. — Our present host, Mr Bolga, was
one of the first who found the corps dashed into pieces amongst the
trees & rocks. — |163|

26th During to day & yesterday there has been a strong breeze from the
SW; the amount of evaporation which a current of air produces in these
countries is very great & in consequence the comparative state of
dryness of the road has been today very remarkable. After dinner I
walked to the Bay & had a good view of the Organ mountains; I was
much struck by the justness of one of Humboldts observations, that
hills in a Tropical country seen from a distance are of a uniform blue tint,
but that contrary to what generally is the case the outline is defined with
the clearest edge. — Few things give me so much pleasure as reading
the Personal Narrative; I know not the reason why a thought which has
passed through the mind, when we see it embodied in words,
immediately assumes a more substantial & true air. — In the same
manner as when we meet in dramatick writings a character which we
have known in life, it never fails to give pleasure. —

27th Walked to the Botanic Garden,[1] this name must be given more out
of courtesy than anything else; for it really is solely a place of amusement. — The chief & great interest it possesses, is the cultivation of
many plants which are notorious from their utility. — There are
some|164| acres covered with the Tea tree. — I felt quite disappointed at
seeing an insignificant little bush with white flowers & planted in

straight rows. — Some leaves being put into boiling water, the infusion
scarcely possessed the proper tea flavour. — There were trees of Cam-
phor, Sago, Cinnamon, Cloves & Pepper, the leaves of all, especially
the Cloves & Cinnamon, had a delightful aromatick taste & smell. —
The Bread-fruit was growing in great luxuriance; the leaves from their
great size & deep divisions were uncommonly handsome. Oh for the
time, when I shall see it in its native Pacific isles. — The Mango &
Jack-fruit were likewise here; I did not before know their names. — The
landscape about Bahia takes its character from these two most beautiful
trees; as for the Mango I had no idea any tree could cast so black a
shadow. — They both bear to the evergreen vegetation of the Tropics
the same ratio which laurels do to our English trees. — In this zone these
three latter, together with the Banana, Orange, Cabbage palm &
Cocoa-nut tree, stand before all others (with the exception perhaps of
the tree fern & some firs) in the beauty of their appearance; At the same
time how remarkably they contribute to the subsistence of mankind: &
in this double respect how far do they surpass those of Europe. — The
Tropics appear the natural birthplace of the human race; but the mind,
like many of its fruits |165|seems in a foreign clime to reach its greatest
perfection.

[1] This piece of land had been bought by the Brazilian Government in the previous century
for use as a gunpowder factory, and for the acclimatization of imported plants. It was
converted into the Royal Botanic Garden by the Regent Dom Pedro I in 1819.

28th   Visited the shore behind the Sugar Loaf & again obtained vast
numbers of insects. — The situation being much the same as that of
Barmouth[1] many of the insects were closely allied; as I watched the
elegant Cicindelæ running on this sand, Barmouth with all its charms
rose vividly before my mind. —

[1] CD had been a member of a reading party at Barmouth in North Wales during July and
August 1828, and had collected many beetles and other insects. He visited Barmouth
again briefly in July 1829, and also escorted his sisters there the following month. See
Correspondence 1: 57–62, 88, 91.

29th   Cloudy greyish day; something like an Autumnal one in England;
without however its soothing quietness. — I wanted to send a note this
morning into the city & had the greatest difficulty in procuring anybody
to take it. — All white men are above it, & every black about here is a
slave. — This, amongst other things, is one great inconvenience of a
slave country. —

30th   Again ascended with Derbyshire the Caucovado & took with me the
Mountain Barometer. — I make it to be 2,225 above level of the sea. —

(*Note in margin:* Real height 2330) During the time we were on the summit we were either in a cloud or rain. — Whilst passing through the woods, I observed the same fact, which I have mentioned about the interior forests, viz the smallness of the trunks of the trees. — Very few reached (I believe not more than 3 or 4) reached seven feet in circumference: & only one 9^{ft} 7^{inch}. — The Caucovado is notorious for Maroon or run-away slaves; the last time we ascended, we met three most villanous|166| looking ruffians, armed up to the teeth. — they were Maticans or slave-hunters, & receive so much for every man dead or alive whom they may take. — In the former case they only bring down the ears. — A slave, who has since voluntarily delivered himself up, run away from Mr Lennons estate on the Macaè & lived in a cave for two years & a half. — So easy is it in these countries for a man to support himself. — Amongst other things which the anti-abolitionists say, it is asserted that the freed slave would not work. I repeatedly hear of run-away ones having the boldness of working for wages in the neighbourhead of their masters. If they will thus work when there is danger, surely they likewise would when that was removed. — Again the blacks, who have been seized by British men of war, are hired out to different tradesmen for seven years, by which time it is supposed they could support themselves. — I have heard many instances from the masters, that they claim their freedom before the expiration of the time: & set up for themselves. — What will not interest or blind prejudice assert, when defending its unjust power or opinion? —

31st Staid at home; the evenings now soon close in; whilst I am lamenting the northern progress of the sun, everybody in England is rejoiced at|167| it: as yet I am no ways accustomed to this reversed order of things. It sounds very good to hear of fruits only ripening at Christmas. —

June 1st Took a long ride, in order to geologize some of the surrounding hills. — After passing for some time through lanes shaded by hedges of Mimosas, I turned off into a track into the forest. — The woods even at this short distance from the city are as quiet & unfrequented as if a civilized man had never entered them. — The path wound up the hill: at the height of 5 or 6000 feet I enjoyed one of those splendid views, which may be met with on every side of Rio. — At this elevation the landscape has attained its most brilliant tint. — I do not know what epithet such scenery deserves: beautiful is much too tame; every form, every colour is such a complete exaggeration of what one has ever beheld before. — If it may be so compared, it is like one of the gayest scenes in the Opera House or Theatre. —

2nd Collected in the neighbourhead of the house: I trust there is a change in the weather: the Hygrometer showed the air to be twice as dry in the middle of the day as in the morning. — There was a good example of what Humboldt says of "the thin vapour, which without changing|168| the transparency of the air, renders its tints more harmonious, softens the effects" &c &c. In one of these days when there is such a profusion of light, the consequent dark shadows are well opposed to the general brightness of the view. —

Sunday 3rd Staid quietly at home, & in the evening walked to the Lagoa. Called on a Mr Roberts, one of the endless nondescript characters of which the Brazils are full. —broken down agents to speculation companies; officers who have served under more flags than one: &c &c to all of whom I am charitable enough to attribute some little peccadillo or another. —

4th Got up at 4 oclock to go out hunting: the person who keeps the hounds is a priest & dean. —the pack only consists of five dogs, their names, Trumpeta, Mimosa, Clariena, Dorena & Champaigna; the huntsman is a black man & performed the other offices of body servant & Clerk. The padre is a very rich man & a great favourite of the last queens; we got to his country house at 5 oclock & found with him another brother priest. — It was very curious to see the miserable manner such men could live in; one sort of shed where dogs, black men & themselves appeared to live together; & the whole place dirty & out of order. — At about seven we arrived at our hunting ground, & put up the horses at a small farm house situated in the middle of the woods. — The hunting consists in all |169|the dogs being turned into the forest & each separately pursues its own game. — The hunters with guns station themselves in the places most likely for the animals, such as small deer & pachas (like guinea pigs) to pass by. — And in the intervals they shoot parrots & Toucans &c.. — I soon found this very stupid & began to hunt my own peculiar game. — The wood contained by far the largest trees I have yet seen. —the average I should think was double of what I have before seen, being about 6 feet in circumference, of course as before there are many larger & smaller trees. — Perhaps in consequence of the greater size this one was much less impenetrable than the generality & might easily be traversed in all directions. — The eldest son of the farmer accompanied us & was a good specimen of the country Brazilian

youths.— His dress consisted of a tattered shirt, pair of trowsers, &
wooden slippers (in keeping on which he showed most singular
dexterity) & no hat & long hair.— He carried with him an old fashioned
gun & an enormous knife.— They use the latter for killing animals & as
they walk along incessantly continue cutting the branches so as to
improve old & make new paths.— This practice is universal, & in
consequence of the habit|170| of carrying the knife, many murders take
place.— It is not at all necessary for them to approach the person as they
can throw the knife to a great distance with force & precision.— The day
before this young man had shot 2 large bearded monkeys & had left
another dead in the tree: these monkeys have prehensile tails, which
when dead by the very tip will support the whole weight of the
animal.— He took with him a mulatto with an axe & to my surprise
proceeded in order to get the monkey, to cut down an enormous tree;
they soon affected this & as it fell with an awful crash it tore up the earth
& broke other trees & itself.— We joined our party, whom we found
shooting beautiful little green parrots; the young Brazilian soon sig-
nalized himself by his hawks eye & steady hand.— We then eat our
dinner & drank wine in the true Don Quixote fashion out of a bag of
goats skin.— After a score of profound bows & with our hands to our
hearts repeating "Monte, Monte, obligado", we took leave of the two
hospitable & intelligent padres proceeded home.— I found on my table
a letter from Shrewsbury dated March 12th.— [1]|171|

I also found King, who had arrived late the evening before in the
Beagle.— He brought the calamitous news of the death of three of our
ship-mates.— They were the three of the Macacù party who were ill
with fever when the Beagle sailed from Rio.— 1st Morgan, an extraordinary powerful man & excellent seaman; he was a very brave man &
had performed some curious feats, he put a whole party of Portugeese
to flight, who had molested the party; he pitched an armed sentinel into
the sea at St Jago; & formerly he was one of the boarders in that most
gallant action against the Slaver the Black Joke.— 2nd Boy Jones one of
the most promising boys in the ship & had been promised but the day
before his illness, promotion.— These were the only two of the sailors
who were with the Cutter, & picked for their excellence.— And lastly,
poor little Musters; who three days before his illness heard of his
Mothers death. Morgan was taken ill 4 days after arriving on board &
died near the Abrolhos, where he was lowered into the sea after
divisions on Sunday—for several days he was violently delirious &
talked about the party.— Boy Jones died two days|172| after arriving at

Bahia, & Musters two days after that. — They were both for a long time insensible or nearly so. — They were both buried in the English burial ground at Bahia; where in the lonely spot are also two other midshipmen. The other five of the party were all slightly attacked; none of them for more than a day or two. — Macacù has been latterly especially notorious for fevers: how mysterious & how terrible is their power.[2] It is remarkable that in almost *every* case, the fever appears to come on several days after returning into the pure atmosphere. — I could quote numbers of such cases: is it the sudden change of life, the better & more stimulating food, which determines the period? — Humboldt & Bonpland,[3] after living for months in the forests, as soon as they returned to the coast, both were seized by violent fevers. —

The Beagle made a very good passage up; being only 5 days, she passed a few miles inside of the Abrolhos. — A French corvette sailed 8 days before & promised our Captain to have dinner ready for him on his arrival at Bahia; as it turned out the case was reversed; such is the advantage of a good knowledge of the winds|173| & coast. — She staid a week at Bahia. — And 12 days back to Rio; she would have been some days shorter on the passage, had she not been becalmed at Cape Frio. —

[1] This was a letter from Caroline Darwin. See *Correspondence* 1: 215–18.
[2] According to the account given by FitzRoy (see *Narrative* 2: 76–7), the Macacú was notorious for malaria; but the mode of infection through the mosquito as a vector was, of course, still unknown.
[3] Aimé Bonplan was the botanist who accompanied Alexander von Humboldt on his travels in South America.

5th   Worked at the produce of yesterdays hunt; in the evening went out geologizing. — Earl has returned (he has been staying for a week with some friends in the city) & brought a good deal of news from the Beagle. —

6th   Went on board & breakfasted with the Captain, spent the day between the city & the Beagle. — Going on board gives in a small degree the comfortable feeling which is always experienced on returning home. — Having lived so long on shore, I have almost forgotten how to stow myself in my own corner. —

7th   Rode with M[r] Bolger to the chapel of Nossa Senhora de Penha; this being one of the sights of the country. — Our road lay through the North & back part of the city, which covers a much greater space than I had imagined. The suburbs are very filthy & are surrounded by marshes covered with the Mangrove; the tide occasionally flows into them, & is sufficient to cause a continual putrefaction of vegetable &

animal matter, which is rendered very perceptible to the nose. — The land surrounding the Bay is generally thus situated for instance Macucù & in consequence unhealthy.|174| As we proceeded in this direction nothing could be more uninteresting than the country. — Nossa Senhora is a gay little chapel built on one of the naked rounded hills of gneiss so frequent in this country. — Some hundreds of steps lead to the summit & there is an extensive view of the harbor & its islands. — On our return we rode to the palace of St Christophe; at a distance, from its large & regular dimensions & from the bright colours of the walls, it has a grand appearance. — I was much struck by the beauty of the right hand side building; I did not expect to see any thing so elegant in the Brazils. — The gate, which the Duke of Northumberland sent as a copy of the one at Sion house,[1] stands on the edge of a hill where there is no path; even under such circumstances it is highly ornamental.

[1] The gate presented to Dom Pedro I by the Duke of Northumberland in 1815 now forms the entrance to the Jardim Zoologico.

8th Collected some Corallines on the rocks, which surround part of Botofogo Bay. —

9th Started at ½ after six with Derbyshire for a very long walk to the Gavia. — This mountain stands near the sea, & is recognised at a great distance by its most singular form. — Like the generality of the hills, it is a precipitous rounded cone, but on the summit is a flat angular mass, whence it takes the name of "table" or *topsail* mountain. —|175| The narrow path wound round its Southern base; the morning was delightful; & the air most fragrant & cool. — I have no where seen liliaceous plants & those with large leaves in such luxuriant plenty; growing on the border of the clear shaded rivulets & as yet glittering with drops of dew, they invited the traveller to rest. — The ocean, blue from the reflected sky, was seen in glimpses through the forest. — Islands crowned with palms varied our horizon. — As we passed along, we were amused by watching the humming birds. — I counted four species — the smallest at but a short distance precisely resembles in its habits & appearance a *Sphinx*. — The wings moved so rapidly, that they were scarcely visible, & so remaining stationary the little bird darted its beak into the wild flowers. — making an extraordinary buzzing noise at the same time, with its wings. — Those that I have met with, frequent shaded & retired forests & may there be seen chasing away the rival butterfly. In vain we attempted to find any path to ascend the Gavia; this steep hill subtends to the coast at an angle of 42°. — We returned home; at our furthest point we had a good view|176| of the coast for many miles. — It was skirted by a band of thick brushwood: behind which was a wide plane of marshes & lakes; which in places were so green, that they looked like meadows. —

Sunday 10th   Like a schoolboy in his holidays, I tremble as I perceive another week completed. —

11th   Rode to the place where I was the other day hunting with the Padre; having put up my horse, I started for the woods. — A mulatto & a little Brazilian boy accompanied me; — the latter was quite a child, but dressed in the same manner as I described the eldest son. — I never saw anything at all equal to his power of perception. — Many of the rarest animals in the most obscure trails were caught by him. — I should have as soon expected a beetle to have turned traitor & been my coadjutor, as to have found so able a one in this little fellow. — It really was like what one reads of the talent of observation which the Indians possess, my eyes with years of practice were not at all on a par with this childs. — I wish the Brazilians, as they advance in age, could keep the pleasant & engaging manners which they possess in youth. — My companions left me & I proceeded on my scramble into the interior of|177| the forest. A profound gloom reigns everywhere; it would be impossible to tell the sun was shining, if it was not for an occasional gleam of light shooting, as it were through a shutter, on the ground beneath; & that the tops of the more lofty trees are brightly illuminated. — The air is motionless & has a peculiar chilling dampness. — Whilst seated on the trunk of a decaying tree amidst such scenes, one feels an inexpressible delight. — The rippling of some little brook, the tap of a Woodpecker, or scream of some more distant bird, by the distinctness with which it is heard, brings the conviction how still the rest of Nature is. —

I returned to the house; where I found several people collected after dinner; this being one of their numerous feast-days. — The many contrivances for catching animals which my large pockets (not the least subject for surprise) contained, afforded ample grounds for curiosity & wonder. In both of which, with a great deal of good-nature they most freely indulged. — They assuredly thought me a greater curiosity than anything their woods contained. —

12th   Worked in the morning at yesterdays produce, a forest is a gold mine to a Naturalist & yesterdays a very rich one. — At one |178|oclock I went to the Admirals for a grand boat race. — The first arriving of the launches, yawls, cutters & other large boats, was an imposing sight. —

It immediately made one understand how powerful a flotilla of such boats would be in war. — The racing was rather too long; especially as the Beagle did not come off quite so triumphantly as might have been wished for. — The evening passed away pleasantly, & by moonlight on the beach several foot races were got up between the officers & the crews of Captains gigs.

13ᵗʰ   Dined with Mʳ Cairnes; who is the only merchant whom I have met with in society. — The generality are little above shopkeepers. I spent an agreeable evening. — Mʳ Price, a merchant from round the Horn & a passenger with Capᵗ Waldegrave, gave a great deal of amusing & interesting information about the plains or what we better designate these the horse & cattle breeding countries. — Mʳ Price married a Spanish lady who is since dead & has with him his two little daughters, Carlotta & Theresa; the Signoritas can speak nothing but Spanish; very pretty, & their motions most exceeding graceful; Theresa, the least about 8 years old, could not help dancing when she heard music, & with a rose in each hand as her partner, danced most exquisitely. — |179|

14ᵗʰ   Dined with Mʳ Aston; a very merry pleasant party; in the evening went with Mʳ Scott (the Attache) to hear a celebrated pianoforte player. — He said Mozarts overtures were too easy. I suppose in the same proportion as the music which he played was too hard for me to enjoy. —

15ᵗʰ   Collected some beautiful Corallines on the rocks at Botofogo bay. — Mʳ Earl does not return to his lodgings here, but remains in town till the Beagle sails. —

16ᵗʰ   Started early in the morning for Tijeuka to see the waterfalls. — Neither the height or the body of water is anything very imposing; but they are rendered beautiful, by the dampness so increasing the vegetation, that the water appears to flow out of one forest & to be received & hidden in another below. — On the road the scenery was very beautiful; especially the distant view of Rio. — As a Sultan in a Seraglio I am becoming quite hardened to beauty. — It is wearisome to be in a fresh rapture at every turn of the road. And as I have before said, you must be that or nothing. —

17ᵗʰ   Took my usual evening stroll to the bay; there to lie down on the coast & watch the setting sun gild the bare sides of the Sugar Loaf. — Wickham & Chaffers paid me a visit. — |180|

18ᵗʰ   King came & spent the day with me; we both on horseback started for the old forest. — He shot some birds & as is generally the case I found many interesting animals of the lower classes. — We found a little Palm tree, only a few inches in circumference, which I believe to be 305 years old. — I judge of this from its number of rings, each of which I imagine marks a year. — On the road home I overtook my old friend the Padre, returning with his dogs from the Gavia. — He presented to me a magnificent specimen of the little once, which after five hours hunting, he had succeeded in shooting.

19ᵗʰ   Spent the day between the city & being on board. — They are very busy in stowing provisions for sea. — The ship looks in same inextricable confusion which she was in in Plymouth. — The Warspite is making sweeps & boarding netting for us. — Our complement of men is increased. — Mʳ Forsyth is removed from the Flag ship into the Beagle & fills the place of poor little Musters. —

20ᵗʰ, 21ˢᵗ & 22ⁿᵈ   During these days have been busily employed with various animals; chiefly however corallines: & my walks in consequence have not been extended far from the house. —

23ʳᵈ   Again I went to the forest, which so often has been proved so fruitful in all kinds of animals. — It is in all probability the last time I shall ever wander in a|181| Brazilian forest. — I find the pleasure derived from such scenes increases, instead of as might have been expected, diminishing. To day instead of the rude tracks, I followed a brook, which in a narrow ravine flowed amongst the huge granitic blocks. — No art could depict so stupendous a scene. — the decaying trunks of enormous trees scattered about, formed in many places natural bridges; beneath & around them the damp shade favoured the growth of the Fern & Palm trees. — & looking upwards the trees in themselves lofty, thus seen, appeared of an almost incredible height. — I soon found even by creeping, I could not penetrate the entangled mass of the living & dead vegetation. — On coming out of the forest, the effect without any exaggeration is that of the full light of the sun breaking on a person who has just left a darkened room. — These woods belong to the government; & the house where I put up the horse is called Chacera o Macâco. — My host the owners name is Antonio da Rocha, & to his hospitality I am indebted for so many delightful walks. — Before going he showed me his garden—where to an European eye there was a singular union of plants. —|182| On one side a fine set of cabbages were growing & joining to these the long stubble of a rice field. — This latter is scarcely to be distinguished from barley; but the ears are different, the seeds being further apart & therefore not having so compact a head. —

After returning home in the evening the quiet neighbourhead of Botofogo was in unusual agitation in celebrating the eve of St Juan. — Round the numerous bonfires there is a continual firing of rockets, guns, crackers, accompanied by shouts of "Viva St Juan". — This is continued during the greater part of the night. — I presume, not having had the luck to have had a gun-powder plot, the Brazilians thus celebrate an innocent saint. —

24th Dined with Mr Cairns; & as far as society goes the pleasantest evening since I left England. — The Captain was there & has announced that the Beagle will sail this day week. — In the evening my little friend Signorita Theresa, whom I find is only 6 years old, gained universal admiration by her dancing & acting. —

25th In the evening took a farewell stroll to the Lagoa, & saw for the last time its waters stained purple by the last rays of twilight. —|183|

26th Rode to the city & went on board in order to make final arrangements for living in the ship after my long absence. — I dread this process nearly as much as I did at Devonport. — There have been several alterations in the ship. —amongst others we have 2 long nine-pounders; this will make us much more independent: several cases occurred during the last war where very small vessels terribly injured large ones, from having one great gun & keeping out of range of the other. — I am sorry to see so many new faces on the deck. —in the whale boat which took me ashore there was not one old-hand. —

27th This is my last day on shore, so I was determined it should not be an idle one. — In the bay I found some fine Corallines; the examination of which occupied me during the whole day. — Upon the whole I am tolerably contented with what I have done at Rio in Natural Hist: several important branches have been cut off: Geology is here uninteresting, Botany & Ornithology too well known. — And the sea totally unproductive excepting in one place in Botofogo Bay. —so that I have been reduced to the lower|184| classes, which inhabit the dry land or fresh water. — The number of species of Spiders which I have taken is something enormous. — The time during these eleven weeks has passed so delightfully, that my feelings on leaving Botofogo are full of regret & gratitude—

28th Removed all my things from shore & am now once again in the intricacy of my own corner, writing this journal. — It is something quite cheering to me to hear the old noises. —the men foreward singing; the centinel pacing above my head & the little creeking of the furniture in the Cabin &c.

29th   We go to sea on next Tuesday, so that I have a nice short time for finishing the collections I made at Botofogo.

The very interesting & important news of the minority of Earl Grey on the reform was brought late last night by the Packet. —[1] The latest information is 20th of May. — The distance of time & space from the events takes from me the keen interest for Politicks & Newspapers. —

[1] The Prime Minister's Reform Bill had been defeated in the House of Lords on 7 May 1832.

30th   Went to the city to purchase several things. — Nothing can be more wearisome than shopping here. — From the length of time the Brazilians detain you & the unreasonable price they at first ask, it is clear that they think both these precious things are equally valueless to an Englishman. —|185|

Sunday, July 1st   Attended divine service on board the Warspite: the ceremony was imposing; especially the preliminary parts such as the "God save the King", when 650 men took off their hats. — Seeing, when amongst foreigners, the strength & power of ones own Nation, gives a feeling of exultation which is not felt at home. — This ship would be in exactly the same state, if she was going to fight another battle of Trafalgar. — It is in the whole & its parts a most splendid piece of mechanism. — Can one wonder at pride in the Captain, when he knows that all & everything bends to his will? When standing on the Quarter deck, in the midst of such a crew, can there be imagined a more lofty situation? — After divisions (the men being all arranged along deck in the two watches), the head officers go the rounds of the whole ship. — I accompanied them, & thus well saw all the store-rooms &c. — Those who have never seen them will form no just idea of their cleanliness & extreme neatness. — After Church I was introduced to two officers who were fond of Nat: History: I was surprised to find in one of their cabins an aviary of Cape-birds & plants in frames. — I dined in the Ward-room & had a very agreeable party. — Coming from a ten-gun Brig into such comforts & luxuries, makes one a little envious. — So many corners unoccupied, appeared to my eyes as great a waste|186| as throwing good food overboard. — After the Kings health & "God save the King" the band played some beautiful music. — It was no common pleasure to hear the Overture to Figaro, Semiramide, Il Barbiere. After so long a fast, the appetite for Music becomes very keen. —

Before I returned to the Beagle I saw all the hammocks carried down out of the nettings. —it is said that this rush of the men surprised Napoleon more than anything else on an English ship.

2nd   Walked to Botofogo & called on the Admiral, Mr Aston & Mr Price. —
The latter I hope we shall again see at Valparaiso: He is afraid 17 years
in Chili has quite unfitted him for any other country, & now on his road,
he is sorry he ever attempted the change. — It will make Valparaiso very
pleasant if we are lucky enough to find him there. —

3rd   Went to the city. On landing, found the Palace Square crowded with
people round the house of two money changers who were murdered
yesterday evening in a more atrocious manner than usual. — It is quite
fearful to hear what enormous crimes are daily committed & go
unpunished. — If a slave murders his master, after being confined for
some-time he then becomes a government one. — However great the
charge|187| may be against a rich man, he is certain in a short time to be
free. — Everybody can here be bribed. — A man may become a sailor or
a physician or any profession, if he can afford to pay sufficiently. — It
has been gravely asserted by Brazilians that the only fault they found
with the English laws was that they could not perceive rich respectable
people had any advantage over the miserable & the poor.
   The Brazilians, as far as I am able to judge, possess but a small share
of those qualities which give dignity to mankind. Ignorant, cowardly,
& indolent in the extreme; hospitable & good natured as long as it gives
them no trouble; temperate, revengeful, but not quarrelsome; contented with themselves & their customs, they answer all remarks by
asking "why cannot we do as our grandfathers before us did". — Their
very appearance bespeaks their little elevation of character. —figures
short, they soon become corpulent; and their faces possessing little
expression, appear sunk between the shoulders. — The Monks differ
for the worse in this latter respect; it requires little physiognomy to see
plainly stamped persevering cunning, sensuality & pride. — One old
man I always stop to look at, the only thing I ever saw like it, is Scoens[1]
Judas Iscariot. — All that I have said about the countenances|188| of the
priests, may be transferred to the voices of the older women. — Being
surrounded by slaves, they become habituated to the harsh tones of
command & the sneer of reproach. — Their manners are seldom softened by terms of endearment: they are born women, but die more like
fiends. — It will be more readily believed, when I state that Mr Earl has
seen the stump of the joint, which was wrenched off in the thumb-
screw which is not unfrequently kept in the house. —
   The state of the enormous slave population must interest everyone
who enters the Brazils. — Passing along the streets it is curious to
observe the numbers of tribes which may be known by the different

ornaments cut in the skin & the various expressions. — From this results the safety of the country. The slaves must communicate amongst themselves in Portugeese & are not in consequence united. — I cannot help believing they will ultimately be the rulers. I judge of it from their numbers, from their fine athletic figures, (especially contrasted with the Brazilians) proving they are in a congenial climate, & from clearly seeing their intellects have been much underrated. —they are the efficient workmen in all the necessary trades. — If the free blacks increase in numbers (as they must) & become discontented at not being equal to white men, the epoch of the general liberation would not be far distant.|189| I believe the slaves are happier than what they themselves expected to be or than people in England think they are. — I am afraid however there are many terrible exceptions. — The leading feature in their character appears to be wonderful spirits & cheerfulness, good nature & a "stout heart" mingled with a good deal of obstinacy. — I hope the day will come when they will assert their own rights & forget to avenge their wrongs. —

[1] Evidently referring to an engraving by Martin Schoen or Schongauer (c. 1430–91) showing the taking of Christ. A faint pencil note 'Who' is written in the margin, probably in Hensleigh Wedgwood's hand.

4th   In the evening unmoored ship; now therefore it is certain we leave Rio in the morning. — I am very glad, as nothing can be more dull than lying in the harbor. — And I always find the interval between sailing & the first day announced hangs heavily on hand. —

5th   A little after 9 oclock we tripped our anchor, & with a gentle breeze stood out of the bay. — Capt⁵ Talbot & Harding accompanied us beyond Santa Cruz. — As we sailed past the Warspite & Samarang (our old Bahia friend) they manned the rigging & gave us a true sailor-like farewell, with three cheers. — The band at the same time striking up "To glory you steer". — The Captain had intended touching at Cape Frio, but as the lightning did so. —we made a direct course for the South. — Near to the Isle de Raza the wind lulled, & we are now becalmed & shall probably remain so during the night:|190| The moon is now shining brightly on the glassy water. —every one is in high spirits at again being at sea & a little more wind is all that is wanted. — The still & quiet regularity of the ship is delightful; at no time is "the busy hum of men" so strongly perceived as when leaving it for the open ocean. —

6th   Scarcely any wind. The Sugar Loaf is still in view & points out the entrance into Rio.

7th, Sunday 8th & 9th   The weather has been most provoking; light variable breezes, a long swell, & I very sick & miserable. — This second attack of sea-sickness has not brought quite so much wretchedness as the former one. But yet what it wants in degree is made up by the indignation which is felt at finding all ones efforts to do anything paralysed. —

10th In the afternoon the calm was broken by a stiff breeze, almost a gale: (i.e. a very heavy one in a Landsmans eyes). — We first lowered the Top-gallant yards, & then struck the masts. — This was the first time that I have been able to look about, when there has been anything of a sea up. — It was a beautiful spectacle to see how gracefully the Beagle glided over the waves, appearing as if by her own choice she avoided the heavy shocks. — As the night came on, the sky looked very dirty, & the waves with their white crests dashed angrily against the|191| ships sides. — In the middle watch however the wind fell & was succeeded by a calm: this is always the worst part of a gale, for the ship not being steadied by the wind pressing on the sails rolls in a most uncomfortable manner between the troughs of the sea. —

11th The day has passed in listless discomfort. — if I had been well several things would have interested me during these latter days. — The vessel has been followed by many sorts of Petrels. — a very elegant one, the Cape-pidgeon, we met as is generally the case on passing the Tropic. — Several Whales have been seen. — I just had a peep at one, but to my jaundiced eyes, it even possessed little interest. —

12th The wind yet continues foul, but light: we are only about 150 miles from Rio, & 700 from Cape St Mary's.

13th A beautiful day; the bright sky & smooth water reminded me of the delightful cruises on the Tropical seas. — But as now we are pressing all sail to the stormy regions of the South, the sooner such scenes are forgotten, the more tolerable will the present be. — Everybody is full of expectation & interest about the undescribed coast of Patagonia. — [1] Endless plans are forming for catching Ostriches, Guanaco, Foxes &c.|192| Already in our day-dreams have we returned heavily loaded with Cavies, Partridges, Snipes &c. — I believe the unexplored course of the Rio Negro will be investigated. — What can be imagined more exciting than following a great river through a totally unknown country? — Every thing shows we are steering for barbarous regions, all the officers have stowed away their razors, & intend allowing their beards to grow in a truly patriarchal fashion. —

[1] The inhabitants of Patagonia were named by Magellan when he first encountered them in 1520 the Patagónes, 'big feet of bears', because of the enormous boots of fur that they wore.

14th   Fine day & a prosperous breeze.

Sunday 15th   From noon of yesterday to the same time to day we had run 160 knots & all congratulated ourselves on soon doubling Cape St Mary's. — On the contrary however we experienced the true uncertainty of a sailors life. — By the evening it blew a gale right in our teeth. — Top-gallant masts were sent on deck. —& with close reefed main top-sail, trysails & fore sails, we beat up against a heavy sea. —

We are about 80 miles from the Morro de St Martha. —It is a curious fact, that often as the different officers have passed this point they have always met a gale. — The Beagle, on her return to England from the last expedition, experienced the heaviest she had had during the whole time. — In the morning I was much interested by|193| watching a large herd of Grampuses, which followed the ship for some time. — They were about 15 feet in length, & generally rose together, cutting & splashing the water with great violence. — In the distance some whales were seen blowing. — All these have been the black whale. — The Spermaceti is the sort which the Southern Whalers pursue. —

16th   There was a good deal of sea up & I in consequence, with my spirits a good deal down. —

17th   My eyes were rejoiced with the sight of studding sails, alow & aloft. —that is wind abaft the beam & favourable.

18th   We are driving along at the rate 8 & 9 knots per hour. — A wonderful shoal of Porpoises at least many hundreds in number, crossed the bows of our vessel. — The whole sea in places was furrowed by them; they proceeded by jumps, in which the whole body was exposed; & as hundreds thus cut the water it presented a most extraordinary spectacle. — When the ship was running 9 knots these animals could with the greatest ease cross & recross our bows & then dash away right ahead. — Thus showing off to us their great strength & activity. — Several flying-fish were skimming over the water; considering time of year & Latitude 31° 37′ S: Long 49° 22′ W, I was surprised to see them.|194|

19th   A calm day.

20th   There is a fine breeze but we can hardly keep our course. — At noon we were 160 miles from Cape St Mary. — We have experienced to day a most complete change of climate. — From the joint cause of shoal water & probably a current from the South, the temperature of the sea at noon was 61°½, it being in the morning 68°¼. — The wind felt quite chilling; the thermometer standing at 59°. — By the time we arrive in harbor, we shall have made a very bad passage & I am sure to me a very tedious one. — The only thing I have been able to do is reading Voyages & Travels. — these are now to me much more interesting than even novels. —

21st The weather to day felt just like an Autumnal day in England. — In the evening the wind freshened & a thick fog came on. — These are very frequent in the neighbourhead of the Plata, & we are only now about 50 miles from the Mouth. — The night was dirty & squally: we were surrounded by Penguins & Seals which made such odd noises that in the middle watch Mr Chaffers went below to report to Mr Wickham that he heard cattle lowing on shore. —

Sunday 22nd   We have had this morning a true specimen of the Plata weather. — The lightning was most vivid, accompanied by heavy rain & gusts of wind. — The day has been exceedingly|195| cold & raw. — We passed through large flocks of different sea-birds. — & some insects & a bird very like a yellow hammer flew on board. — We are about 50 miles from Cape St Marys. — I have just been on deck. — the night presents a most extraordinary spectacle. — the darkness of the sky is *interrupted* by the most vivid lightning. — The tops of our masts & higher yards ends shone with the Electric fluid playing about them. — the form of the vane might almost be traced as if it had been rubbed with phosphorus. — (*Note in margin:* St Elmo's fire.) To complete these natural fireworks. — the sea was so highly luminous that the Penguins might be tracked by the stream of light in their wake. — As the night looked dirty & there were heavy squalls of rain & wind, we have dropped our anchor. —

23rd All day we have been beating up the river, & now at night we are come to an anchor. — We were generally at the distance of four or five miles from the Northern shore. — Thus seen, it presents a most uniform appearance. — a long straight line of sandy beach was surmounted by a sloping bank of green turf. — On this viewed through a glass were large herds of cattle feeding. — Not a tree broke the continuity of outline: & I only observed one hut, near to which was the Corral or enclosure of stakes, so frequently mentioned by all travellers in the Pampas. — I am afraid we shall not even tomorrow reach M. Video. — |196|

24th The wind yet continues dead on end against us, & as there is a strong current setting out we make scarcely any progress. — The same line of low & green coast is to be seen as yesterday, only not quite so near. — It is quite curious, how much I have suffered from the cold. — The thermometer stands above 50°, & I am loaded with clothes; yet judging from my feelings I should have thought it a very cold English winter day. — Others in the vessel have not experienced this so strongly, so that I presume my constitution in a shorter time becomes habituated to a warm climate. — & therefore on leaving it more strongly feels the contrary extreme. —

I procured this evening a Watch-bill & as most likely our crew will for rest of the voyage remain the same. — I will copy it. — Boatswains mates, J. Smith & W. Williams: — Quarter-Masters, J. Peterson, White, Bennett, Henderson: — Forecastle Men, J. Davis; Heard: Bosworthick (Ropemaker); Tanner; Harper (sailmaker); Wills (armourer); — Foretop-men, Evans; Rensfrey; Door; Wright; Robson; MacCurdy; Hare; Clarke; — Main top-men Phipps; J. Blight; Moore; Hughes; Johns B.; Sloane; Chadwick; Johns; Williams; Blight, B.; Childs; — Carpenters crew, Rogers; Rowe; J. May; James; Idlers, Stebbing (instrument men-der); Ash gunroom steward; Fuller, Captains do; R Davis, boy do; Matthews, missionary; E Davis, Officers cook; G Phillips, ships cook; Lester, cooper; Covington, fiddler & boy to Poop-cabin; Billet, gunroom boy; Royal Marines. — Beazeley, sergeant; Williams, Jones, Burgess, Bute, Doyle, Martin, Middleton, Prior (midshipmens steward); — Boatswain, Mr Sorrell; Carpenter, Mr May; |197| Midshipmen, M^{rss} Stewart, Usborne, Johnson, Stokes, Mellersh, King, Forsyth. — Hell-yar, Captains clerk. — Mr Bino, acting surgeon — Mr Rowlett, purser. — Mr Chaffers, Master. — Mr Sulivan, 2^d Lieutenant; Mr Wickham, 1st Lieutenant; R. FitzRoy, Commander. — There are (including Earl, the Fuegians & myself) 76 souls on board the Beagle. —

I hear the cable rattling through the Hawse-hole so we have come to anchor for the night.

25th A fine breeze has carried us to an anchor within six or seven miles of Monte Video. — At about noon we passed between Maldonado & the little island of Lobos covered with seals. — At some future time we shall lay in the harbor at Maldonado. — the country in the neighbourhead is more uneven than in the other parts of the coast, but from the sandy hillocks has a dreary uninteresting appearance. — To day the water from its calmness & reddish muddy colour looked like that of a river: of course however, the Southern bank is far beyond the reach of vision. — The fresher discoloured water from its less specific gravity floats on the surface of the salt. — this was curiously shown by the wake of the vessel, where a line of blue might be seen mingling in little eddies with the adjoining fluid. — in this case instead of stirring up the mud, it was the reverse & stirred up the clear water. — |198|

26th We entered the bay about 9 oclock: just as we were coming to an anchor, signals were made from the Druid, a frigate lying here; which (to our utter astonishment & amusement) ordered us to "Clear for action" & shortly afterward "Prepare to cover our boats". We set sail again & the latter part of order was shortly explained by the arriving of 6 boats heavily armed with Carronades & containing about 40 marines, all ready for fighting, & more than 100 blue-jackets. — Captain Hamilton came on board & informed us that the present government is a military usurpation. —& that the head of the party had seized upon 400 horses, the property of a British subject; & that in short the flotilla of boats went to give weight to his arguments. — The revolutions in these countries are quite laughable; some few years ago in Buenos Ayres, they had 14 revolutions in 12 months. —things go as quietly as possible; both parties dislike the sight of blood; & so that the one which appears the strongest gains the day. — The disturbances do not much affect the inhabitants of the town, for both parties find it best to protect private property. — The present governor has about 260 Gaucho cavalry & about same number of Negro infantry. —the opposite party is now collecting a force & the moment he enters the town the others will scamper out. — Mr Parry (a leading merchant here) says he is quite certain a 150 men from the|199| Frigate could any night take M: Video. The dispute has terminated by a promise of restitution of the horses; but which I do not think is very clear will be kept. — I am afraid, it is not impossible that the consequences will be very unpleasant to us: The Druids officers have not for some weeks been allowed to go on shore, & perhaps we shall be obliged to act in the same manner. — How annoying will be the sight of green turf plains, whilst we are performing a sort of quarantine on board. —

27th I had no opportunity of taking a long walk. —so that I went with the Captain to Rat island. — Whilst he took sights I found some animals & amongst them there was one very curious. —at first sight every one would pronounce it to be a snake: but two small hind legs or rather fins marks the passage by which Nature joins the Lizards to the Snakes. —[1]

[1] This was a skink or legless lizard.

28th Landed early in the morning on the Mount, This little hill is about 450 feet high & being by far the most elevated land in the country gives the name Monte Video. — The view from the summit is one of the most uninteresting I ever beheld. — Not a tree or a house or trace of cultivation give cheerfulness to the scene. — An undulating green plain & large herds of cattle has not even the charm of novelty. — Whoever has seen Cambridgeshire, |200| if in his mind he changes arable into pasture ground & roots out every tree, may say he has seen Monte Video. — Although this is true, yet there is a charm in the unconfined feeling of walking over the boundless turf plain: Moreover if your view is limited to a small space, many objects possess great beauty. — Some of the smallest birds are most brilliantly coloured; much more so than those in Brazil. — The bright green turf being browsed short by the cattle, is ornamented by dwarf flowers; amongst which to my eyes the Daisy claimed the place of an old friend. — The only other plants of larger size are tall rushes & a thistle resembling much the Acanthus; this latter with its silvery foliage covers large spaces of ground. — I went on board with a party of midshipmen; who had been shooting & had killed several brace of Partridges & wild Ducks, & had caught a large Guano about 3 feet long. — These lizards at certain times of the year are reckoned excellent food. — The evening was calm & bright, but in the middle of night it blew a sudden gale. — All hands were piped up to send Top-gallant masts on deck & to get in the Cutter: In such scenes of confusion, I am doubtful whether the war of the elements or shouts of the officers be most discordant. —

*Sunday 29*th This morning we are pitching heavily, & occasionally a sea breaks over us. — The weather is yet boisterous & the rain very cold. —|201|

*30*th I was busily employed with the collections of Saturday. — The Captain this morning procured information of some old Spanish charts of Patagonia, which are now at Buenos Ayres. — He immediately determined to run up there to see them. —

*31*st At one oclock we stood out of the Bay with a light fair wind. — As we passed the Druid, we picked up Mr Hammond, a midshipman belonging to her who has now joined the Beagle. — Mr Hammond is a connection of poor little Musters. — Before sailing, I went ashore to the Town with the Captain; the appearance of the place does not speak much in its favor; it is of no great size; possesses no architectural beauties, & the streets are irregular & filthily dirty. — It is scarcely credible that any degree of indolence would permit the roads to be in such a bad state as they are. — The bed of a torrent with blocks of stone lying in mud is an exact resemblance. — It was distressing to see the efforts of the Bullocks, as harnessed by their horns to the clumsy carts they managed to stumble on amongst the stones. — As far as regards the inhabitants, they are a much finer set than at Rio de Janeiro. — Many of the men have handsome expressive faces & athletic figures; either of which it is very rare to meet with amongst the Portugeese. — I believe in about a weeks time we|202| shall return to M. Video & complete our equipment. —

August 1ˢᵗ We have had a famous breeze & are now at anchor about 12 miles from Buenos Ayres. — At one time to day it was just possible to see both the Northern & Southern shores of the river at the same time. — A river of such great size & dimensions possesses no interest or grandeur. —

2ⁿᵈ We certainly are a most unquiet ship; peace flies before our steps. On entering the outer roadstead, we passed a Buenos Ayres guard-ship. — When abreast of her she fired an empty gun; we not understanding this sailed on, & in a few minutes another discharge was accompanied by the whistling of a shot over our rigging. Before she could get another gun ready we had passed her range. — When we arrived at our anchorage, which is more than three miles distant from the landing place; two boats were lowered, & a large party started in order to stay some days in the city. — Wickham went with us, & intended immediately going to Mʳ Fox, the English minister, to inform him of the insult offered to the British flag. — When close to the shore, we were met by a Quarantine boat which said we must all return on board, to have our bill of health inspected, from fears of the Cholera. — Nothing which we could|203| say about being a man of war, having left England 7 months & lying in an open roadstead, had any effect. — They said we ought to have waited for a boat from the guard-ship & that we must pull the whole distance back to the vessel, with the wind dead on end against us & a strong tide running in. — During our absence, a boat had come with an officer whom the Captain soon despatched with a message to his Commander to say "He was sorry he was not aware he was entering an uncivilized port, or he would have had his broardside ready for answering his shot". — When our boats & the health one came alongside. — the Captain immediately gave orders to get under weigh & return to M Video. — At same time sending to the Governor, through the Spanish officer, the same messuages which he had sent to the Guard-ship, adding that the case should be throughily investigated in other quarters. — We then loaded & pointed all the guns on one broardside, & ran down close along the guard-ship. Hailed her, & said that when we again entered the port, we would be prepared as at present & if she dared to fire a shot we would send our whole

The mole at Montevideo, by T. A. Prior after A. Earle (*Narrative* 1: 105).

broardside into her rotten|204| hulk. — We are now sailing quietly down the river. — From M Video the Captain intends writing to Mr Fox & to the Admiral; so that they may take effective steps to prevent our Flag being again insulted in so unprovoked a manner. — From what I could see of the city of Buenos Ayres it appears to be a very large place & with many public buildings. — Its site is very low & the adjoining coast is elevated but a few feet above the level of the water. —

3rd   In the morning watch, before it was daylight, the Beagle stood too close in-shore & stuck her stern fast about a foot in the mud. — With a little patience & mæneuvering they got her off, & two whale boats being lowered to sound the bank ahead, we soon gained the channel. —

The navigation of the Plata is difficult, owing to there being no landmarks, the water generally shoal & running in currents & the number of banks in the whole course. — We saw several old wrecks which now serve as buoys to guide other ships. — "It is an ill wind which blows nobody any good". — We arrived at M Video after sunset, & the Captain immediately went on board the Druid. — He has returned & brings the news, that the Druid will tomorrow|205| morning sail for Buenos Ayres, & demand an apology for their conduct to us. — Oh I hope the Guard-ship will fire a gun at the Frigate; if she does, it will be her last day above water.

4th   We altered our anchorage, & stood much closer in. — we found an excellent berth amongst the merchant ships. — After dinner went with Wickham to Rat island & collected some animals. — In the evenings the greater length of twilight is very pleasant: it is quite a new phenomenon to watch the purple clouds of the Western sky gradually to fade into the leaden hue of night. — This is a beauty of which the equinoctial regions can seldom boast. — And to an Europeans eyes it is a great loss. —

5th   This has been an eventful day in the history of the Beagle. — At 10 oclock in the morning the Minister for the present military government came on board & begged for assistance against a serious insurrection of some black troops. — Cap FitzRoy immediately went ashore to ascertain whether it was a party affair, or that the inhabitants were really in danger of having their houses ransacked. — The head of the Police (Damas) has continued in power through both governments, & is considered as entirely neutral; being applied to, he gave it as his opinion that it would be doing a service to the state to land our force. — |206| Whilst this was going on ashore, the Americans landed their boats & occupied the Custom house. — Immediately the Captain arrived at the mole, he made us the signal to hoist out & man our boats. In a very few minutes, the Yawl, Cutter, Whaleboat & Gig were ready with 52 men heavily armed with Muskets, Cutlasses & Pistols. After waiting some time on the pier Signor Dumas arrived & we marched to a central fort, the seat of government. During this time the insurgents had planted artillery to command some of the streets, but otherwise remained quiet. They had previously broken open the Prison & armed the prisoners. — The chief cause of apprehension was owing to their being in possession of the citadel which contains all the ammunition. — It is suspected that all this disturbance is owing to the mæneuvering of the former constitutional government. — But the politicks of the place are quite unintelligible: it has always been said that the interests of the soldiers & the present government are identical. —& now it would seem to be the reverse. — Capt. FitzRoy would have nothing to do with all this: he would only remain to see that private property was not attacked. — If the National band were not rank cowards, they might at once seize the citadel & finish the business; instead of|207| this, they prefer protecting themselves in the fortress of St. Lucia. — Whilst the different parties were trying to negociate matters, we remained at our station & amused ourselves by cooking beefsteaks in the Courtyard. — At sun-set the boats were sent on board & one returned with warm clothing for the men to bivouac during the night. — As I had a bad headache, I also came & remained on board. — The few left in the Ship under the command of Mr Chaffers have been the most busily engaged of the whole crew. — They have triced up the Boarding netting, loaded & pointed the guns. —& cleared for action. — We are now at night in a high state of preparation so as to make the best defence possible, if the Beagle should be attacked. — To obtain ammunition could be the only possible motive. —[1]

[1] FitzRoy wrote: 'Scarcely had the Druid disappeared beneath the horizon, when the chief of the Monte Video police and the captain of the port came on board the Beagle to request assistance in preserving order in the town, and in preventing the aggressions of some mutinous negro soldiers. I was also requested by the Consul-general to afford the British residents any protection in my power; and understanding that their lives, as well as property, were endangered by the turbulent mutineers, who were more than a match for the few well-disposed soldiers left in the town, I landed with fifty well-armed men, and remained on shore, garrisoning the principal fort, and thus holding the mutineers in check, until more troops were brought in from the neighbouring country, by whom they were surrounded and reduced to subordination. The Beagle's crew were not on shore more than twenty-four hours, and were not called upon to act in any way; but I was told by the principal persons whose lives and properties were threatened, that the presence of these seamen certainly prevented bloodshed.' See Narrative 2: 95.

6th The boats have returned. — Affairs in the city now more decidedly show a party spirit, & as the black troops are enclosed in the citadel by double the number of armed citizens, Capt FitzRoy deemed it advisable to withdraw his force. — It is probable in a very short time the two adverse sides will come to an encounter: under such circumstances, Capt FitzRoy being in possession of the central fort, would have|208| found it very difficult to have preserved his character of neutrality. — There certainly is a great deal of pleasure in the excitement of this sort of work. —quite sufficient to explain the reckless gayety with which sailors undertake even the most hazardous attacks. — Yet as time flies, it is an evil to waste so much in empty parade. —

7th To my great grief it is not deemed prudent to walk in the country. —so I was obliged to go ashore to the dirty town of M: Video. — After dinner went out collecting to Rat Island. —

8th There has been a good deal of wind & rain. — In the evening the barometer fell, so the Captain determined immediately to strike Top-masts & let go another anchor. At sunset it blew a full gale of wind, but with our three anchors & no hamper aloft, we snugly rode whilst the breeze heavily whistled through the rigging. —

9th A merchant ship has drifted some way from her anchorage. — The Captain in middle of day went to her & found that at first she had only veered out 30 fathoms of cable. —(whilst we were riding with 70). A length of cable is a great security, as it takes away any sudden stress & by its friction does not strain so much on the anchor. — It|209| is quite curious how negligent all merchant vessels are. —yesterday very few struck Top-gallant masts. — Some years ago 14 vessels at Buenos Ayres went on shore & were lost, out of which only three had taken any & none sufficient precautions.

The Captain managed to go to the town to day, & brought back news that the disturbances increase in violence. — There has been some skirmishing with the black-troops, & a fresh party seems to have risen for the head of government. In the paltry state of Monte Video, there are actually about 5 contending parties for supremacy. — It makes one ask oneself whether Despotism is not better than such uncontrolled anarchy. — The weather yet continues wet & boisterous: it is a consolation, although a poor one, that the two distinct causes which prevent us from going ashore should come together.

10th During the whole of the night there have been several vollies of musketry fired in the city, & we all thought there must have been some heavy fighting.

11th   But this morning we hear not even one has been wounded. — in fact both parties are afraid of coming within reach of musket range of each other. Yesterday Lavalleja, the military governor, entered the town & was well received by everybody|210| excepting his former black troops. These he threatened to expel from the citadel & planted some guns to command the gate. — To revenge this the Blacks last night made a sally, & hence arose the firing. — This morning the news comes that Lavelleja, who was unanimously but yesterday received, has been obliged to fly the city, & that it is now certain that Signor Frutez & the constitutional government will gain the day. — One is shocked at the bloody revolutions in Europe, but after seeing to what an extent such imbecile changes can proceed, it is hard to determine which of the two is most to be dreaded. The weather for these last days has been wet & uncomfortable in the extreme.

Sunday 12th   The utter consternation of the civic guard during the other nights skirmish has given general amusement. — Large bodies immediately threw away their white cross belts that they might not be recognised in the dark: & the impetuosity with which they rushed down the streets, if it could have been directed to a charge, would have been most imposing. — In evening dined with Mr Parry.

13th   At last the unsettled politicks & weather have permitted us to walk in the country: Wickham, Sulivan, Hammond & myself went out shooting & if our sport was not very good the exercise was most delightful. — |211| Hammond & myself walked in a direct line for several miles to some plains covered with thistles, where we hoped to find a flock of Ostriches. — We saw one in the distance; if I had been by myself, I should have said it was a very large deer running like a race-horse. —as the distance increased it looked more like a large hawk skimming over the ground. —the rapidity of its movements were astonishing. — As the breeze was rather too stiff for boats, it had been determined to walk from the Mount round the bay to the town. — When far distant from it, Wickham & Sullivan found themselves so tired, that they declared they could move no further. — By good luck a horseman came up, whom we hired to carry them by turns till another horse was found; & thus we arrived just before the city gates were closed for the night. —

14th Signor Frutez entered the Town in full parade & was saluted by the forts. — He was accompanied by 1800 wild Gaucho cavalry; many of them were Indians. — I believe it was a magnificent spectacle; the beauty of the horses, & the wildness of their dresses & arms were very curious. — |212|

15th As the boat was landing me at the Mount, we surprised a large Cabra or Capincha on the rocks. After a long & animated chace in a little bay I succeeded in shooting it through the head with a ball. — These animals abound in the Orinoco & are not uncommon here, but from their shyness & powers of swimming & diving are difficult to be obtained. — It is like in its structure a large guinea-pig; in its habits a water rat. — it weighed 98 pounds. — Having sent my game on board in triumph, I collected great numbers of different animals: some beautiful snakes & lizards & beetles. Under stones were several scorpions about 2 inches long; when pressed by a stick to the ground, they struck it with their stings with such force as very distinctly to be heard. —

The Druid has returned from Buenos Ayres & brought from its government a long apology for the insult offered to us. — The Captain of the Guard-ship was immediately arrested & it was left to the British consuls choice whether he should any longer retain his commission. —

16th Spent the day in examining the rich produce of yesterdays labor. — The Beagle goes to sea the day after tomorrow for her first cruize. — |213|

17th All day & night it has blown a stiff breeze from the South. There have been several hail-storms, which forms our first introduction to frozen water. — A sea soon gets up in the river & from its little depth the waves become so muddy that they look like mountains of mud. — This riding with our head to wind shakes the very foundation of my stomach. —

18th Several officers are on shore & cannot yet come off. — The Captain however ventured to sail to Rat Island to obtain sights. It was beautiful to see how the whale-boat hops over the sea. — In returning he carried away the yard of his sail. —

Sunday 19th In the morning there was a fresh breeze from the NW—A wind in this direction soon emptys the river; at night we had 18 feet under our stern, in the morning only 13. From this cause, independently of intending to sail in the course of the day, it was advisable to move our anchorage. — The instant we had tripped our anchor the wind drifted us within a few yards of the buoy which marks the old wreck. Then is the time to watch sailors working: one foul rope & we should have been on shore. — The sailors in the city were saying, A dios Barca Inglese, A Dios. — A merchant ship certainly would have had no chance of escaping: but with our body of men it is the work of a second|214| to set sail & get way on the ship. This has been for me the first specimen of working off a lee-shore with a stiff breeze blowing. — During the morning we tacked about, waiting for the weather to moderate & at last again anchored. — In the afternoon we sent on board the Packet some parcels &c & my box of specimens, & the boats returning from the shore, we made sail. — A fine breeze carried us 40 miles from the Mount, where we anchored for the night. — In such shoal water as in the Plata the sea is very short; I have never seen so much spray break over the Beagle & I have not often felt a more disagreeable sensation in my stomach. —

20th In the afternoon we anchored 8 miles off Point Piedras on the Southern shore of the river. — At this distance there were only 18 feet water. — The Captain intends at present verifying the leading points in the coast. — The Spaniards on shore having already filled up the details. — Any minute knowledge of an almost uninhabited coast where shipping cannot approach, will never be of any great value. —

21st In middle of day anchored near Point Piedras, & sent our boats to sound. — Shortly after getting under way, the water suddenly shoaled & we grazed the bottom rather too sensibly. — |215| In calm weather this is of little consequence, but when there is any sea, it does not take long to knock a hole in the bottom. — The coast was very low, & covered with thickets. — the extreme similarity of different parts of the banks is the chief cause of the difficulty of navigating this river. — The weather has been beautifully clear during these last two days. — I do not believe there has been one single cloud in the heavens. — Several land-birds took refuge in the rigging, such as larks, fly-catchers, doves & butcher-birds & all appeared quite exhausted. — To night we have anchored North of Cape St Antonio. — as soon as we double this we shall be in the open ocean. — Already the water has lost its ugly muddy colour. —

22nd All day we have been sailing within two or three miles off the coast. — For 40 miles it has been one single line of sandy hillocks, without any break or change.[1] The country within is uninhabited, & ships never frequent this track, so that it is the most desolate place I have ever visited. — At sunset, before anchoring, we came rather suddenly on a bank, & were obliged instantly to put the ship up to the

wind. — This fine weather is of the greatest importance to the surveying
& as long as it lasts, sailing slowly along the coast is sufficient for all
purposes. — |216|

[1] According to FitzRoy: 'farther inshore are thickets affording shelter to numbers of
jaguars'. See *Narrative* **2**: 97.

23rd The weather continues most beautiful: a bank of clouds in the SW
frightened us in the morning but now at night we are at an anchor with
a calm. — No people have such cause anxiously to watch the state of
weather as Surveyors. — Their very duty leads them into the places
which all other ships avoid & their safety depends on being prepared
for the worst. — Every night we reef our top-sails so as to lose no time if
a breeze should force us to move. — Yesterday morning getting up the
anchor & securing it & setting all sail only took us five minutes. — We
have not made much progress during the day; for we have tacked all the
time parallel to the coast. —

24th We have made a good run; at last North of Cape of Corrientes the
coast in a small degree has altered its appearance: instead of the
undulating chain of sand-hillocks the horizon is bounded by low table
land. This being divided by broard gaps or vallies, presents so many
square masses. — We have seen during the day the smoke from several
large fires within the country: it is not easy to guess how they arise. — It
is too far North for the Indians & the country is uninhabited by the
Spaniards. — The sun set in a cloudless sky; & there is every prospect of
the Northerly wind lasting; if so tomorrow we shall double Corrientes
& if we can, land in the boats on the promontory.|217|

25th We have made an excellent run of 70 miles to day. — Indeed the
breeze to my taste was much too good, as it prevented us from
attempting to land at Cape Corrientes, which we doubled at Noon. —
We sailed very close to the shore, & it was very interesting viewing the
different countries as we rapidly passed on. — North of Corrientes, a
dead level line of cliff takes the place of the sand hillocks. — The cliff is
perpendicular & about 30 feet high, & with a few exceptions is continued all the way South of the Cape. — From the mast-head a great
extent of flat Pampas was seen without any break or elevation. — To
every ones astonishment there was near the promontory of Corrientes
an Estancia. — Cattle were ·ery abundant near the house, & the place
looked prosperous. — (*Note in margin:* We have heard they have 50000
head.) Two or three men on horseback were watching us with great
interest: so we hoisted our pennant & colours, & doubtless for the first

time they had ever been seen in this sea. — This farm must be about 200 miles from any town, & the greater part of the interval consists in desert salt plains. — There cannot easily be imagined a more desolate habitation for civilized man. —

Sunday 26th Torrents of rain & the atmosphere was so thick that it was impossible to continue the survey. — We remained therefore at anchor. — The bottom was rocky & in consequence plenty of fish: almost every man in the|218| ship had a line overboard & in a short time a surprising number of fine fish were caught. — I also got some Corallines which were preeminently curious in their structure. — We had to day a beautiful illustration how useful the Barometer is at sea. — During the last three or four fine days it has been slowly falling. — the Captain felt so sure that shortly after it began to rise we should have the wind from the opposite quarter, the South, that when he went to bed he left orders to be called when the Barometer turned.

27th Accordingly at one oclock it began to rise, & the Captain immediately ordered all hands to be piped up to weigh anchor. — In the course of an hour from being a calm it blew a gale right on shore, so that we were glad enough to beat off. — By the morning we were well out at sea; so with snug sail cared little for the breeze or the heavy swell. — If we had not a Barometer, we probably should have remained two hours longer at anchor, & then if the gale had been a little harder we should have been in a most dangerous situation. — As it was, the sea was very heavy & irregular. — it fairly pitched our Howitzer out of the slide into the sea. — This was not our only misfortune, as in weighing ship we tore our anchor into pieces & quite|219| disabled it for use. — During the night the weather moderated &

28th this morning we stood in again for the shore. — By the time we got within a few miles of the land it was almost calm, but the swell from the ocean was extraordinarily great. — This is what might be expected from the gradual shoaling of the water. — The surf on the beach was proportionally violent: for ¼ of a mile the sea was white with foam & a cloud of spray traced for many miles the line of coast. As it was impossible to take observations, we are this evening again standing out to sea, patiently to wait till the elements are quiet. —

29th The morning was thick with rain: but in the afternoon in spite of the remaining swell, some miles of the coast were traced. — at night the weather looked dirty & we have stood out to sea. —

This day last year I arrived home from N. Wales & first heard of this Voyage. — During the week it has often struck me how different was my situation & views then to what they are at present: it is amusing to imagine my surprise, if anybody on the mountains of Wales had whispered to me, this day next year you will be beating off the coast of Patagonia: — And yet how common & natural an occurrence|220| it now appears to me. — Nothing has made so vivid an impression on my mind as those days of painful uncertainty: the clearness with which I recollect the most minute particulars, gives to the period of an year the appearance of far shorter duration. — But if I pause & in my mind pass from month to month, the time fully grows proportional to the many things which have happened in it. —

30th Very wet day: about noon it fell calm, & we could hear the surf roaring although about six miles distant from the beach. — The weather looked exceedingly threatening; but after all it did not blow more than a stiff breeze during the night. —

31st By the middle of the day we got within surveying distance of the coast. — We let go the anchor: but the sky clearing we soon had a regular dry SW wind. — The anchor would not hold in the sand and we were forced again to stand out. — To night it has lulled, & we have anchored. — Tomorrow I trust we shall be enabled to continue the survey, which has been interrupted for a week. — At last I find myself decidedly much less afraid of sea-sickness, although during two of the days I was on my "beam ends".|221|

September 1st The breeze freshened during the night & in the morning there was a good deal of sea. In heaving up the anchor, a sudden pitch in the vessel broke it off just above the flues. — It has been a cloudless day; but with a strong breeze right in our teeth. — To night we have anchored & to our universal joy the wind has chopped round to the North. —

Sunday 2nd This day will always be to me a memorable anniversary; in as much as it was the first in which the prospect of my joining the voyage wore a prosperous appearance. — Again in heaving up the anchor (one of the best & largest) it broke off like the former ones: — it is supposed that the bottom consists of a clay so stiff as nearly to resemble rock, & that during the night the flue of anchor works into it, so that no power is able to wrench it out. — So early in the voyage it is a great loss. The wind blew a gale; but under close reefed topsails we ran along about 70 miles of coast. — Out of all this range scarcely two parts could be distinguished from each other: nothing interrupts the line of sand hillocks. — Tomorrow we shall be near to Baia Blanca;[1] where I hope we shall remain some time. — This last week, although lost for surveying, has produced several animals; the examination of which has much interested me. —|222|

[1] The modern spelling, later also used by CD, is Bahia Blanca.

3rd The weather has been tolerably fair for us; but in the evening the breeze was fresh & a good deal of sea. — At this time, the situation of the vessel was for a few minutes very dangerous. — We came suddenly on a bank where the water was very shoal. — It was a startling cry, when the man in the chains sang out, "& a half, two". Our bottom was then only two feet from the ground. —if we had struck, it is possible we should have gone to the bottom; & the long swell of the open ocean would soon dash the strongest timber into pieces. —(*Note in margin:* We have since had reason to believe it was a mistake of the Leadsman.) It is beautiful to see the quiet calm alertness of the sailors on such occasions. — We soon deepened our water when we altered our course. — At present we are riding in a wild anchorage, waiting for the morning. —

4th We have remained all day at our anchorage: the weather has been cloudy for some days past & it is almost necessary to obtain observations of the sun to ascertain our situation. — I am throughily tired of this work, or rather no work; this rolling & pitching about with no end gained. — Oh for Baia Blanca; it will be a white day for me, when we gain it. —|223|

5th We ran along 40 miles of coast & then anchored near to the mouth of the Bay: During the day I took several curious marine animals. —

6th In the morning we stood into the bay; but soon got entangled in the midst of shoals & banks; we came again to an anchor. — At this time a small Schooner passed near to us. —an officer was sent on board to procure information about the bay &c: The Schooner was a Sealer, bound from the settlement at Baia Blanca to the Rio Negro; south of which she intended fishing for the Seals. — Mr Harris, a half partner & Captain, volunteered piloting us into the bay on condition of being carried up in a boat to the Settlement; where there was another Schooner bound for the same port, & in which he intended taking a passage. — By Mr Harris's assistance we arrived in the evening at a fine bay; where sheltered from all bad weather, we moored ship. — Mr Harris gave us a great deal of useful information about the country. — Baia Blanca has only been settled within the last six years: previous to which even the existence of the bay was not known. — It is designed as a frontier fort against the Indians & thus to connect Buenos Ayres to Rio Negro. — In the time of the old Spaniards, before the independence, the latter was purchased from the native chief of the place. — The settlers at Baia Blanca did not follow this|224| just example, & in consequence ever since a barbarous & cruel warfare has been carried on: — But I shall mention more about this presently. —

7th In the morning the Captain, Rowlett the pilot & myself started with a pleasant breeze for the Settlement: it is distant about twenty miles. — Instead of keeping the middle channel, we steered near to the Northern shore: from this cause & from the number of similar islands, the pilot soon lost his reckoning. — We took by chance the first creek we could find: but following this for some miles, it gradually became so narrow that the oars touched on each side & we were obliged to stop. — These Islands rather deserve the name of banks; they consist of mud which is so soft that it is impossible to walk even the shortest distance; in many the tops are covered by rushes; & at high water the summits of these are only visible. — From our boat nothing within the horizon was to be seen but these flat beds of mud; from custom an horizontal expanse of water has nothing strange in it; but this had a most unnatural appearance, partaking in the character of land & water without the advantages of either. — The day was not very clear & there was much refraction, or as the sailors expressed it, "things loomed high", the only thing within our view which|225| was not level was the horizon; rushes looked like bushes supported in the air by nothing, & water like mud banks & mud-banks like water. — With difficulty the boat was turned in the little creek; & having waited for the tide to rise, we sailed straight over the mud banks in the middle of the rushes. By heeling the boat over, so that the edge was on a level with the water, it did not draw more than a foot of water. — Even with this we had much trouble in getting her along, as we stuck several times on the bottom.

    In the evening we arrived at the creek which is about four miles distant from the Settlement. — Here was a small schooner lying & a mud-hut on the bank. — There were several of the wild Gaucho cavalry waiting to see us land; they formed by far the most savage picturesque group I ever beheld. — I should have fancied myself in the middle of Turkey by their dresses. — Round their waists they had bright coloured shawls forming a petticoat, beneath which were fringed drawers. Their boots were very singular, they are made from the hide of the hock joint of horses hind legs, so that it is a tube with a bend in it; this they put on fresh, & thus drying on their legs is never again removed. — The spurs are|226| enormous, the rowels being from one to two inches long. — They all wore the Poncho, which is [a] large shawl with a hole in the middle for the head. — Thus equipped with sabres & short muskets they were mounted on powerful horses. — The men themselves were far more remarkable than their dresses; the greater number were half Spaniard & Indian. —some of each pure blood & some black. — The Indians, whilst gnawing bones of beef, looked, as they are, half recalled wild beasts. — No painter ever imagined so wild a set of expressions. — As the evening was closing in, it was determined not to return to the vessel by the night. —so we all mounted behind the Gauchos & started at a hand gallop for the Fort. —

Our reception here was not very cordial. The Commandante was inclined to be civil; but the Major, although second in rank, appears to be the most efficient. He is an *old* Spaniard, with the old feelings of jealousy. — He could not contain his surprise & anxiety at a Man of War having arrived for the first time in the harbor. He asked endless questions about our force &c., & when the Captain, praising the bay, assured him he could bring up even a line of battle ship, the old gentleman was appalled & in his minds eye saw the British Marines taking his fort. — These ridiculous suspicions made it|227| very disagreeable to us. —so that the Captain determined to start early in the morning back to the Beagle. —[1]

The Settlement is seated on a dead level turf plain, it contains about 400 inhabitants; of which the greater number are soldiers: The place is fortified, & good occasion they have for it: The place has been attacked several times by large bodies of Indians. — The War is carried on in the most barbarous manner. The Indians torture all their prisoners & the Spaniards shoot theirs. — Exactly a week ago the Spaniards, hearing that the main body of their armies were gone to Northward, made an excursion & seized a great herd of horses & some prisoners. Amongst these was the head chief, the old Toriano who has governed a great district for many years. — When a prisoner, two lesser chiefs or Caciques came one after the other in hopes of arranging a treaty of liberation: It was all the same to the Spaniards, these three & 8 more were lead out & shot. — On the other hand, the Commandante's son was taken some time since; & being bound, the children (a refinement in cruelty I never heard of) prepared to kill him with nails & small knives. — A Cacique then said that the next day more people would be present, &|228| there would be more sport, so the execution was deferred, & in the night he escaped.

A Spanish friend of Mr Harris received us hospitably. — His house consisted in one large room, but it was cleaner & more comfortable than those in Brazil. — At night I was much exhausted, as it was 12 hours since I had eaten anything. —

[1] FitzRoy wrote: 'Leaving the boat's crew to bivouac, as usual, I accepted a horse offered to me, and took the purser up behind; Mr. Darwin and Harris being also mounted behind two gaucho soldiers, away we went across a flat plain to the settlement. Mr. Darwin was carried off before the rest of the party, to be cross-questioned by an old major, who seemed to be considered the wisest man of the detachment, and he, poor old soul, thought we were very suspicious characters, especially Mr. Darwin, whose objects seemed most mysterious.

In consequence, we were watched, though otherwise most hospitably treated; and when I proposed to return, next morning, to the boat, trifling excuses were made about the want of horses and fear of Indians arriving, by which I saw that the commandant wished to detain us, but was unwilling to do so forcibly; telling him, therefore, I should walk back, and setting out to do so, I elicited an order for horses, maugre the fears and advice of his major, who gave him all sorts of warnings about us. However, he sent an escort with us, and a troop of gaucho soldiers were that very morning posted upon the rising grounds nearest to the Beagle, to keep a watch on our movements.

We afterwards heard, that the old major's suspicions had been very much increased by Harris's explanation of Mr. Darwin's occupation. "Un naturalista" was a term unheard of by any person in the settlement, and being unluckily explained by Harris as meaning "a man that knows every thing," any further attempt to quiet anxiety was useless.' See *Narrative* 2: 103–4.

8th We rode to the boat early in the morning; & with a fresh breeze arrived at the ship by the middle of the day. — It was then reported to the Captain that two men on horseback had been reconnoitring the ship. The Captain well knowing that so small a party of Spaniards would not venture so far, concluded they were Indians. — As we intended to wood & water near to that spot it was absolutely necessary for us to ascertain whether there was any camp there. — Accordingly three boats were manned & armed; before reaching the shore, we saw five men gallop along the hill & then halt. The Captain upon seeing this sent back the other two boats, wishing not to frighten them but to find out who they were. — When we came close, the men dismounted & approached the beach, we immediately then saw it was a party of cavalry from Baia Blanca. — After landing & conversing with them, they told us they had been sent down to look after the Indians; this to a certain degree was true, for we found marks of a|229| fire; but their present purpose evidently was to watch us; this is the more probable as the officer of the party steadily kept out of sight, the Captain having taxed them with being so suspicious; which they denied. — The Gauchos were very civil & took us to the only spot where there was any chance of water. — It was interesting seeing these hardy people fully equipped for an expedition. — They sleep on the bare ground at all times & as they travel get their food; already they had killed a Puma or Lion; the tongue of which was the only part they kept; also an Ostrich, these they catch by two heavy balls, fastened to the ends of a long thong. — They showed us the manner of throwing it; holding one ball in their hands, by degrees they whirl the other round & round, & then with great force send them both revolving in the air towards any object. — Of course the instant it strikes an animals legs it fairly ties them together. — They gave us an Ostrich egg & before we left them, they found another nest or rather depositary in which were 24 of the great eggs. — It is an undoubted fact that many female Ostriches lay in the same spot, thus forming one of their collections. |230| Having given our friends some dollars, they left us in high good humor & assured us they would some day bring a live Lion. — We then returned on board. — During the last two days the Captain has formed a plan which will materially affect the rest of our voyage. — Mr Harris is connected with two small schooners employed in sealing & now at Rio Negro. He & the other Captain is well ackquainted with the adjoining coast. The Captain thought this so fine an opportunity that he has hired them both by the Month & intends sending officers in each who will survey this intricate coast whilst the Beagle (after returning to M Video) will proceed to the South. — By this means the time spent on the Eastern coast will be much shorter & this is hailed with joy by everybody. — Mr Harris will immediately go to Rio Negro to bring the vessels & soon after that we shall return to the Rio Plata.[1]

[1] In order to be able to survey the shallow coastal waters and inlets between Baia Blanca and the Rio Negro as thoroughly as possible, FitzRoy took a decision that later proved sadly expensive to him. He wrote: 'At last, after much anxious deliberation, I decided to hire two small schooners—or rather decked boats, schooner-rigged—from Mr. Harris, and employ them in assisting the Beagle and her boats. Mr. Harris was to be in the larger, as pilot to Lieutenant Wickham—and his friend Mr. Roberts, also settled at Del Carmen, on the river Negro, was to be Mr. Stokes's pilot in the smaller vessel. These small craft, of fifteen and nine tons respectively, guided by their owners, who had for years frequented this complication of banks, harbours, and tides, seemed to me capable of fulfilling the desired object—under command of such steady and able heads as the officers mentioned—with this great advantage; that, while the Beagle might be procuring supplies at Monte Video, going with the Fuegians on her first trip to the southward, and visiting the Falkland Islands, the survey of all those intricacies between Blanco Bay and San Blas

might be carried on steadily during the finest time of year. One serious difficulty, that of my not being authorized to hire or purchase assistance on account of the Government, I did not then dwell upon, for I was anxious and eager, and, it has proved, too sanguine. I made an agreement with Mr. Harris, on my own individual responsibility, for such payment as seemed to be fair compensation for his stipulated services, and I did hope that if the results of these arrangements should turn out well, I should stand excused for having presumed to act so freely, and should be reimbursed for the sum laid out, which I could so ill spare. However, I foresaw and was willing to run the risk, and now console myself for this, and other subsequent mortifications, by the reflection that the service entrusted to me did not suffer.' See *Narrative* 2: 110.

In personal letters to Captain Beaufort at the Admiralty, dated 10 May, 7 June, 16 July, 26 October, 16 November and 5 December 1833, FitzRoy expressed his anxiety as to whether his action would be supported officially, and his eventual frustration when it was not (see *Beagle Record* pp. 131–3, 142–5, 162–3 amd 170–1). His total expenditure amounted in the end to £1680 (see *Narrative* 2, Appendix pp. 97–8).

Sunday 9th   In the morning divine service was read on the lower deck. — After dinner a large party of officers went on shore to see the country. — For the first two miles from the beach, it is a succession of sand hillocks thickly covered with coarse herbage; then comes the Pampas, which extend for many miles & in the distance is the Sierra de Ventana, a chain of mountains which we imagine to be lofty. — The ground was in every direction tracked by the Ostriches|231| & deer. — One large one of the latter bounded up close to me. — Excepting these, death appeared to reign over all other animals. — I never saw any place before so entirely destitute of living creatures.

10th   All hands have been busily employed to day; some surveying: some digging a well for water & others cutting up an old wreck for fire wood. — I took a long walk with a rifle, but did not succeed in shooting anything. I saw some deer & Ostriches, the latter made an odd deep noise; I also found a warren of the Agouti, or hare of the Pampas; it is about the size of two English ones, but in its habits resembles a rabbit.

In the evening the merchant Schooner arrived from the Settlement; bringing with it Mr Harris, bound for Rio Negro; & our Spanish host who was invited to pay us a visit. — Mr Harris tells us that the Majors fears are not yet quieted, & that no one in the place, excepting our host, would venture to pay us a visit. —

When the schooner sailed, Mr Rowlett accompanied her, in order at Rio Negro to try to procure fresh provisions for the ship. —

11th   Having proved to our Spanish friends that we were not Pirates, the Captain with two boats started for the Settlement. — Nearly all the men were employed on|232| shore; so that the ship was left in as unusual as delightful a state of quietness.

12th   Went out shooting with Mr Wickham with our rifles:—to my great delight I succeeded in shooting a fine buck & doe.— The Captains servant shot three more.— We were obliged to send a boats crew to carry them to the shore.— One of mine however was previously disposed of.— I left it on the ground a substantial beast, but in the evening the Vultures & hawks had picked even the bones clean.— In our walk I found also an Ostriches nest; it contained only one egg.—

13th   The ships anchorage was removed a few miles up the harbor; in order to be nearer a newly discovered watering place.— Here we shall remain some weeks; if the present clear dry weather lasts, the time will pass very pleasantly.—

14th   I am spending September in Patagonia, much in the same manner as I should in England, viz in shooting; in this case however there is the extra satisfaction of knowing that one gives fresh provisions to the ships company.— To day I shot another deer & an Agouti or Cavy.— The latter weighs more than 20 pounds; & affords the very best meat I ever tasted.— Whilst shooting I walked several miles within the interior; the general features of the country remain the same, an undulating sandy plain covered with coarse herbage &|233| which as it extends, gradually becomes more level.— The bottoms of some of the vallies are green with clover: it is by cautiously crawling so as to peep into these that the game is shot.—

If a deer has not seen you stand upright; generally it is possessed with an insatiable curiosity to find out what you are; & to such an extent that I have fired several times without frightening it away.—

15th   The Spaniards, whom we some time since thought were Indians, have been employed hunting for us & have generally bivouacced near the coast.— They offered to lend me a horse to accompany them in one of their excursions; of this I gladly accepted.— The party consisted of 9 men & one woman; the greater number of the former were pure Indians, the others most ambiguous; but all alike were most wild in their appearance & attire.— As for the woman, she was a perfect non descript; she dressed & rode like a man, & till dinner I did not guess she was otherwise.— The hunters catch everything with the two or three balls fastened to the thongs of leather; the manner of proceeding is to form themselves into a sort of crescent, each man less than a quarter of a mile apart; one goes some way ahead & endeavours to drive the animals towards the others & thus in a manner encircling them.— I saw one most beautiful chace;|234| a fine Ostrich tried to escape; the

Gauchos pursued it at a reckless pace, each man whirling the balls round his head; the foremost at last threw them, in an instant the Ostrich rolled over & over, its legs being fairly lashed together by the thong. — Its dying struggles were most violent. — The men then formed a ring & drove to the centre several cavies; they only killed one; but their riding was most excellent, especially in the quickness & precision with which they turn. — The horses are soon fatigued from such violent exercise & it is necessary often to change them & pick out fresh ones from the herd which always accompanies a party. — At this time of year, the eggs of the ostrich is their chief prize. — In this one day they found 64, out of which 44 were in two nests; the rest scattered about by ones or twos. — They also catch great numbers of Armadilloes. — In the middle of the day they lighted a fire & soon roasted some eggs & some Armadilloes in their hard cases: — They had neither water, salt or bread; of the two latter for weeks together they never taste; so that it makes little difference to them where they live. —

Like to snails, all their property is on their backs & their food around them. — It is very interesting to watch, whilst seated|235| round the fire, the swarthy but expressive countenances of my half-savage hosts. The creature of a woman flirted & actually was affected; she pretended to be frightened of my gun & screamed out "no est cargado"? We returned to the beach in the evening, where the same scene of eggs & Armadilloes occurred again and I went on board. — My feet were a good deal tired, the stirrups being so small that even without shoes I had difficulty in getting in the two first toes. — The Gauchos always have these uncovered & separate from the other three. —

Sunday 16ᵗʰ   The party who went out to shoot fresh provisions brought home 2 deer, 3 Cavies & an ostrich. — With the net also a most wonderful number of fish were caught; in one drag more than a tun weight were hauled up; — including ten distinct species. —

17ᵗʰ, 18ᵗʰ   Have been employed during these two days with various marine animals which I procured from the beach & by dredging. — What we had for dinner to day would sound very odd in England. — Ostrich dumpling & Armadilloes; the former would never be recognised as a bird but rather as beef. — The Armadilloes when unlike to the Gauchos' fashion, cooked without their cases, taste & look like a duck. — Both of them are very good. — |236|

19ᵗʰ   Walked to the plains beyond the sand hillock & shot some small birds for specimens. — It is a complete puzzle to all of us, how the

Ostriches, Deer, Cavies, &c which are so very numerous, contrive to get water. Not one of us has seen the smallest puddle (excepting the well which is 8 feet deep) & it is scarcely credible they can exist without drinking. I should think this sandy country in the summer time must be a complete desert; even now in spring & all the flowers in bud the sun is very powerful, there being no shelter & the heat being reflected from the sand hillocks.

20th   Staid on board: —

21st   In the morning there was a good deal of wind; so that I did not leave the ship. —

22nd   Had a very pleasant cruize about the Bay with the Captain & Sulivan. — We staid sometime on Punta Alta about 10 miles from the ship; here I found some rocks. — These are the first I have seen, & are very interesting from containing numerous shells & the bones of large animals.[1] The day was perfectly calm; the smooth water & the sky were indistinctly separated by the ribbon of mud-banks: — the whole formed a most unpicturesque picture. — It is a pity such bright clear weather should be wasted on a country, where half its charms do not appear. — We got on board just in time to escape a heavy squall & rain. — |237|

[1] In FitzRoy's words (see *Narrative* 2: 106–7), 'My friend's attention was soon attracted to some low cliffs near Point Alta, where he found some of those huge fossil bones, described in his work; and notwithstanding our smiles at the cargoes of apparent rubbish which he frequently brought on board, he and his servant used their pick-axes in earnest, and brought away what have since proved to be most interesting and valuable remains of extinct animals'. This was truly a red-letter day for biology, marking the initial discovery of the first of the lines of evidence that eventually led CD to question and ultimately to reject the doctrine of the fixity of species. In his private diary for 1837 he wrote: 'In July opened first note book on "Transmutation of Species"—had been greatly struck from about month of previous March on character of S. American fossils, and species on Galapagos Archipelago. These facts origin (especially latter) of all my views.' Here in the cliffs at Punta Alta he unearthed for the first time the fossilized remains of enormous animals that he recognised at once as being similar in all respects but size to living counterparts, although the full implications of this momentous finding probably did not strike him until his specimens had been examined by Richard Owen soon after the return of the *Beagle* to England in October 1836.

He was immediately concerned to establish the exact origin of the strata in which the fossils occurred. The entry in his notebook runs as follows: 'Sep. 22. Entrance of creek, dark blue sandy clay much stratified dipping to NNW or N by W at about 6°. On the beach a succession of thin strata dipping at 15° to W by S—conglomerate quartz and jasper pebbles—with shells—vide specimens. On the coast about 12 feet high, and in the conglom. teeth and thigh bone. Proceeding to NW there is a horizontal bed of *earth* containing much fewer shells—but armadillo—this is horizontal but widens gradually, hence I think conglomerate with broken shells was deposited by the action of tides—*earth* quietly. Is this above the clay which is seen a short time previously? Covered by diluvium and sand hillocks as earthy bank—thickened & cropped out in direction NNW, it probably overlies the clay.' See Down House Notebook 1.10, and *CD and the Voyage* p. 166.

Sunday 23rd   A large party was sent to fish in a creek about 8 miles distant; great numbers of fish were caught. — I walked on to Punta alta to look after fossils; & to my great joy I found the head of some large animal, imbedded in a soft rock. — It took me nearly 3 hours to get it out: As far as I am able to judge, it is allied to the Rhinoceros. — I did not get it on board till some hours after it was dark. —

24th   Employed in carefully packing up the prizes of yesterday. — In the morning one of the Schooners arrived & the other is shortly expected. They have had a very bad passage of 6 days. — Mr Rowlett brings back an excellent account of Rio Negro. — Nothing could exceed the civility of the Governor & the inhabitants. — It was rendered the more striking from the contrast of our reception at the fort of Baia Blanca. —

25th   The Schooner has been taken to the Creek. — Mr Wickham & a party of men have erected tents on shore & are living there during the refit of the vessel. — I accompanied the little settlement & whilst they were rigging the tents I walked to Punta alta & again obtained several fossils. — I came quite close to an Ostrich on her nest; but did not see her till she rose up & with her long legs stretched across the country. — |238|

26th   The weather is most beautiful. — Passing from the splendor of Brazil to the tame sterility of Patagonia has shown to me how very much the pleasure of exercise depends on the surrounding scenery. —

27th   That no time may be lost during the altering of the Schooner, we have changed our anchorage & stood further out, so as to survey some of the outer banks. —

28th, 29th & *Sunday 30th*   We have been for these three days cruizing about the mouth of the harbor. — The two latter were boisterous, & there was a considerable swell on the sea. — I, as usual very sick & miserable; my only comfort is, that two or three of the officers are but very little better & that like to myself they always feel the motion when first going out of harbor. —

October 1st   The morning threatened us with heavy weather; but it blew over in a hail storm. We have anchored near to a cliff, upon which the Captain intends to erect some land mark as a guide on entering the harbor. —

2nd   Early in the morning the Captain with a large party landed in the four whale-boats. — Dinner for all hands was taken, as it was intended to work at the land-mark all day & return in the evening. — King & I went in one direction to geologize & Mr Bynoe in another to shoot. — During our walk I observed the wind had freshened & altered its point; but I paid no further attention to it. — When we returned|239| to the beach, we found two of the boats hauled up high & dry & the others gone on board. — The Captain two hours previously had had some difficulty in getting off & now the line of white breakers clearly showed the impossibility. — It was an unpleasant prospect, to pass the night with thin clothes on the bare ground; but it was unavoidable, so we made the best of it. — Mr Stokes & Johnson were left in command & made what arrangements they could. — At night no supper was served out; as we were 18 on shore & very little food left. — We made a sort of tent or screen with the boats sails & prepared to pass the night. — It was very cold, but by all huddling in a heap, we managed pretty well till the rain began, & then we were sufficiently miserable. —

3rd   At day-break things wore a very bad appearance. — The sky looked dirty & it blew a gale of wind; a heavy surf was roaring on the beach; & what was the worst of all the men thought this weather would last. — The Beagle was pitching very deeply & we thought it not impossible she would be forced to slip cable & run out to sea. — We afterwards heard she rode it out well, but that some of the seas went right over her, although having 120 fathoms of cable out. — It was now time to look after our provisions: we breakfasted on some|240| small birds & two gulls, & a large hawk which was found dead on the beach. — Our dinner was not much better, as it consisted in a fish left by the tide & the bones of the meat, which we were determined to keep for the next day. — In the evening however to our great joy & surprise the wind lulled & the Captain in his boat was able to come within some hundred yards of the coast; he then threw over a cask with provisions which some of the men swam out to & secured. — This was all very well; but against the cold at night there was no remedy. — Nothing would break the wind, which was so cold that there was snow in the morning on the Sierra de Ventana. — I never knew how painful cold could be. I was unable to sleep even for a minute from my body shivering so much. The men also who swam for the provisions suffered extremely, from not being able to get warm again. —

4th   By the middle of the next day we were all on board the Beagle & most throughily after our little adventure did we enjoy its luxuries. — In the evening we moved our anchorage and stood in towards our old place. —

5th   Some of the men felt rather unwell, but none of us are made at all ill by it. — The wind has been very light all day, & we have made little progress.|241|

6th   We beat up the channel against a strong breeze & anchored at night in the old place opposite the well. The sand-hillock here is christened "Anchor-stock hill". —

Sunday 7th I walked to the creek where the tents are pitched for preparing the Schooners, & slept there during the night. Wickham has established quite a comfortable little town: — An encampment in the open air always has something charming about it. Even a Gypsies hut in England makes me rather envious; but here, in the wide plain, the little establishment made quite a picture. — This creek has been very useful for the vessels; the larger one is nearly ready for sea, & the other will be so in a few days. —

8th The Captain had bought from the Gaucho soldiers a large Puma or South American lion, & this morning it was killed for its skin. — These animals are common in the Pampas, I have frequently seen their footsteps in my walks: it is said they will not attack a man; though they evidently are quite strong enough. — The Gauchos secured this one; by first throwing the balls & entangling its front legs, they then lassoed or noosed him, when, by riding round a bush & throwing other lassos, he was soon lashed firm and secure. — |242|

After breakfast I walked to Punta Alta, the same place where I have before found fossils. — I obtained a jaw bone which contained a tooth: by this I found out that it belongs to the great ante-diluvial animal the Megatherium.[1] This is particularly interesting as the only specimens in Europe are in the Kings collection at Madrid, where for all purposes of science they are nearly as much hidden as if in their primæval rock. — I also caught a large snake, which at the time I knew to be venemous; but now I find it equals in its poisonous qualities the Rattle snake. In its structure it is very curious, & marks the passage between the common venemous & the rattle snakes. Its tail is terminated by a hard oval point, & which, I observe, it vibrates as those possessed with a more perfect organ are known to do. —

[1] In Down House Notebook 1.10, CD wrote: 'Megatherium like Armadillo case, teeth: —'; and in letters sent soon afterwards to his sister Caroline and to Henslow (see *Correspondence* 1: 276–82), he wrongly identifies as belonging to *Megatherium* the fossils he had found with a 'curious osseous coat' like that of a huge armadillo, this being a pardonable error originating from Cuvier's misleading account of the specimen in Madrid. The mistake was rectified in due course when Richard Owen classified the *Beagle* material, and showed that the osseous coat actually belonged to a glyptodont *Hoplophorus*, and that CD had brought back not only *Megatherium*, but also other new edentates such as *Toxodon*, *Mylodon* and *Glossotherium*.

The cliffs at Punta Alta are now irretrievably submerged beneath the naval base of Puerto Belgrano, but further along the coast in the vicinity of Monte Hermoso it is still possible to find pieces of glyptodont armour and other fossils.

9th   Staid on board. —

10th   In the morning there was a fresh breeze, & I did not go on shore. —

11th   Took a long walk in a straight line into the interior; uninteresting as the country is, we certainly see it in by far the best time. It is now the height of Spring; the birds are all laying their eggs & the flowers in full blossom. — In places the ground is covered with the pink flowers of a Wood Sorrell & a wild pea, & dwarf Geranium. — Even with this &|243| a bright clear sky, the plain has a dreary monotomous aspect. —

12th   To day I walked much further within the country; but all to no use; every feature in the landscape remains the same. — I found an Ostriches nest which contained 27 eggs. — Each egg equals in weight 11 of a common hens; so that the quantity of food in this nest was actually the same as 297 hens eggs. — We had some difficulty in getting on board; as there was a very fresh breeze right in our teeth. —

13th, Sunday 14th, 15th   On Sunday the Schooners came down from the creek & anchored alongside. — Their appearance is much improved by their refit; but they look very small. — "La Paz" is the largest, carrying 17 tuns; La Lievrè only 11 & ½. — Between the two they have 15 souls. — Mr Stokes & Mellersh are in La Paz; Mr Wickham & King in the other. — They sail on Wednesday; I look forward to our separation with much regret; our society on board can ill afford to lose such very essential members. — I am afraid the whole party will undergo many privations; the cabin in the smaller one is at present only 2 & ½ feet high! Their immediate business will be to survey South of B. Blanca: & at the end of next month we meet them at Rio Negro, in the bay of St Blas. —|244|

16th   Again I walked to Punta alta to look for fossil bones: on the road I crossed the track of a large herd of the Guanaco or American Camel. — the marks were as large as a cow, but more cloven. We laid in a good stock of fresh provisions for sea; as 6 deer were shot & great numbers of fish caught. —

17th   The Beagle & the two Schooners, forming a little fleet sailed together & anchored at night in the entrance of the Bay. —

18th   We continued to sound. — At noon the Schooners made sail to the South; we gave them three hearty & true cheers for a farewell. —

19th   The Captain landed for half an hour at Monte Hermoso, (or *Starvation* point as we call it) to take observations. — I went with him & had the good luck to obtain some well preserved fossil bones of two or three sorts of Gnawing animals. — One of them must have much resembled the Agouti but it is smaller. — We are now at night pressing on for the Rio Plata. —

20th The wind is very light. —

Sunday 21st & 22nd During these two days it has been a thick fog, with light breezes: We are all getting anxious for the moment of receiving letters to arrive. — Moreover, there is another substantial reason; our bread fails us on next Sunday, at present all hands are on a 2/3 allowance. The detainement from the Schooners is the cause of the miscalculation in the stores. —|245|

23rd The fog cleared away, only to disappoint us with an unfavourable breeze: —

24th The night was pitch dark, with a fresh breeze. — The sea from its extreme luminousness presented a wonderful & most beautiful appearance; every part of the water, which by day is seen as foam, glowed with a pale light. The vessel drove before her bows two billows of liquid phosphorus, & in her wake was a milky train. — As far as the eye reached, the crest of every wave was bright; & from the reflected light, the sky just above the horizon was not so utterly dark as the rest of the Heavens. — It was impossible to behold this plain of matter, as it were melted & consuming by heat, without being reminded of Miltons description of the regions of Chaos & Anarchy. —[1]

[1] CD wrote: 'Milton's *Paradise Lost* had been my chief favourite, and in my excursions during the voyage on the *Beagle*, when I could take only a single small volume, I always chose Milton' (see *Autobiography* p. 85). In describing to Henslow (see *Correspondence* 1: 280, and also *Journal of Researches* pp. 114–15) a possibly new species of toad coloured black and vermilion, he says 'Milton must allude to this very individual when he talks of "squat like a toad"' (*Paradise Lost*, Book 4, line 800). In an entry in Down House Notebook 1.7, written at Coquimbo in May 1835 before setting out for Copiapó, CD reminds himself not to leave the volume of Milton behind.

25th A fair breeze, right aft; we have not for the last 24 hours gone less than 6 knots an hour. — It may sound strange, but it is necessary for a person to be some time in a ship, before he understands how to enjoy a favourable wind; it is something like the pleasure of riding fast, although with no particular end in view; & this pleasure must be solely derived from habit. — In the same manner, during a fair breeze nothing can be more delightful than the general cheerfulness which pervades the whole ship. —|246|

26th The day has been very cloudy: but what are clouds & gloom to those who have just heard from their friends at home. My letters from Shrewsbury are dated May 12th & June 28th. —[1] Receiving letters unfits one for any occupation; so that I have done nothing but read the Newspapers; it is rather a laborious undertaking & to make it tolerable it requires the high interest of the present politicks of England. —

[1] These were letters from Susan and Caroline Darwin. See *Correspondence* 1: 234–6 and 241–3.

27th Went to the city to purchase some things.

Sunday 28th Rode with Mr Hammond to dine with a friend of his who has an Estancia in the country. — The town is built on a promontory & for two or three miles behind it an irregular suburb extends. — It is in this neighbourhead alone, that the ground is enclosed. — On each side of us the hedges were composed of enormous Agaves & in the vacant places were large Cacti. — I have seldom seen anything more strange to an Europæan eye than the appearance which, from this cause, the fields presented. — The house of the gentleman (Mr Grenville) with whom we were going to dine, was situated in the open camp; but from the large orchards surrounding it, the place had an unusually cheerful air. — In the garden Peaches, Quinces, Apples, Vines, Figs, Lemons & Oranges flourished with great luxuriance: the two latter formed most delightfully shady walks. Numerous Olive trees were in flower, these|247| very much resemble the Ilex, their leaves are however narrower & longer. — After a very pleasant dinner we returned to the ship. — Mr Grenville is one of the few Englishmen who has served under the Brazilian flag & who is a gentleman. — He is of a poor but good family & was, as a very young man (amongst many others) enticed out by Lord Cochrane when he served the Chilians. Subsequently to this, Mr Grenville had the command of a large Brazilian frigate, & in it fought some gallant actions. — He is now married to a very pleasant, & what is very rare, domestic Spanish lady. — With her he got the Estancia, where he is now living. —

29th Walked round the fortifications; & entered the country through the gate by which the English took M: Video. — The degree to which the ground near to the city is strewed with the bones of cattle & horses is truly astonishing & quite corresponds to the annual vast export of hides. — In the evening dined with Mr Parry & met there Cap. Paget of the Samarang. — Our old friend the Samarang came here a few weeks after we sailed to the South. — the Druid having gone to England. —

30th We got under weigh early in the morning for Buenos Ayres, but a fresh breeze right in our teeth lasted the whole day; so that when we anchored at night, we had not made much progress. — |248|

31ˢᵗ A beautiful day: but the wind has been steadily against us. — In the evening all the ropes were coated & fringed with Gossamer web. — I caught some of the Aeronaut spiders which must have come at least 60 miles. How inexplicable is the cause which induces these small insects, as it now appears in both hemispheres, to undertake their aerial excursions. —

November 1ˢᵗ A calm delightful day. — I know not the reason why such days always lead the mind to think of England and home. — It would seem as if the serenity of the air allowed the thoughts with greater ease to pass & repass the long interval. —

2ⁿᵈ Passing the Guard-ship (who this time treated us with greater respect) we anchored at noon in the outer roads. — The boats were lowered & a large party of officers went on shore; the landing is very awkward; from the shoalness of the water a cart is obliged to come a long way out to meet the boat. — We immediately went out riding: there is no way of enjoying the shore so throughily as on horseback: after being for some months in a ship, the mere prospect of living on dry land is very pleasant, & we were all accordingly in high spirits. — It is from this cause, I suppose, that most Foreigners believe that English sailors are all more or less mad. — |249|

3ʳᵈ The city of Buenos Ayres is large, & I should think one of the most regular in the world. — Every street is at right angles to the one it crosses; so that all the houses are collected into solid squares called "quadras". — On the other hand the houses themselves are like our squares, all the rooms opening into a neat little court. — They are generally only one story high, with flat roofs; which are fitted with seats & are much frequented by the inhabitants in Summer. In centre of the town is the Plaza, where all the public offices, Fortress, Cathedral &c are. — It was here that the old Viceroys lived, before the revolution. — The general assemblage of buildings possesses considerable architectural beauty, although none individually do so.

In the evening went out riding with Hammond & in vain tried to reach the camp; in England any one would pronounce the roads quite impassible; but the bullock waggons do contrive to crawl slowly on, a man however generally goes ahead to survey which is the best part to be attempted. — I do not suppose they travel one mile per hour, & yet with this the bullocks are much jaded: it is a great mistake to imagine with the improved roads and increased velocity of travelling that in the same proportion the cruelty towards the animals becomes greater. —

|250| For some miles round the town the country is enclosed by ditches & hedges of Agave or Aloes with Fennel.— One ride is sufficient to account for the horror which the few English gentlemen who reside here express for Buenos Ayres.— In our ride we passed the public place for slaughtering the cattle: the beasts were all lassoed in the Corral; so that there was no skill shown, the only thing which surprised me is the wonderful strength of horses compared to bullocks. After being caught round the horns, one horse dragged them to any distance; the poor beast after vainly in its efforts ploughing up the ground to resist the force, would dash at full speed to one side; the horse immediately turns to receive the shock, & stands so firmly as almost to throw the bullock down when he comes to the end of the Lasso.— When brought to the spot for killing, the matador with great caution cuts the hamstrings & then being disabled sticks them; it is a horrible sight: the ground is made of bones, & the men, horses & mud are stained by blood.—

Sunday 4th   Walked into several of the Churches & admired the brilliancy of the decorations for which the city is celebrated.— It is impossible not to respect the fervor which appears to reign during the Catholic service as compared with the Protestant.— The effect is heightened|251| by the equality of all ranks.— The Spanish lady with her brilliant shawl kneels by the side of her black servant in the open aisle.—

I visited the Museum, which is attached to the only remaining convent; although esteemed as second to none by the inhabitants it is very poor. In the evening went out riding with Hamond; we saw the first starting of a troop of waggons for Mendoza.— Changing the bullocks, they travel day & night, but even with this it takes 50 days. These waggons are very narrow & long, they are thatched with reeds & stand on wheels the diameter of which is 10 feet.— They are drawn by 6 bullocks, which are urged on by a goad at least twenty feet long.—it is suspended within the roof, so that it can be easily used.—the point is sharp & for the intermediate bullocks a small point projects downwards thus

For the wheel bullocks a short goad is kept in the waggon.— All this apparatus at first looks like implements of war.—

5th   Rode about 6 leagues into the camp to an English Estancia.— The country is very level & in places from Willows & Poplars being planted by the ditches much resembled Cambridgeshire.— Generally it is open

& consists either of bright green turf or large tracts of a very tall
Sow-thistle (8 or 9 feet high). — |252| Even the very roads were burrowed
by the Viscache. — This animal is nocturnal in its habits; in structure it
is allied to the Cavies, having gnawing teeth & only three toes to its hind
legs; it differs in having a tail. — The holes made by this animal yearly
cause the death of many of the Gauchos. — As Head mentions, every
burrow is tenanted by a small owl, who, as you ride past, most gravely
stares at you. —

6th Spent the day in shopping & in gaining information relative to the
geology of the country. — I trust when the Beagle returns for the winter
to the Rio Plata I shall be able to make some long excursions in this
unpicturesque but curious country. — Buenos Ayres is an excellent
place for making purchases; there are many shops kept by Englishmen
& full of English goods. — Indeed the whole town has more of an
Europæan look than any I have seen in S. America. One is called back
to the true locality, both by the Gauchos riding through the streets with
their gay coloured Ponchos & by the dress of the Spanish ladies. — This
latter, although not differing much from an English one, is most elegant
& simple. — In the hair (which is beautifully arranged) they wear an
enormous comb; from this a large silk shawl folds round the upper part
of the body. Their walk is most graceful, & although|253| often disap-
pointed, one never saw one of their charming backs without crying out,
"how beautiful she must be". —[1]

[1]In a letter to Caroline Darwin, CD wrote: 'We [himself and Hamond] were generally
companions on shore: our chief amusement was riding about & admiring the Spanish
Ladies. — After watching one of these angels gliding down the streets; involuntarily we
groaned out, "how foolish English women are, they can neither walk nor dress". — And
then how ugly Miss sounds after Signorita; I am sorry for you all; it would do the whole
tribe of you a great deal of good to come to Buenos Ayres. —' Later in the same letter he
says: 'I am glad the journal arrived safe; as for showing it, I leave that entirely in your
hands. — I suspect the first part is abominaly childish, if so do not send it to Maer. — Also,
do not send it by the Coach, (it may appear **ridiculous** to you) but I would as soon loose
a piece of my memory as it. — I feel it is of such consequence to my preserving a just
recollection of the different places we visit. — When I get another good opportunity I will
send some more. —' See *Correspondence* 1: 276–9.

7th We expected to have gone on board to day, but from bad weather &
other causes the sailing of the Beagle has been deferred for a few
days. — In the evening Capt. FitzRoy & myself dined at Mr Gores, the
English Charge d'affaires. We had a very pleasant evening: we met
there Colonel Harcourt Vernon, one of the most rare instances of a
tourist leaving the beaten tracks of Europe. — He has already travelled
in Agypt & having a strong wish to see Tropical scenery came to Rio de

Janeiro. And as he says, one walk amidst the glories of Brazil well repays the trouble of crossing the Atlantic. — Colonel Vernon is now going to undertake a most laborious journey, namely to cross the Pampas to Lima, from whence to Mexico & so home. —

8th In the evening went to the Theatre; I did not understand one word; yet, & which I should think was different from other languages, it sounded most distinct & energetic. — We saw here the universal custom amongst the Spaniards of separating the women from the men. — In the boxes they are together, but the pit is full of men & the gallery of women. The price for the boxes is about 14 pence or two *paper dollars*; for the rest of the house it is only one, or seven pence: of English money. — |254|

9th Called with Capt. FitzRoy on Donna Clara or Mrs Clarke. — The history of this woman is most strange. — She was originally a handsome young woman, transported for some atrocious crime. — On board the convict ship on its passage outwards, she lived with the Captain: some time before coming to the Latitude of Buenos Ayres she planned with the rest of the convict women to murder all on board excepting a few sailors. — She with her own hands killed the Captain, & by the help of a few sailors brought the ship into Buenos Ayres. — After this she married a man of considerable property & now inherits it. — Everybody seems to have forgotten her crimes, from the extraordinary labours she underwent in nursing our soldiers after the disastrous attempt (our flags are now in the Cathedral) to take this city. — Mrs Clarke is now an old decrepid woman: with a masculine face, & evidently even yet a most ferocious mind. — Her commonest expressions are "I would hang them all Sir", "I would kill him Sir," for smaller offences, "I would cut their fingers off". — The worthy old lady looks as if she would rather do it, than say so. —

10th Breakfasted with Mr Gore & at noon went on board: in the evening made sail for Monte Video; but as the night was dirty came to an anchor. — |255|

Sunday 11th The wind is unfavourable & we do not make much progress. — Every day is now of consequence, as it is one out of the summer. —

12th & 13th The wind continues dead in our teeth & although carrying on night & day we get on very slowly. — In the evening it blew hard & we dropped the anchor. —

14ᵗʰ This morning we entered the harbor at noon; after having fairly conquered as foul a wind as ever blew. — I received letters dated July 25, August 15ᵗʰ & 18ᵗʰ. —[1]

[1] These were letters from Catherine, Susan and Erasmus Darwin. See *Correspondence* 1: 253–9.

15ᵗʰ Spent the whole day in the city.

16ᵗʰ The dilatory method of doing business in this place again detained me all morning; in the evening enjoyed with Hamond a delightful gallop over the grassy plains. — We called on our way back on a Spanish family. Here I first saw the well known & universal custom of the young ladies giving to any gentlemen present a rose; the Signoritas make their little present with much grace & elegance. — The Signora at the same time, tells you with due formality, to consider the house as your own. —

17ᵗʰ Boisterous weather; glad should I be if the day for taking an everlasting farewell of the Rio Plata was near at hand. —

Sunday 18ᵗʰ After divine service on board I took a quiet ride over the open plains which border the river. — |256|

19ᵗʰ Employed in packing up specimens of Nat: History for England. —

20ᵗʰ Went out collecting on the Mount. — In the course of my walk I came quite close to two of the great lizards of this country. — From the nose to end of tail the length must have been at least 3 feet. —

21ˢᵗ All day long provisions & stores are hoisting in; never, without excepting Plymouth, have I seen the ship, even the quarter deck crowded with all sorts of things. — I am glad of it, for I am impatient to be again at sea. — I suspect however before our return there will oftener be occasion for patience than for the contrary extreme. —

22ⁿᵈ Rode with Mʳ Hamond to the Rio St Lucia. — the distance is about 12 miles & the path lies over an undulating plain of turf. — On our return we were obliged to go some miles round to avoid one of the great beds of thistles. These are quite impassible, as they are armed with long prickles, & grow close together to the height of six feet. — Riding is the only source of enjoyment in this country. —

23ʳᵈ At night there was a grand ball given in order to celebrate the reestablishment of the President. — It was a much gayer scene than I should have thought this place could have produced. — the desire which the inhabitants have on such occasions of appearing splendidly dressed is excessive: & to gratify it |257| the ladies will spare no sacrifices. The music was in very slow time & the dancing, although most formal, possessed much gracefulness. — The ball was given in the Theatre; nothing surprised me so much as the arrangements of the house; every part not actually occupied by the dancers was entirely open to the lowest classes of Society. —so that all the passages to the boxes, back parts of the pitt, were filled by any people who liked to look on. — And nobody ever seemed even to imagine the possibility of disorderly conduct on their parts. How different are the habits of Englishmen, on such Jubilee nights!

24th   Went to the Theatre & heard the opera of Cenerentola.[1]

[1] *Cenerentola* by Rossini had been performed at Covent Garden in 1830.

Sunday 25th   Rode with M[r] Parry to Las Pietras; a pretty village so called from some rocks of a singular shape. — One calls a village pretty in this country, if it possesses a dozen fig trees & is situated a hundred feet above the general level. —

26th   The ship got under weigh at noon, but we anchored at night without leaving M: Video. — The occasion of this delay caused a painful scene on board. — During the morning the heat on shore was excessive, & far more intolerable than that of the Tropics. I fully felt the truth of what M[r] Daniell|258| has ascertained to be the fact; namely that the difference between the heat of the suns rays & temperature of the atmosphere increases as the latitude becomes higher & in a greater ratio than the Temp. decreases. Hence it happens that the thermometer would actually rise higher when exposed to the sun in London than under the Equator; also it proves how completely all the effects of climate depend on mean temperature. — The day had been beautiful, but the barometer foretold a change, so that in a *calm* we anchored & struck our top gallant masts. — It was not in vain, a little after 10 oclock the squall struck us & it blew heavily all night.

27th   The morning was dirty, but the afternoon was fair & we ran up the river about 30 miles in order to pump in fresh water. — Anchored off the cliffs called Santa Maria. —

28th   A beautiful day; but fair wind of yesterday is now foul. — We sail direct for the bay St. Blas, where we appointed to meet the Schooners by the 20th of this month. — After meeting them we push directly onwards to Terra del Fuego, so that we may not loose any more of these precious long days. — I thank our good fortune that the Mount is at last out of sight; & I sincerely trust we may not see its outline for several months to come. —|259|

29th, 30th Beautiful days, calm sea, & a fine breeze; what can the heart of man desire more?

December 1st In the evening the weather looked threatening; & during the first watch there was a strong breeze. —it died away in a baffling calm; which the sailors call the "Doldrums". —

Sunday 2nd A cloudy day with a strong breeze.

3rd We anchored at night not far from the entrance of St Blas. — Within a few miles the two Schooners were at anchor. — Mr Wickham came on board & reports all well in the vessels. — They had a fine passage from Bahia Blanca; but during the month they have been surveying these coasts, there has been much dirty weather; & a little wind soon raises a great sea. — The report of the Bay of San Blas is so bad, that I suppose we shall not enter it. —

4th We ran down alongside the Schooners; & all the necessary business between them & the Beagle was carried on with the greatest activity: — The morning passed away most merrily in hearing & relating everything which has happened since we parted. — The coast, however, on which the Schooners have been employed seems to be even more uninteresting than that of Bahia Blanca. — The instructions for the next three months are as follows: — Mr Wickham, after cauking La Lievrè at R. Negro, runs up B. Blanca; returns immediately & joins Mr Stokes, who will be employed in this neighbourhead. — They then|260| in company sail for Port Desire;[1] & from that point, these little vessels will survey the coast up to Rio Negro. — The Beagle will meet them there in March; which month being very boisterous, our whole fleet intend lying snug in the river. — All the Officers dined together in the Gun-room; soon after which the Beagle made sail. — We are now with a rattling breeze & a bright moon scudding for Nassau Bay, behind Cape Horn. —

[1] Puerto Deseado on a modern map.

5th & 6th During these two delightful days we have been gliding onwards; but at a very slow pace. — I have been employed in examining some small Crustacea; most of which are not only of new genera, but very extraordinary ones. —

7th & 8th Fine, light weather.

Sunday 9th From the high irregular swell, there must have been bad weather to the South, so that we are lucky in escaping it. —

10th A strong breeze; At noon we were a little to the South of Port Desire. —

11th The Barometer had given good warning of a change of weather: it is the anniversary of our first attempt to get out of the English channel, & as on that day we were met by a heavy breeze from the SW. — With me the association was perfect though not very satisfactory, between the two days: my stomach plainly declared it was of terrestrial origin & did not like the sea. — |261|

12th It continued to blow fresh & in the middle of the day suddenly freshened into the heaviest squall I have ever seen. Luckily it gave us good notice, so that every thing was furled & the ship put before the wind; it is always interesting to watch the progress of a squall; the black cloud with its rising arch which gives passage to the wind; then the line of white breakers, which steadily approaches till the ship heels over & the squall is heard whistling through the rigging. — The climate during the few last days has undergone a complete change. — The Temp. varies from 45° to 50°, & the air has the bracing *feel* of an English winter day: But the most curious thing is to see the hammocks piped down at ½ after seven & the sun some way above the horizon. — it is a spectacle we have not beheld for the last 15 months.

13th In the evening the wind veered round & became fair: we are however some leagues further North than we were two days ago — so much for those unlucky South Westers. —

14th Light variable wind, generally against us. —

15th Very foggy. — every thing conspires to make our passage long. This evening the low land South of the St^s of Magellan was just visible from the deck. — |262|

Sunday 16th We made the coast of Tierra del Fuego[1] a little to the South of Cape St Sebastian & then altering our course ran along a few miles from the shore. — The Beagle had never visited this part before; so that it was new to every body. — Our ignorance whether any natives lived here, was soon cleared up by the usual signal of a smoke. — & shortly by the aid of glasses we could see a group & some scattered Indians evidently watching the ship with interest. — They must have lighted the fires immediately upon observing the vessel, but whether for the purpose of communicating the news or attracting our attention, we do not know. — The breeze was fresh & we ran down about 50 miles of coast & anchored for the night. — The country is not high, but formed of horizontal strata of some modern rock, which in most places forms abrupt cliffs facing the sea. — It is also intersected by many sloping vallies, these are covered with turf & scattered over with thickets & trees, so as to present a cheerful appearance. The sky was gloomy & the atmosphere not clear, otherwise the views would in some places have been pretty. — At a great distance to the South was a chain of lofty mountains, the summits of which glittered with snow. — We are at anchor to the South of St Pauls head. — |263|

[1] When in 1520 Magellan was making his way in the *Trinidada* through the Straits that he had discovered, he is said to have seen many fires lit by the natives in the hills to the south, and named the country Tierra del Fuego, 'Land of Fire'.

17th The Ship rolled so much during the night from the exposed anchorage, that there was no comfort to be obtained. — At daylight which is about 3 oclock we got under weigh & with a fair breeze stood down the coast. At Port St Policarpo, the features of the country are changed. — high hills clothed in brownish woods take the place of the horizontal formations. — A little after noon we doubled C. St. Diego & entered the famous Straits Le Maire. — We had a strong wind with the tide; but even thus favoured it was easy to perceive how great a sea would rise were the two powers opposed to each other. — The motion from such a sea is very disagreeable; it is called "pot-boiling", & as water boiling breaks irregularly over the ships sides. — We kept close to the Fuegian shore; the outline of the rugged inhospitable Staten Land was visible amidst the clouds. — In the afternoon we anchored in the bay of Good Success, here we intend staying some days. — In doubling the Northern entrance, a party of Fuegians were watching us, they were perched on a wild peak overhanging the sea & surrounded by wood. — As we passed by they all sprang up & waving their cloaks of skins sent forth a loud sonorous shout. —this they continued for a long time. — These people followed the ship up the harbor & just before dark we|264| again heard their cry & soon saw their fire at the entrance of the Wigwam which they built for the night. — After dinner the Captain went on shore to look for a watering place; the little I then saw showed how different this country is from the corresponding zone in the Northern Hemisphere. — To me it is delightful being at anchor in so wild a country as Tierra del F.; the very name of the harbor we are now in, recalls the idea of a voyage of discovery; more especially as it is memorable from being the first place Capt. Cook anchored in on this coast; & from the accidents which happened to Mr Banks & Dr Solander. —[1] The harbor of Good Success is a fine piece of water & surrounded on all sides by low mountains of slate. — These are of the usual rounded or saddle-backed shape, such as occur in the less wild parts of N: Wales. — They differ remarkably from the latter in being clothed by a very thick wood of evergreens almost to the summit. The last time Cap. FitzRoy was here it was in winter; he says the landscape was of the same brownish green tint & but little more snow on the hills. — The Barometer had been very low & this evening it suddenly rose $\frac{3}{10}$ of an inch, & now at night it is blowing a gale of wind & rain & heavy squalls sweep down upon us from the mountains:|265| Those who know the comfortable feeling of hearing the rain & wind beating against the windows whilst seated round a fire, will understand our feelings: it would have been a very bad night out at sea, & we as well as others may call this Good Success Bay. —

[1] Joseph Banks and Daniel Carl Solander had sailed with Captain Cook in the *Endeavour* from 1768 to 1771. While climbing at Bay of Good Success on 16–17 January 1769, they were caught in a snowstorm and two members of their group perished. See *Journal of the right hon. Sir Joseph Banks during Captain Cook's first voyage in H.M.S. Endeavour in 1768–71 to Terra del Fuego, Otahite, New Zealand, Australia, the Dutch East Indies, etc.* Edited by Sir Joseph D. Hooker. London, 1896.

18th   The Captain sent a boat with a large party of officers to communicate with the Fuegians. —[1] As soon as the boat came within hail, one of the four men who advanced to receive us began to shout most vehemently, & at the same time pointed out a good landing place. — The women & children had all disappeared. — When we landed the party looked rather alarmed, but continued talking & making gestures with great rapidity. — It was without exception the most curious & interesting spectacle I ever beheld. — I would not have believed how entire the difference between savage & civilized man is. — It is greater than between a wild & domesticated animal, in as much as in man there is greater power of improvement. — The chief spokesman was old & appeared to be head of the family; the three others were young powerful men & about 6 feet high. — From their dress &c &c they resembled the representations of Devils on the Stage, for instance in Der Freischutz. — The old man had a white feather cap; from under|266| which, black long hair hung round his face. — The skin is dirty copper colour. Reaching from ear to ear & including the upper lip, there was a broard red coloured band of paint. —& parallel & above this, there was a white one; so that the eyebrows & eyelids were even thus coloured; the only garment was a large guanaco skin, with the hair outside. — This was merely thrown over their shoulders, one arm & leg being bare; for any exercise they must be absolutely naked. — Their very attitudes were abject, & the expression distrustful, surprised & startled: — Having given them some red cloth, which they immediately placed round their necks, we became good friends. — This was shown by the old man

A Fuegian at Portrait Cove, by T. Landseer after C. Martens (*Narrative* **2:** frontispiece).

patting our breasts & making something like the same noise which people do when feeding chickens. — I walked with the old man & this demonstration was repeated between us several times: at last he gave me three hard slaps on the breast & back at the same time, & making most curious noises. — He then bared his bosom for me to return the compliment, which being done, he seemed highly pleased: — Their language does not deserve to be called articulate: Capt. Cook says it is like a man clearing his throat; to which may be added another|267| very hoarse man trying to shout & a third encouraging a horse with that peculiar noise which is made in one side of the mouth. — Imagine these sounds & a few gutterals mingled with them, & there will be as near an approximation to their language as any European may expect to obtain. Their chief anxiety was obtain knives; this they showed by pretending to have blubber in their mouths, & cutting instead of tearing it from the body. —they called them in a continued plaintive tone Cochilla, —probably a corruption from a Spanish word. — They are excellent mimics, if you cough or yawn or make any odd motion they immediately imitate you. — Some of the officers began to squint & make monkey like faces; — but one of the young men, whose face was painted black with white band over his eyes was most successful in making still more hideous grimaces. — When a song was struck up, I thought they would have fallen down with astonishment; & with equal delight they viewed our dancing and immediately began themselves to waltz with one of the officers. — They knew what guns were & much dreaded them, & nothing would tempt them to take one in their hands. — Jemmy Button came in the boat with us; it was interesting to watch their conduct|268| to him. — They immediately perceived the difference & held much conversation between themselves on the subject. — The old man then began a long harangue to Jemmy; who said it was inviting him to stay with them: —but the language is rather different & Jemmy could not talk to them. — If their dress & appearance is miserable, their manner of living is still more so. — Their food chiefly consists in limpets & muscles, together with seals & a few birds; they must also catch occasionally a Guanaco. They seem to have no property excepting bows & arrows & spears: their present residence is under a few bushes by a ledge of rock: it is no ways sufficient to keep out rain or wind. —& now in the middle of summer it daily rains & as yet each day there has been some sleet. — The almost impenetrable wood reaches down to high water mark. —so that the habitable land is literally reduced to the large stones on the beach. —& here at low water, whether it may be night or day, these wretched looking beings pick up a livelihood. — I believe if the world was searched, no lower grade of man could be found. — The Southsea Islanders are civilized compared to them, & the Esquimaux, in subterranean huts may enjoy some of the comforts of life. —

After dinner the Captain paid the Fuegians|269| another visit. — They received us with less distrust & brought with them their timid children. — They noticed York Minster (who accompanied us) in the same manner as Jemmy, & told him he ought to shave, & yet he has not 20 hairs on his face, whilst we all wear our untrimmed beards. — They examined the color of his skin; & having done so, they looked at ours. — An arm being bared, they expressed the liveliest surprise & admiration. — Their whole conduct was such an odd mixture of astonishment & imitation, that nothing could be more laughable & interesting. — The tallest man was pleased with being examined & compared with a tall sea-man, in doing this he tried his best to get on rather higher ground & to stand on tip-toes: He opened his mouth to show his teeth & turned his face en profil; for the rest of his days doubtless he will be the beau ideal of his tribe. — Two or three of the officers, who are both fairer & shorter than the others (although possessed of large beards) were, we think, taken for Ladies. — I wish they would follow our supposed example & produce their "squaws". — In the evening we parted very good friends; which I think was fortunate, for the dancing & "sky-larking" had occassionally bordered on a trial of strength. —|270|

[1] For FitzRoy's account of this first meeting with the Fuegians see *Narrative* 2: 120–2.

19th   I determined to attempt to penetrate some way into the country. — There is no level ground & all the hills are so thickly clothed with wood as to be quite impassable. — The trees are so close together & send off their branches so low down, that I found extreme difficulty in pushing my way even for gun-shot distance. — I followed therefore the course of a mountain torrent; at first from the cascades & dead trees, I hardly managed to crawl along; but shortly the open course became wider, the floods keeping clear the borders. — For an hour I continued to follow the stream, & was well repaid by the grandeur of the scene. — The gloomy depth of the ravine well accorded with the universal signs of violence. — in every direction were irregular masses of rock & uptorn trees, others decayed & others ready to fall. — To have made the scene perfect, there ought to have been a group of Banditti. — in place of it, a seaman (who accompanied me) & myself, being armed & roughly dressed, were in tolerable unison with the surrounding savage Magnificence. We continued ascending till we came to what I suppose must have been the course of a water-spout, & by its course reached a considerable elevation. — The view was imposing but not very pictures-que: the whole wood is composed of the antarctic Beech (the Winters bark & the Birch are comparatively |271| rare). This tree is an evergreen, but the tint of the foliage is brownish yellow: Hence the whole lanscape has a monotomous sombre appearance; neither is it often enlivened by the rays of the sun. — At this highest point the wood is not quite so thick, —but the trees, though not high are of considerable thickness. — Their curved & bent trunks are coated with lichens, as their roots are with moss; in fact the whole bottom is a swamp where nothing grows except rushes & various sorts of moss. —the number of decaying & fallen trees reminded me of the Tropical forest. — But in this still solitude, death instead of life is the predominant spirit. — The delight which I experienced, whilst thus looking around, was increased by the knowledge that this part of the forest had never before been traversed by man. —

20th I was very anxious to ascend some of the mountains in order to collect the Alpine plants & insects. — The one which I partly ascended yesterday was the nearest, & Capt. FitzRoy thinks it is certainly the one which Mr Banks ascended, although it cost him the lives of two of his men & very nearly that of Dr Solander. — I determined to follow a branch of the watercourse, as by this means all danger of losing yourself even in the case of a|272| snow storm is removed. — The difficulty of climbing was very great: as the dead & living trunks were so close, that in many places it was necessary to push them down to make a path. — I then gained a clearer place & continued following the rivulet. — This at last dwindled away, but having climbed a tree I took the bearing of the summit of the hill with a compass & so steered a straight course. — I had imagined the higher I got, the more easy the ascent would be, the case however was reversed. From the effects of the wind, the trees were not above 8 or 10 feet high, but with thick & very crooked stems; I was obliged often to crawl on my knees. At length I reached what I imagined to be green turf; but was again disappointed by finding a compact mass of little beech trees about 4 or 5 feet high. — These were as thick as Box in the border of a flower garden. — For many yards together my feet never touched the ground. I hailed with joy the rocks covered with Lichens & soon was at the very summit. — The view was very fine, expecially of Staten Land & the neighbouring hills; Good Success Bay with the little Beagle were close beneath me. In ascending the bare summit, I came close to two Guanaco & in the course of my walk saw several more.— These beautiful|273| animals are truly alpine in their habits, & in their wildness well become the surrounding landscape.— I cannot imagine anything more graceful than their action: they start on a canter & when passing through rough ground they dash at it like a thorough bred hunter.— The noise they make is very peculiar & somewhat resembles the neighing of a colt.[1] A ridge connected this hill with one several miles distant & much more lofty, even so that snow was lying on it; as the day was not far advanced I determined to walk there & collect on the road.— Some time after I left this hill (Banks Hill, Capt: FitzR) a party of 6 from the ship reached it, but by a more difficult path; but in descending they found an easier.— After 2 hours & a half walking I was on the top of the distant peak.—it was the highest in the immediate neighbourhead & the waters on each side flowed into different seas.— The view was superb, & well was I repaid for the fatigue.— I could see the whole neck of land which forms the East of Strait Le Maire.— From Cape St Diego as far as the eye could reach up the NW coast; & what interested me most, was the whole interior country between the two seas.— The Southern was mountainous & thickly wooded; the Northern|274| appeared to be a flat swamp & at the extreme NW part there was an expanse of water, but this will be hereafter examined. It looked dirty in the SW & I was afraid to stay long to enjoy this view over so wild & so unfrequented a country.— When Sir J. Banks ascended one of these mountains it was the middle of January which corresponds to our August & is certainly as hot as this month, & even with the occurrence of a snow-storm the misfortunes they met with are inexplicable. The snow was lying on the ESE side of the hills, & the wind was keen.—but on the lee side the air was dry & pleasant. Between the stony ridges & the woods there is a band of peat bogs & over this the greater part of my track lay.—but nearly all the difficulty was avoided by following a regular path which the Guanacos frequent; by following this I reached in much shorter time the forest & began the most laborious descent through its entangled thickets.— I collected several alpine flowers, some of which were the most diminutive I ever saw; & altogether most throuighly enjoyed the walk.—

---

[1] FitzRoy wrote: 'Soon after daylight this morning, some very large guanacoes were seen near the top of Banks Hill. They walked slowly and heavily, and their tails hung down to their hocks. To me their size seemed double that of the guanacoes about Port Desire. Mr. Darwin and a party set off to ascend the heights, anxious to get a shot at the guanacoes and obtain an extended view, besides making observations. They reached the summit, and saw several large animals, whose long woolly coats and tails added to their real bulk, and gave them an appearance quite distinct from that of the Patagonian animal; but they could not succeed in shooting one.' See *Narrative* 2: 122.

21st The Beagle got under weigh at 4 AM. — & doubtless to the grief of the Fuegians: The same evening we were with them they departed in a body, but yesterday they returned with a reinforcement of natives|275| who most likely came to beg for "Cochillas". — We doubled Cape Good Success, then the wind fell light & it became misty. — So calm a sea & atmosphere would have surprised those who think that this is the region where winds & waters never cease fighting. —

22nd In the morning watch it freshened into a fine Easterly wind. — which is about as lucky & rare an event as getting a prize ticket in a lottery. We soon closed in with the Barnevelts; & running past Cape Deceit with its stony peaks, about 3 oclock doubled the old-weather-beaten Cape Horn. — The evening was calm & bright & we enjoyed a fine view of the surrounding isles. — The height of the hills varies from 7 or 800 to 1700, & together they form a grand irregular chain. — Cape Horn however demanded his tribute & by night sent us a gale right in our teeth. —

23rd With close-reefed sail the Beagle made good weather of it; & much to her credit fell nothing to leeward. —

24th In the morning of the 24th Cape Horn was on our weather bow. — We now saw this notorious point in its proper form, veiled in a mist & its dim outline surrounded by a storm of wind & water:|276| Great black clouds were rolling across the sky & squalls of rain & hail swept by us with very great violence: so that the Captain determined to run into Wigwam cove. — This harbor is a quiet little basin behind Cape Spencer & not far from Cape Horn. — And here we are in quite smooth water; & the only thing which reminds us of the gale which is blowing outside. — is the heavy puffs or Whyllywaws, which every 5 minutes come over the mountains, as if they would blow us out of the water. —

25th This being Christmas day, all duty is suspended, the seamen look forward to it as a great gala day; & from this reason we remained at anchor. — Wigwam Cove is in Hermit Island; its situation is pointed out by Katers Peak, a steep conical mountain 1700 feet high which arises by the side of, & overlooks the bay:— Sulivan Hamond & myself started after breakfast to ascend it:—the sides were very steep so [as] to make the climbing very fatiguing, & parts were thick with the Antarctic Beech. From the summit a good geographical idea might be obtained of the surrounding isles & distant main land. — These islands would appear to be the termination of the chain of the Andes; the mountain tops only being raised above the ocean. — Whilst looking round on this inhospitable region|277| we could scarcely credit that man existed in it. — On our return on board, we were told we had been seen from the ship: this we knew to be impossible, as the Beagle is anchored at the mouth of the harbor & close under a lofty peak, behind which is Katers. As it was certain men had been seen crawling over the rock on this hill, they must have been Fuegians. — From their position, all our parties were in view. — & what must have been their feelings of astonishment — the whole of wigwam cove resounded with guns fired in the Caverns at the Wild fowl; we three also screaming to find out echos, Sulivan amusing himself by rolling down the precipes huge stones, & I impetuously hammering with my geological tools the rocks. They must have thought us the powers of darkness; or whatever else, fear has kept them concealed. — Wigwam Cove has frequently been visited: it was named by Mʳ Weddell. The Chanticleer, with Capt. Forster remained here some months; the remains of the tent where he swung the Pendulum exist yet. —

The sky looked ominous at sunset & in the middle of the night the hands were turned up to let go another anchor, for it blew a tremendous gale.

26ᵗʰ   The weather continues unsettled & most|278| exceedingly unpleasant; on the hills snow falls, & in the vallies continued rain & wind. — The temperature in day-time is about 45° & at night it falls to 38° or 40°; from the continued cloudy state of the atmosphere, the suns rays seldom have much power. — Considering this is the middle of the summer & that the Latitude is nearly the same as Edinburgh, the climate is singularly uncongenial. Even on the fine days, there is a continual succession of rain or hail storms; so that on shore there is not a dry spot. —

27ᵗʰ, 28ᵗʰ & 29ᵗʰ   To our great loss, the weather during these three days has been very bad, with much rain & violent squalls from the SW. — Yesterday the Captain went to reconnoitre the bays formed by the many islands at the back of Hermits. — I accompanied him, but the weather is so bleak & raw as to render boating rather disagreeable. — We ascended some of the hills, which as usual showed us the nakedness of the land. —

In most of the coves there were wigwams; some of them had been recently inhabited. The wigwam or Fuegian house is in shape like a cock of hay, about 4 feet high & circular; it can only be the work of an hour, being merely formed of a few branches & imperfectly thatched with grass, rushes &c. As shell fish, the chief source of subsistence, are soon exhausted in any one place, |279| there is a constant necessity for migrating; & hence it comes that these dwellings are so very miserable. It is however evident that the same spot at intervals, is frequented for a succession of years. —the wigwam is generally built on a hillock of shells & bones, a large mass weighing many tuns. — Wild celery, Scurvy grass, & other plants invariably grow on this heap of manure, so that by the brighter green of the vegetatioι the site of a wigwam is pointed out even at a great distance. —

The sea is here tenanted by many curious birds, amongst which the Steamer is remarkable; this [is] a large sort of goose, which is quite unable to fly but uses its wings to flapper along the water; from thus beating the water it takes its name. Here also are many Penguins, which in their habits are like fish, so much of their time do they spend under water, & when on the surface they show little of their bodies excepting the head. —their wings are merely covered with short feathers. So that there are three sorts of birds which use their wings for more purposes than flying; the Steamer as paddles, the penguin as fins, & the Ostrich spreads its plumes like sails to the breeze. —|280|

30ᵗʰ Remained at anchor.

31ˢᵗ The sun having at last shown itself at the proper time, observations were obtained & as the weather did not look quite so bad we put to sea.

January 1ˢᵗ For this & the following day we had a moderate wind from the old quarter SW; we all thought that after so much bad weather we should at least have a few fine days; the wind lulled & we hailed with joy a light air from the East; but in a couple of hours it veered to the North & then blew a strong gale from the SW. —

2ⁿᵈ, 3ʳᵈ This is always accompanied by constant rain & a heavy sea; & now after four days beating we have scarcely gained a league. Can there be imagined a more disagreeable way of passing time? — Whilst weathering the Diego Ramirez rocks, the Beagle gave an unusual instance of good sailing; with closed reefed topsails & courses & a great sea running, close hauled to the wind she made 7 & ½ knots.

4ᵗʰ–9ᵗʰ During all these precious days we have been beating day & night against the Westerly winds. The cause of our slow progress is a current which is always setting round the coast & which counterbalances the little which can be gained by beating up against strong winds & a heavy sea. — After passing the Il Defonsos rocks, it blew strong & in 24 hours we|281| were rather to leeward of them. — After this the wind was steady from the NW with much rain, & we drifted down to the Latitude of 57° 23'. — On the 8th it blew what Sailors term a strong gale (it is the first we have had) the Beagle is however so good a sea-boat, that it makes no great difference.

9th   To day the weather has been a little better, but now at night the wind is again drawing to the old quarter. — We doubled Cape Horn on the 21st, since which we have either been waiting for good or beating against bad weather & now we actually are about the same distance, viz. a hundred miles, from our destination. — There is however the essential difference of being to the South instead of the East. — Besides the serious & utter loss of time & the necessary discomforts of the ship heavily pitching & the miseries of constant wet & cold, I have scarcely for an hour been quite free from sea-sickness: How long the bad weather may last, I know not; but my spirits, temper, & stomach, I am well assured, will not hold out much longer. —

10th   A gale from the SW.

11th   A very strong breeze, with heavy squalls; by carrying a press of sail, we fetched within a mile of Christmas Sound. — This|282| rough precipitous coast is known by a mountain which from its castellated form was called by Capt. Cook York Minster. We saw it only to be disappointed, a violent squall forced us to shorten sail & stand out to sea. — To give an idea of the fury of the unbroken ocean, clouds of spray were carried over a precipice which must have been 200 feet high. —

12th   A gale with much rain, at night it freshened into a regular storm. — The Captain was afraid it would have carried away the close reefed main topsail. — We then continued with merely the trysails & storm stay sail.

Sunday 13th   The gale does not abate: if the Beagle was not an excellent sea-boat & our tackle in good condition, we should be in distress. A less gale has dismasted & foundered many a good ship. The worst part of the business is our not exactly knowing our position: it has an awkward sound to hear the officers repeatedly telling the look out man to look well to leeward. — Our horizon was limited to a small compass by the spray carried by the wind:—the sea looked ominous, there was so much foam that it resembled a dreary plain covered by patches of drifted snow. — Whilst|283| we were heavily labouring, it was curious to see how the Albatross with its widely expanded wings, glided right up the wind. —

Noon. At noon the storm was at its height; & we began to suffer; a great sea struck us & came on board; the after tackle of the quarter boat gave way & an axe being obtained they were instantly obliged to cut away one of the beautiful whale-boats.—the same sea filled our decks so deep, that if another had followed it is not difficult to guess the result. — It is not easy to imagine what a state of confusion the decks were in from the great body of water.— At last the ports were knocked open & she again rose buoyant to the sea. — In the evening it moderated & we made out Cape Spencer (near Wigwam Cove), & running in, anchored behind false Cape Horn. — As it was dark there was difficulty in finding a place; but as the men & officers from constant wet are much tired, the anchor was "let go" in the unusual depth of 47 fathoms.— The luxury of quiet water after being involved in such a warring of the elements is indeed great.— It could have been|284| no ordinary one since Capt. FitzRoy considers it the worst gale he was ever in. —[1] It is a disheartening reflection; that it is now 24 days since doubling Cape Horn, since which there has been constant bad weather, & we are now not much above 20 miles from it.

[1] FitzRoy noted: 'It was well that all our hatchways were thoroughly secured, and that nothing heavy could break a-drift. But little water found its way to the lower deck, though Mr. Darwin's collections, in the poop and forecastle cabins on deck, were much injured . . . The roller which hove us almost on our beam ends, was the highest and most hollow that I have seen, excepting one in the Bay of Biscay, and one in the Southern Atlantic; yet so easy was our little vessel that nothing was injured besides the boat, the netting (washed away), and one chronometer.' See *Narrative* 2: 126.

14ᵗʰ The winds certainly are most remarkable; after such a storm as yesterdays, it blew a heavy gale from the SW.— As we are in smooth water it does not so much signify.— We stood to the North to find an harbor; but after a wearying search in a large bay did not succeed. I find I have suffered an irreparable loss from yesterdays disaster, in my drying paper & plants being wetted with salt-water.— Nothing resists the force of an heavy sea; it forces open doors & sky lights, & spreads universal damage.— None but those who have tried it, know the miseries of a really heavy gale of wind.— May Providence keep the Beagle out of them. —

15ᵗʰ Standing to the East, we found a most excellent anchorage in Goree Sound & moored ship, secured from wind & sea: We shall probably remain here some weeks as the Fuegians & Matthews are to be settled here|285| & there will be some boat expeditions. The object of our disastrous attempt to get to the Westward was to go to the Fuegian York Minsters, country.— Where we now are is Jemmy Buttons & most luckily York Minster from his free choice intends to live here with Matthews & Jemmy. — Goree Sound is situated by Lennox Island & near to the Eastern entrance of Beagle channell. —

16th The Captain took two boats to search for a good place for the settlement. — We landed & walked some miles across the country. — It is the only piece of flat land the Captain has ever met with in Tierra del F & he consequently hoped it would be better fitted for agriculture. — Instead of this it turned out to be a dreary morass only tenanted by wild geese & a few Guanaco. — The section on the coast showed the turf or peat to be about 6 feet thick & therefore quite unfit for our purposes. We then searched in different places both in & out of the woods, but nowhere were able to penetrate to the soil; the whole country is a swamp. — The Captain has in consequence determined to take the Fuegians further up the country. — This place seems to be but sparingly inhabited. — In one place we found recent traces; even so that the fire where limpets had been roasted|286| was yet warm. — York Minster said it had only been one man; "very bad man" & that probably he had stolen something. — We found the place where he had slept. — it positively afforded no more protection than the form of a hare. — How very little are the habits of such a being superior to those of an animal. — By day prowling along the coast & catching without art his prey, & by night sleeping on the bare ground. —

17th, & 18th Spent in preparing for a long excursion in the boats. — In consequence, the Captain determined to take the whole party to J. Buttons country in Ponsonby Sound.

19th In the morning, three whale-boats & the Yawl started with a fair wind. — We were 28 in number & the yawl carried the outfit given to Matthews by the Missionary society. — The choice of articles showed the most culpable folly & negligence. Wine glasses, butter-bolts, tea-trays, soup turins, mahogany dressing case, fine white linen, beavor hats & an endless variety of similar things shows how little was thought about the country where they were going to. The means absolutely wasted on such things would have purchased an immense stock of really useful articles. — Our course lay towards the Eastern en-trance|287| of the Beagle channel & we entered it in the afternoon. — The scenery was most curious & interesting; the land is indented with numberless coves & inlets, & as the water is always calm, the trees actually stretch their boughs over the salt water. In our little fleet we glided along, till we found in the evening a corner snugly concealed by small islands. — Here we pitched our tents & lighted our fires. — nothing could look more romantic than this scene. — the glassy water of the cove & the boats at anchor; the tents supported by the oars & the smoke curling up the wooded valley formed a picture of quiet & retirement. —

20th We began to enter to day the parts of the country which is thickly inhabited. — As the channel is not generally more than three or 4 miles broard, the constant succession of fresh objects quite takes away the fatigue of sitting so many hours in one position. — The Beagle channel was first discovered by Cap FitzRoy during the last voyage,[1] so that it is probable the greater part of the Fuegians had never seen Europæans. — Nothing could exceed their astonishment at the apparition of our four boats: fires were lighted on every point to attract our attention & spread the news. — |288| Many of the men ran for some miles along the shore. — I shall never forget how savage & wild one group was. — Four or five men suddenly appeared on a cliff near to us. — they were absolutely naked & with long streaming hair; springing from the ground & waving their arms around their heads, they sent forth most hideous yells. Their appearance was so strange, that it was scarcely like that of earthly inhabitants. —

We landed at dinner time; the Fuegians not at first inclined to be friendly, for till one boat pulled in before the others, they kept their slings in readiness: — We soon delighted them by trifling presents such as tying red tape round the forehead; it is very easy to please but as difficult to make them content; the last & first word is sure to be "Yammerschooner" which means "give me". —[2] At night we in vain endeavoured to find an uninhabited cove; the natives being few in number were quiet & inoffensive:

[1] The Beagle Channel was discovered by M. Murray, Master of the *Beagle*, in April 1830. See *Narrative* 1: 429.

[2] According to Thomas Bridges on p. 639 of *Yamana–English: a dictionary of the speech of Tierra del Fuego*, edited by F. Hestermann and M. Gusinde and privately printed at Mödling, Austria, in 1933, the phrase 'yamašk-ūna' does indeed mean 'Do be liberal to me'.

21st In the morning however, a fresh party having arrived, they became troublesome, some of the men picked up stones & the women & children retreated; I was very much afraid we should have had a skirmish; it would have been shocking to have fired on such naked|289| miserable creatures. — Yet their stones & slings are so destructive that it would have been absolutely necessary. — In treating with savages, Europæans labor under a great disadvantage, untill the cruel lesson is

taught how deadly firearms are. — Several times when the men have
been tired & it was growing dark, all the things have been packed up to
remove our quarters; & this solely from our entire inability to frighten
the natives. One night the Captain fired double barrelled pistols close
to their faces, but they only rubbed their heads & when he flourished
his cutlass they were amused & laughed. They are such thieves & so
bold Cannabals that one naturally prefers separate quarters. —

The country on each side of the channel continues much the same,
slate hills thickly clothed by the beech woods run nearly parallel to the
water; the low point of view from a boat & the looking along one valley
& thus loosing the beautiful succession of ridges, is nearly destructive
to picturesque effect. —

22nd After an unmolested night in what would appear to be neutral
ground, between the people we saw yesterday & Jemmys. — we
enjoyed a delightful pull through the calm water. — The Northern
mountains have become more lofty & jagged. — their summits are
partially|290| covered with snow & their sides with dark woods: it was
very curious to see as far as the eye ranged, how *exact* & truly horizontal
the line was at which the trees ceased to grow. — it precisely resembled
on a beach the high-water mark of drift sea-weed. —

At night we arrived at the junction with Ponsonby Sound; we took up
our quarters with a family belonging to Jemmys or the *Tekenika*[1]
people. — They were quiet & inoffensive & soon joined the seamen
round a blazing fire; although naked they streamed with perspiration at
sitting so near to a fire which we found only comfortable. — They
attempted to join Chorus with the songs; but the way in which they
were always behind hand was quite laughable. — A canoe had to be
despatched to spread the news & in the morning a large gang arrived. —[2]

[1] The name 'Yapoo Tekeenica' used by FitzRoy and others to describe the inhabitants of
this part of Tierra del Fuego seems to have arisen from the difficulty of communicating
with them. According to E. L. Bridges (*Uttermost part of the earth*, Hodder & Stoughton,
London, 1948), who was born and brought up with the Fuegians at the Mission
established by his father, Thomas Bridges, at Ushuaia on the Beagle Channel, the word
'iapooh' means an otter, while 'teke uneka' means 'I do not understand you'. FitzRoy's
original informant must therefore have been conveying a lack of understanding rather
than the name of the place. The canoe-using natives of the Beagle Channel called
themselves 'Yamana' ('People'), while the particular group who lived on the shores of the
Murray Narrow were the 'Yahgashagalumoala' or 'People from Mountain Valley Channel', which Thomas Bridges shortened to Yahgans.
[2] Note in the margin in ink and in very small writing runs: 'Not [*illeg*] boat. Bougainville
simple things such as skins scarlet cloth, a nail & Linen'.

Woollya, by T. Landseer after R. FitzRoy (*Narrative* 2: 208).

23rd Many of them had run so fast that their noses were bleeding, & they talked with such rapidity that their mouths frothed, & as they were all painted white red & black they looked like so many demoniacs who had been fighting. — We started, accompanied by 12 canoes, each holding 4 or 5 people, & turning down Ponsonby, soon left them far behind. — Jemmy Button now perfectly knew the way & he guided us|291| to a quiet cove where his family used formerly to reside. We were sorry to find that Jemmy had quite forgotten his language, that is as far as talking, he could however understand a little of what was said. It was pitiable, but laughable, to hear him talk to his brother in English & ask him in Spanish whether he understood it.[1] I do not suppose, any person exists with such a small stock of language as poor Jemmy, his own language forgotten, & his English ornamented with a few Spanish words, almost unintelligible. — Jemmy heard that his father was dead; but as he had had a "dream in his head" to that effect, he seemed to expect it & not much care about it. — He comforted himself with the natural reflection "me no help it". — Jemmy could never find out any particulars about his father, as it is their constant habit, never to mention the dead. — We believe they are buried high up in the woods. —anyhow Jemmy will not eat land-birds, because they live on dead men. — This is one out of many instances where his prejudices are recollected, although language forgotten.

When we arrived at Woolliah[2] (Jemmys cove) we found it far better suited for our purposes, than any place we had hitherto seen. — There was a considerable space of cleared & rich ground, & doubtless Europæan vegetables would flourish well. — |292| We found a strange family living there, & having made them friends, they, in the evening, sent a canoe to Jemmys relations. — We remained in this place till the 27th, during which the labors of our little colony commenced. — On the 24th the Fuegians began to pour in; Jemmys mother, brother, & uncle came; the meeting was not so interesting as that of two horses in a field. — The most curious part was the astonishing distance at which Jemmy recognized his brothers voice.[3] To be sure, their voices are wonderfully powerful. — I really believe they could make themselves heard at treble the distance of an Englishmen. — All the organs of sense are highly perfected; sailors are well known for their good eyesight, & yet the Fuegians were as superior as another almost would be with a glass. — When Jemmy quarrelled with any of the officers, he would say "me see ship, me no tell". — Both he & York have invariably been in the right; even when objects have been examined with a glass.

Everything went on very peacibly for some days. — 3 houses were built, & two gardens dug & planted. — & what was of most consequence the Fuegians were very quiet & peacible; at one time there were about 120 of them. — the men sat all day long watching our proceedings & the poor women working like slaves for their subsistence. The men did not manifest much surprise at anything & never even appeared to look|293| at the boats. — Stripping for washing & our white skins seemed most to excite their attention. — They asked for every thing they saw & stole what they could. — Dancing & singing absolutely delighted them. — Things thus remained so quiet, that others & myself took long walks in the surrounding hills & woods. — On the 27th however suddenly every woman & child & nearly all the men removed themselves & we were watched from a neighbouring hill. — We were all very uneasy at this, as neither Jemmy or York understood what it meant; & it did not promise peace for the establishment. — We were quite at a loss to account for it.[4] Some thought that they had been frightened by our cleaning & firing off our fire-arms the evening before. — perhaps it had some connection with a quarrell between an old man & one of our sentries. — the old man being told not to come so close, spat in the seamans face & then retreating behind the trench, made motions, which it was said, could mean nothing but skinning & cutting up a man. He acted it over a Fuegian, who was asleep, & eyed at the same time our man, as much as to say, this is the way I should like to serve you: — Whatever might have been the cause of the retreat of the Fuegians, the Captain thought it advisable not to sleep another night there. All the|294| goods were therefore moved to the houses, & Matthews & his companions prepared to pass rather an aweful night. — Matthews behaved with his usual quiet resolution: he is of an eccentric character & does not appear (which is strange) to possess much energy & I think it very doubtful how far he is qualified for so arduous an undertaking. — In the evening we removed to a cove a few miles distant & in the morning returned to the settlement.

[1] Note in ink in margin 'Man violently crying along side'.
[2] See engraving of Woollya (now spelt Wulaia) by T. Landseer after R. FitzRoy in *Narrative* 2; facing p. 208.
[3] There are notes in pencil in the margin near here about the Fuegians that I cannot decipher.
[4] Note in pencil in margin: 'Reason upon effects of shooting'.

28th We found everything quiet; the canoes were employed in spearing fish & most of the people had returned. — We were very glad of this &

now hoped everything would go on smoothly. — The Captain sent the Yawl & one Whale boat back to the ship; & we in the other two re-entered the Beagle channel in order to examine the islands around its Western entrance. To everyones surprise the day was overpowringly hot, so much so that our skin was burnt; this is quite a novelty in Tierra del F. — The Beagle channel is here very striking, the view both ways is not intercepted, & to the West extends to the Pacific. — So narrow & straight a channell & in length nearly 120 miles, must be a rare phenomenon. — We were reminded, that it was an arm of the sea, by the number of Whales, which were spouting in different directions: the water is so deep that one morning two monstrous whales were swimming within stone |295|throw of the shore. — In the evening having pitched our tents, unfortunately a party of Fuegians appeared. — If these barbarians were a little less barbarous, it would have been easy, as we were superior in numbers, to have pushed them away & obliged them to keep beyond a certain line. —but their courage is like that of a wild beast, they would not think of their inferiority in number, but each individual would endeavour to dash your brains out with a stone, as a tiger would be certain under similar circumstances to tear you. — We sailed on till it was dark & then found a quiet nook; the great object is to find a beach with pebbles, for they are both dry & yield to the body, & really in our blanket bags we passed very comfortable nights. — It was my watch till one oclock; there is something very solemn in such scenes; the consciousness rushes on the mind in how remote a corner of the globe you are then in; all tends to this end, the quiet of the night is only interrupted by the heavy breathing of the men & the cry of the night birds. —the occasional distant bark of a dog reminds one that the Fuegians may be prowling, close to the tents, ready for a fatal rush. —

29th In the morning we arrived at the point where the channel|296| divides & we entered the Northern arm. The scenery becomes very grand, the mountains on the right are very lofty & covered with a white mantle of perpetual snow: from the melting of this numbers of cascades poured their waters through the woods into the channel. — In many places magnificent glaciers extended from the mountains to the waters edge. — I cannot imagine anything more beautiful than the beryl blue of these glaciers, especially when contrasted by the snow: the occurrence of glaciers reaching to the waters edge & in summer, in Lat: 56° is a most curious phenomenon: the same thing does not occur in Norway under Lat. 70°. — From the number of small ice-bergs the channel represented in miniature the Arctic ocean. — One of these glaciers placed us for a minute in most imminent peril; whilst dining in a little bay about ½ a mile from one & admiring the beautiful colour of its vertical & overhanging face, a large mass fell roaring into the water; our boats were on the beach; we saw a great wave rushing onwards & instantly it was evident how great was the chance of their being dashed into pieces. — One of the seamen just got hold of the boat as the curling breaker reached it: he was knocked over & over but not hurt & most fortunately our boat received no damage. — If they|297| had been washed away; how dangerous would our lot have been, surrounded on all sides by hostile Savages & deprived of all provisions. —[1]

[1] FitzRoy's account of this episode gives special credit to CD for his part in saving the boats. It runs: 'Our boats were hauled up out of the water upon the sandy point, and we were sitting round a fire about two hundred yards from them, when a thundering crash shook us—down came the whole front of the icy cliff—and the sea surged up in a vast heap of foam. Reverberating echoes sounded in every direction, from the lofty mountains which hemmed us in; but our whole attention was immediately called to great rolling waves which came so rapidly that there was scarcely time for the most active of our party to run and seize the boats before they were tossed along the beach like empty calabashes. By the exertions of those who grappled them or seized their ropes, they were hauled up again out of reach of a second and third roller; and indeed we had good reason to rejoice that they were just saved in time; for had not Mr. Darwin, and two or three of the men, run to them instantly, they would have been swept away from us irrecoverably. Wind and tide would soon have drifted them beyond the distance a man could swim; and then, what prizes they would have been for the Fuegians, even if we had escaped by possessing ourselves of canoes. At the extremity of the sandy point on which we stood, there were many large blocks of stone, which seemed to have been transported from the adjacent mountains, either upon masses of ice, or by the force of waves such as those which we witnessed. Had our boats struck these blocks, instead of soft sand, our dilemma would not have been much less than if they had been at once swept away . . . The following day (30th) we passed into a large expanse of water, which I named Darwin Sound—after my messmate, who so willingly encountered the discomfort and risk of a long cruise in a small loaded boat.' A mountain overlooking the Sound whose height was estimated as 6,800 ft, roughly the same as that of Sarmiento, was also named after CD (see *Narrative* 2: 216–17). In a letter to Catherine Darwin, CD later described Sarmiento as 'the highest mountain in the South, excepting M.!!Darwin!!' (see *Correspondence* 1: 381), but a modern atlas shows Sarmiento as 2,300 m, while Darwin is only 2,135 m.

30th The scenery was very grand, we were sailing parallel, as it were to the backbone of Tierra del; the central granitic ridge which has determined the form of all the lesser ones: It was a great comfort finding all the natives absent; the outer coast during the summer is on account of the seals, their chief resort. — At night we had miserable quarters, we slept on boulders, the intervals being filled up with putrefying seaweed; & the water flowed to the very edge of the tent. —

31st The channel now ran between islands; & this part was entirely unknown; it rained continually & the weather well became its bad character. —

February 1st & 2nd The countryside was most desolate, barren, & unfrequented: we landed on the East end of Stuart island, which was our furthest point to the West being about 150 miles from the ship:

3rd Miserable weather: we proceeded by the outside coast to the Southern entrance or arm of the Beagle Ch. & thus commenced our return.

4th & 5th Nothing happened till the evening before reentering Ponsonby Sound. — We met a large body of Fuegians, & had a regular auction to purchase fish; by the means of old|298| buttons, & bits of red cloth we purchased an excellent supper of fish.

6th Arrived at the Settlement. — Matthews gave so bad an account of the conduct of the Fuegians that the Captain advised him to return to the ship. —[1] From the moment of our leaving, a regular system of plunder commenced, in which not only Matthews, but York & Jemmy suffered. Matthews had nearly lost all his things; & the constant watching was most harassing & entirely prevented him from doing anything to obtain food &c. Night & day large parties of the natives surrounded his house. — (*Note in margin:* They tryed to tire him out by making incess noises.) One day, having requested an old man to leave the place, he returned with a large stone in his hand: Another day, a whole party advanced with stones & stakes, & some of the younger men & Jemmys brother were crying. — Matthews thought it was only to rob him & he met them with presents. — I cannot help thinking that more was meant. — They showed by signs they would strip him & pluck all the hairs out of his face & body. — I think we returned just in time to save his life. — The perfect equality of all the inhabitants will for many years prevent their civilization: even a shirt or other article of clothing is immediately torn into pieces. — Until some chief rises, who by his power might be able to keep to himself such presents as animals &c &c, there must be an end to all hopes of bettering their|299| condition. — It would not have been so bad if all the plunder had remained in one family or tribe. — But there was a constant succession of fresh canoes, & each one returned with something. — Jemmy's own relations were absolutely so foolish & vain, as to show to strangers what they had stolen & the method of doing it. —

It was quite melancholy leaving our Fuegians amongst their barbarous countrymen: there was one comfort; they appeared to have no

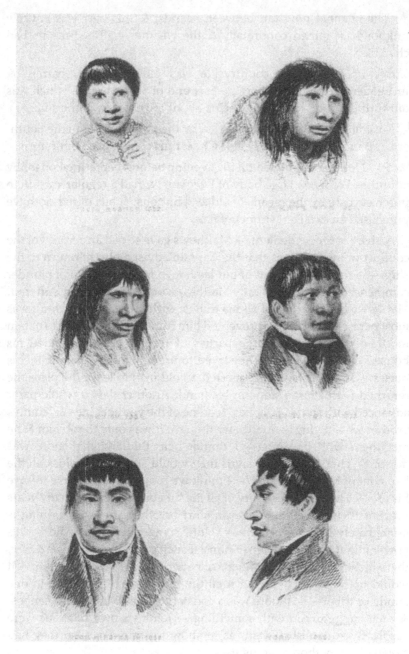

FitzRoy's Fuegians, by T. Landseer (*Narrative* **2:** 324).

personal fears. — But, in contradiction of what has often been stated, 3 years has been sufficient to change savages, into, as far as habits go, complete & voluntary Europæans. — York, who was a full grown man & with a strong violent mind, will I am certain in every respect live as far as his means go, like an Englishman. — Poor Jemmy, looked rather disconsolate, & certainly would have liked to have returned with us; he said "they were all very bad men, no 'sabe' nothing". — Jemmys own brother had been stealing from him as Jemmy said, "what fashion do you call that". — I am afraid whatever other ends their excursion to England produces, it will not be conducive to their happiness. — They have far too much sense not to see the vast superiority of civilized over uncivilized habits; & yet I am afraid to the latter they must return. — |300|

We took Matthews & some of the clothes, which he had buried in the boat & made sail: The Captain to save time determined to go to the South & outside of Navarin Island, instead of our returning by the Beagle channel. We slept at night in the S. entrance of Ponsonby Sound, & in the morning started for the ship.

[1]Matthews was subsequently taken in the *Beagle* to New Zealand, where he joined his brother and continued to work as a missionary.

There was a fresh breeze & a good deal of sea, rather more than is pleasant for a boat. So that on reaching in the evening the Beagle, there was the pleasure of smooth water joined to that of returning after 20 days absence. — The distance we have run in the boats has been about 300 miles & as it was in a East & West direction it afforded an excellent geological section of the country. —

8th & 9th   The ship remained in Goree Sound. —

Sunday 10th   Removed to a bay North of Orange Bay. —

11th–15th   The Captain, in his boat, paid the Fuegians a visit, & has brought back a very prosperous account of them. — Very few of the things belonging to Jemmy, York or Fuegia had been stolen & the conduct of the natives was quite peacible. — If the garden succeeds, this little settlement may be yet the means of producing great good & altering the habits of the truly savage inhabitants: — On the 13th, a party of eight under the command of Mr Chaffers crossed Hardy peninsula so as to reach & survey the West coast. The distance was not great; but from the|301| soft swampy ground was fatiguing. — This peninsula, although really part of an island, may be considered as the most Southern extremity of America: it is terminated by False Cape Horn. — The day was beautiful, even sufficiently so as to communicate part of its charms to the surrounding desolate scenery. — This & a view of the Pacific was all that repaid us for our trouble. —

16th   The same party started again & for the same object, but our course was rather different: Having ascended a more lofty hill, we enjoyed a most commanding view of the two oceans & their islands. — The weather was beautiful; indeed ever since being in harbour Tierra del has been doing its best to make up for the three miserable weeks at sea. —

Sunday 17th   Divine service & a quiet day. —

18th & 19th   The Ship moved to Woollaston Island & during these days, the Northern part has been surveyed.

20th   It blew very hard, & in consequence the Captain has run across the bay to our old quiet place in Goree Road. — The thermometer was only 38° with much rain & hail. —

21st   The weather prevented our returning to Woollaston island & from touching at Acquirre bay, so we made a clean run for good Success Bay.

22nd   To night it is blowing furiously: the water is fairly torn up, & thick bodies of spray are whirled across the Bay. — |302|

23rd   Last nights gale was an unusually heavy one. — We were obliged to let go three anchors. — The Boats were unable to bring off the wooding party, so they were obliged to make it out as well as they could during the night. —

Sunday 24th & 25th   After waiting for fine weather, on Monday I ascended Banks Hill to measure its height & found it 1472 feet. — The wind was so strong & cold; that we were glad to beat a retreat. — If we had been an hour later, the boats could not have reached the shore for us. — This was one of the hills I went up during our last visit, I was surprised that nine weeks had not effaced our footsteps so that we could recognize to whom they belonged. —

26th   Put to sea & steered for the Falkland islands: at night it blew heavily with a great sea: the history of this climate is a history of its gales. —

27th & 28th   Strong breezes. —

March 1st   We arrived early in the morning at Port Louis, the most Eastern point of the Falkland Islands.[1] The first news we received was to our astonishment, that England had taken possession of the Falklands Islands & that the Flag was now flying. — These Islands have been for some time uninhabited, untill the Buenos Ayres Government, a few years since claimed them & sent some colonists. — Our government remonstrated|303| against this, & last month the Clio arrived here with orders to take possession of the place. — A Buenos ayrean man of war was here at the time, with some fresh colonists. — Both they & the vessel returned to the Rio Plata. — The present inhabitants consist of one Englishman, who has resided here for some years, & has now the charge of the British Flag, 20 Spaniards & three women, two of whom are negresses. — The island is abundantly stocked with animals. — there are about 5000 wild oxen, many horses, & pigs. — Wild fowl, rabbits, & fish in the greatest plenty. — Europæan vegetables will grow. — And as there is an abundance of water & good anchorage; it is most surprising that it has not been long ago colonized, in order to afford provisions for Ships going round the Horn. — At present it is only frequented by Whalers, one of which is here now. —

We received all this intelligence from a French boat, belonging [to] a Whaler, which is now lying a wreck on the beach. Between the 12th & 13th of January, the very time when we suffered from the gale off Cape Horn, this fine ship parted from three anchors & drove on shore. — They describe the gale as a perfect hurricane. — They were glad to see us, as they were at a loss what to do. — all the stores are saved & of course plenty of food. — |304| Capt: FitzRoy has offered to take them 22 in number in the Beagle & to purchase on account of the owners, any stores which we may want. The rest must be sacrificed. —

[1] A full account of the chequered history of the Falkland Islands from their discovery by John Davis on 14 August 1592 to the assertion of British sovereignty and hoisting of the Union Jack by HMS *Tyne* on 2 January 1833 is given by FitzRoy in *Narrative* 2: 228–40.

2nd   Mr Dixon, the English resident, came on board. — What a strange solitary life his must be: it is surprising to see how Englishmen find their way to every corner of the globe. I do not suppose there is an inhabited & civilized place where they are not to be found. —

3rd   Took a long walk; this side of the Island is very dreary: the land is low & undulating with stony peaks & bare ridges; it is universally covered by a brown, wiry grass, which grows on the peat. — In this tract, very few plants are found, & excepting snipes & rabbits scarcely any animals. — The whole landscape from the uniformity of the brown color, has an air of extreme desolation. —

4th   A grievous accident happened this afternoon in the death of Mr Hellyer. — One of the residents brought the news that he had found some clothes & a gun on the sea coast. — We made all haste to the place

& in a short time discovered the body, not many yards from the shore, but so entangled in the Kelp, that it was with difficulty it was disengaged. — It was quite evident he had shot a bird & whilst swimming for it, the strong stalks of the sea weed had caught his legs|305| & thus caused his death. —

5th Mr Hellyer was buried on a lonely & dreary headland. — The procession was a melancholy one: in front a Union jack half mast high was carried, & over the coffin the British ensign was thrown; the funeral, from its simplicity was the more solemn, & suited all the circumstances. —

6th–9th Several ships have arrived; we are now five sail in the harbor: An English schooner has agreed to carry the Frenchman & all his stores (which we could not have done) to Monte Video & to receive 20 per cent at the auction. — During these days I have been wandering about the country, breaking rocks, shooting snipes, & picking up the few living productions which this Island has to boast of. — It is quite lamentable to see so many casks & pieces of wreck in every cove & corner: we know of four large ships in this one harbor. One of these was the L'Uranie a French discovery ship who had been round the world. — The weather generally has been cold & very boisterous. —

Sunday 10th In the evening it blew a tremendous gale of wind. — I should never have imagined it possible for such a sea to get up in so few minutes. — The Barometer had given most excellent warning that something uncommon was coming: in the |306|middle of the day it looked like a clear; but at dinner the Captain said the glass says we have not had the worst: about an hour afterwards it reached us in all its fury: The French Brig let go four anchors; the English schooner drove; & a little more would have added another wreck. — At night our Yawl was swamped at her moorings; she did not sink, but was towed on shore & emptied. — some of her gear & sails are lost: —

10th to Sunday 17th This is one of the quietest places we have ever been to. — Nearly all the Ships are gone; & no one event has happened during the whole week: The boats are employed in surveying. —

I walked one day to the town, which consists in half a dozen houses pitched at random in different places. — In the time of the old Spaniards when it was a Botany Bay for Buenos Ayres, it was in a much more flourishing condition. — The whole aspect of the Falkland Islands, were however changed to my eyes from that walk; for I found a rock abounding with shells; & these of the most interesting geological æra. —[1]

[1] Although CD mentions here only the geology of the Falkland Islands, it seems clear from a crucial series of entries in his notebook made at this time that he was already deeply concerned with the questions raised by differences in the geographical distribution of plants and animals as between isolated islands and the mainland. Interspersed with descriptions of the geology, he wrote:

'March 2.
To what animals did the dung beetles in S. America belong — Is not the closer connection of insects and plants as well as this fact point out closer connection than Migration.
Scarcity of Aphidians?
Vide Annales des Sciences for Rio Plata.
The peat not forming at present & but little of the Bog Plants of Tierra del F; no moss; perhaps decaying vegetables may slowly increase it. —beds ranging from 10 to one foot thick.
Great scarcity in Tierra del of Corallines, supplanted by Fuci: Clytra prevailing genus.
Procure Trachaea of Upland Goose.
Tuesday 12th —
Examine Balanus in fresh water beneath high water mark.
Horses fond of catching cattle—aberration of instinct.
Snipes rather [illeg].
Examine pits for Peat. Specimen of do— Have there been any bones ever found &c or Timber.
Are there any reptiles? or Limestone?
21st
Saw a cormorant catch a fish & let it go 8 times successively like a cat does a mouse or otter a fish; & extreme wildness of shags.
Read Bougainville.
In 1784, from returns of Gov. Figueroa, buildings amounted to 34, population including 28 convicts, 82 persons, & cattle of all kinds 7,774.
22nd
East of basin, peat above 12 feet thick resting on clay, & now eaten by the sea. — Lower parts very compact, but not so good to burn as higher up; small bones are found in it like Rats—argument for original inhabitants: from big bones must be forming at present, but very slowly: *Fossils in Slate*: opposite points of dip: & mistake of stratification: What has become of lime?
It will be interesting to observe differences of species & proportionate Numbers: what also appear characters of different habitations.
Migration of Geese in Falkland Islands as connected with Rio Negro?
March 28th —
Emberiza in flocks.
Send watch to be mended.
Enquire period of flooding of R. Negro and Plata.
Is the cleavage of M: Video (an untroubled country) very generally vertical, or what is the dip?—'

See Down House Notebook 1.14, and CD and the Voyage pp. 178–9.

Sunday 24th We have never before stayed so long at a place & with so little for the Journal. — For the sake of the fossil shells, I paid a visit of three days to the town. In a long ride I found the country no ways different from what it is in the neighbourhead|307| of the Ship. — The same entire absence of trees & the same universal covering of brown wiry grass growing on a peat soil. — The inhabitants are a curious mixed race; their habitations are in a miserable condition & deficient in almost every accomodation. The place bespeaks what it has been, viz a bone of contention between different nations. —

On Friday a sealing vessel arrived commanded by Capt. Lowe; a notorious & singular man who has frequented these seas for many years & been the terror to all small vessels. — It is commonly said, that a Sealer, Slaver & Pirate are all of a trade; they all certainly require bold energetic men; & amongst Sealers there are frequently engagements for the best "rookerys". & in these affrays Capt Lowe has gained his celebrity. — In their manners habits &c I should think these men strikingly resembled the old Buccaneers. Capt Lowe brought with him the people belonging to a vessel which was wrecked on the SW coast of Tierra del by the great gale of the 13th of Jan. — Thus we already know of the loss of two vessels & a third which was got off shore. — Capt Lowe considers this Summer to have been the most boisterous he has ever seen. It is satisfactory to have felt the very worst weather, in one of the most notorious places in the world. & that in|308| a class of vessel, which is generally thought unfit to double the Horn. — Few vessels would have weathered it better than our little "diving duck". —

26th   A short time after our arrival here, a small American Sealing vessel came in; Capt. FitzRoy entered into terms for buying it, on condition of its return by the 25th. — As the vessel did not keep her appointment, we supposed she had failed to find her consort, & the Captain therefore purchased Low's Schooner. — She is a fine vessel of 170 tuns, drawing 10 feet of water, and an excellent sea-boat. If the Admiralty sanction the provisioning & payment of men, this day will be an important one in the history of the Beagle. — Perhaps it may shorten our cruize, anyhow it will double the work done; & when at sea, it is always pleasant to be sailing in company; the consort affords an object of attention to break the monotomous horizon of the ocean.[1]

[1] FitzRoy's purchase of Lowe's schooner at his own expense, necessarily made like the chartering of La Paz and La Lièvre without prior approval from London, gave rise later, notwithstanding its important contribution to the success of the surveys that he was required to undertake, to serious troubles with the officials at the Admiralty. He wrote: 'I had often anxiously longed for a consort, adapted for carrying cargoes, rigged so as to be easily worked with few hands, and able to keep company with the Beagle; but when I saw the Unicorn, and heard how well she had behaved as a sea-boat, my wish to purchase her

was unconquerable. A fitter vessel I could hardly have met with, one hundred and seventy tons burthen, oak built, and copper fastened throughout, very roomy, a good sailer, extremely handy, and a first-rate sea-boat; her only deficiencies were such as I could supply, namely, a few sheets of copper, and an outfit of canvas and rope. A few days elapsed, in which she was surveyed very thoroughly by Mr. May, and my mind fully made up, before I decided to buy her, and I then agreed to give six thousand dollars (nearly £1,300) for immediate possession. Being part owner, and authorized by the other owners to do as he thought best with the vessel in case of failure, Mr. Low sold her to me, payment to be made into his partners' hands at Monte Video. Some of his crew being "upon the lay," that is, having agreed to be paid for their work by a small proportion of the cargo obtained, preferred remaining at the Falklands to seek for employment in other vessels, others procured a passage in the *Rapid*, and a few were engaged by me to serve in their own vessel which, to keep up old associations, I named '*Adventure*'. Mr. Chaffers and others immediately volunteered to go in her temporarily (for I intended to place Mr. Wickham in her if he should be willing to undertake the responsibility), and no time was lost in cleaning her out thoroughly, loading her with stores purchased by me from M. le Dilly and from Mr. Bray (lately master of the *Transport*), and despatching her to Maldonado, to be prepared for her future employment.' See *Narrative* 2: 274–5.

FitzRoy wrote to the Admiralty announcing his action as follows: 'I believe that their Lordships will approve of what I have done, but if I am wrong, no inconvenience will result to the public service, since I alone am responsible for the agreement with the owner of the vessels, and am able and willing to pay the stipulated sum.' However, their Lordships' Minute across the corner ran: 'Do not approve of hiring vessels for this service, and therefore desire that they may be discharged as soon as possible.' See *Admiralty Records*, Record Office, ADM/1/1819.

29ᵗʰ   The English Schooner will not conveniently carry all the Frenchmen of the wreck; The Captain offered to carry some, & to day three of her officers came on board. —

April 4ᵗʰ   Our Schooner sailed for Rio Negro, in order if possible to catch Mʳ Wickham before he & Mʳ Stokes set out in their little vessels on a surveying cruize. — Mʳ Chaffers has at present the command. — Mʳ Wickham will have it eventually. — The chief cause of the Beagles|309| present delay is the Captain having purchased what remained of the Frenchmans wreck for refitting the schooner. During this time I have been very busy with the Zoology of the Sea; the treasures of the deep to a naturalist are indeed inexhaustible. —[1]

[1] A pencil note here seems to be concerned with the productivity of the waters in this part of the coast, but it is only partly legible.

6ᵗʰ   After cruizing about the mouth of the Sound to complete the survey, we stood out to sea on our way to the Rio Negro.

Sunday 7ᵗʰ   Our usual luck followed us in the shape of a gale of wind; being in the right direction we scudded before it; by this means we run a long distance, but it was miserable work; every place dark wet & the very picture of discomfort. —

9th The weather to day is beautiful; it is the first time for three months that studding sails have been set. — We attribute all this sun-shine & blue sky to the change in latitude; small although it be. — We are at present 380 miles from the Rio Negro. —

12th We expect to arrive at our destination tomorrow morning. — the weather latterly has been tolerably good but there was too much sea to allow me to be comfortable. —

13th In the morning we were off the mouth of the Rio Negro. Nothing was to be seen of the Schooner. In vain we endeavoured, by firing a gun & hoisting a pilot signal to procure intelligence from the shore. — We suppose the sea on the bar prevented the|310| pilot from coming out. — Thus during the whole day we continued to cruize backwards & forwards. — It was exceedingly annoying; as every one was most anxious to hear that Wickham & his party were all well. — The coast is like, what we saw so much of, about Bahia Blanca, either sandy dumes or a horizontal line of low cliffs. —

Sunday 14th In the middle of the day, a Sail was seen a long way off in the SW. — We immediately made chace & soon found to our joy it was the Schooner. — Mr Chaffers came on board & reported that the Schooner had made good weather of it; but that the gales had been very heavy. — The Captain altered his plans & ordered Mr Chaffers to proceed directly on to Maldonado in the Rio Plata, & there wait our arrival. —

15th Whilst we were beating up to our station at the mouth of the Rio Negro; a small Schooner was seen beating down to us. — Every one immediately declared, they knew by the cut of her sails, that she was Wickham's. It turned out differently; she was a trading vessel to Rio Negro & brought news of our little Schooners. They were all well a week since & were then ready to sail to the South to the Bay of St Joseph. They had suffered one loss in Williams, the marine, who fell overboard in the river & was drowned. As the distance at present is under 100 miles, the Captain determined to run down & pay the Schooners a|311| visit. Mr Wickham will go in the Beagle to Maldonado & Mr Stokes will remain in command. — This arrangement has materially affected me as the Captain has offered that one of the little Schooners, should take me up to the Rio Negro, after staying a few days in the Bay of St Joseph. — For the sake of the geology this is of the highest interest to me; otherwise the passage in so small a vessel will be sufficiently uncomfortable. —

16th We have been standing, during the day, across the great Bay of St Matthias; as the place is unsurveyed we heave to at night:— The weather has been beautiful but too light; the mild warm climate & blue sky is most throughily enjoyed by all of us after our boisterous cruize in the South. What we saw of the coast consisted entirely in horizontal cliffs; in these, the divisions of the strata run for miles together exactly parallel to the surface of the sea.— It looks an El Dorado to a Geologist; such modern formations must contain so many organic remains.—

17th We reached St Josephs Bay, this is a grand circular expanse of water, opening by a narrow mouth into St Matthias. the crook of land which forms it is a remarkable feature in a chart of the coast of Patagonia.— It was expected that Mr Wickham would have been here, but|312| to our sorrow, & more especially to the French passengers, who are very anxious to arrive at M. Video, the little Schooners were not to be seen. The wind being very light & a strong tide setting into the bay, we were obliged to let go a stream anchor. This gave me a most delightful opportunity of taking a glimpse at the cliffs.— They abounded with fossil shells & were in many respects very curious & interesting. My visit was so short that there was only time to see how much was missed. At night, as soon as the tide turned, the anchor was weighed & we proceeded in pursuit of Mr Wickham.—

18th The climate here is quite celestial; cloudless blue skys, light breezes & smooth water.— We hear that this has been a very fine season; how strange it is, that the short distance as compared to the whole surface of the globe of this country from T. del Fuego, should make so much difference.—so that those rapid currents in the atmosphere, which have attained a velocity of from 60 to 100 miles per hour, should not even here be felt.— As the wind is too light, every one is grumbling at this fine weather; we have been slowly working up the bay of St Matthias to Port St Antonio, where we yet hope to find the Schooners.—|313|

19th All our plans have undergone a complete revolution. During the night the soundings were very irregular & in the same proportion dangerous, so that we were obliged to heave to and in consequence of this a current set us far to the South. In the morning a fresh NW breeze sprung up; from these various disadvantages the Captain gave up the attempt to find Mr Wickham or of landing me at Rio Negro, & made sail for Maldonado.— If the wind, that omnipotent & overbearing master, permits it, the Beagle will touch at Maldonado & proceed on to M. Video & Buenos Ayres. — I intend stopping at the former place, as it possesses the two great advantages of retirement & novelty. —

20th It blew half a gale of wind; but it was fair & we scudded before it. — Our decks fully deserved their nickname of a "half tide rock"; so constantly did the water flow over them. —

Sunday 21st At noon 300 miles from Maldonado, with a foul wind.

22nd & 23rd Our usual alternation of a gale of wind & a fine day.

24th We are off the mouth of the Plata. At night there was a great deal of lightning; if a hurricane had been coming, the sky could not have looked much more angry. — Probably we shall hear there has been at M. Video a tremendous Pampero. Our Royal mast head shone with St Elmos fire & therefore according to all good sailors no ill luck followed. — It is curious how the R. Plata seems to form a nucleus for thunder storms; phenomena which both to the South & North of it are comparatively rare. — |314|

25th & 26th At daybreak we found a current had set us several miles to leeward of Maldonado; as the breeze was both strong & fair the Captain determined to run on to M: Video. — We arrived there a little after noon. I went on shore & saw Mr Earl; he remained at this place, during our whole cruize, in hopes of recovering his health, in which respect, however, I am afraid he has had little success. — In the evening received letters from home dated Sept. 12th, Octob 14th, Novem. 12th, & Decr 15th. —[1]

During our absence, things have been going on pretty quietly, with the exception of a few revolutions. —

[1] These were letters from Caroline, Catherine and Susan Darwin (see *Correspondence* 1: 268–71, 273–5 and 283–5), that for 15 December having apparently been lost.

27th Having landed our French passengers & having received all parcels & letters; after dinner weighed anchor & made sail, with a fresh breeze for Maldonado. —

Sunday 28th By noon we arrived at the anchorage of Maldonado & found there, our schooner, all safe & snug. —

29th I took up my residence on shore, & procured lodgings at a well-know[n] old lady, by name Donna Francisca. — The day was spent in vain efforts to make any sort of comfortable arrangements. — The rooms are very high & large; they have but very small windows & are almost destitute of furniture. — They are all on the ground floor & open into each other. — The very existence of what an Englishman calls comfort never passed through the builders mind. —

30th   I rode a few miles round the town; the country is exceedingly similar to that of|315| M: Video, but rather more hilly. — We here have the same fine grass plain, with its beautiful flowers & birds, the same hedges of Cactus & the same entire absence of all trees. After pacing for some weeks the planck decks, one ought to be grateful for the pleasure of treading on the green elastic turf, although the surrounding view in both cases is equally uninteresting.

May 1st   The day has been miserably spent in attempts to transact business by the aid of vilely bad Spanish. — The Beagle sails tomorrow for M. Video & will return in about a fortnight. —

2nd & 3rd   The torrents of rain almost entirely prevent me doing anything. It is impossible to go any distance into the country; as all the rivers are unusually full, & a bridge is an invention scarcely known in these parts. —

The city of Maldonado is in reality only a small village; as is universally the case in [a] Spanish town, all the streets run in parallel lines cutting each other at right angles. — & in the centre is the Plaza with its Church. — I never saw so quiet, so deserted a looking place; it has scarcely any trade, & none by water untill these few last years; it appears only to be a collection of land-owners & a few of the necessary tradesmen, such as blacksmiths & joiners, who do all the business for a circuit of 50 miles round. Nearly the only produce of the country is cattle & horses. — These are both in wonderful|316| numbers. — Every person, even it is said to the beggars, rides; it is thought quite out of the question to walk ever so short a distance. — As a proof, how very common horses are, the price of a saddle will buy three good ones. — It is a most beautiful exhibition to see the boys riding on bare-backed colts & chasing each other over hill & dale, & twisting about in a manner which no one till he has seen it would believe a horse capable of. — Their method of riding is certainly the most perfect & graceful, for showing the full power of a horse in all its actions. —

4th   I rode about four leagues into the Camp to the head of a large fresh water lake called Laguna del Potrero. The object of my ride was to see a white marble, from which lime is manufactured. — The day was beautiful & it was a pleasant ride over hill & dale of turf & surrounded by endless flocks of cattle, sheep & horses. —

Sunday 5ᵗʰ–8ᵗʰ During the greater number of these days, there has been torrents of rain & heavy thunder storms. — The whole country is in a state of inundation, even so that many lives have been lost. — the oldest inhabitants have never seen such weather before. — It has necessarily prevented me from making a trip into the country which I had intended to have almost finished by this time. In consequence of these delays & the bad weather I have scarcely been able to set about anything. It anyhow has afforded me some good lessons|317| in being patient & in speaking Spanish. —

9ᵗʰ The weather being fine I persuaded my two guides & companions to start on our ride. — Don Francisco Gonzales, & Morante, a sort of servant of his, were both well armed, & having plenty of friends & relations in the country were just the people for my purpose. We drove before us a troop of fresh horses; a very luxurious way of travelling as there is then no danger of having a tired or lame one. — I agreed to pay 2 dollars a day (about 8ˢ..6ᵈ) & all expences on the road. — Such is the hospitality in this country, that the latter for 12 days only cost me about 16 dollars. — As the rivers were very full we only went a short distance; a little beyond the head of the Laguna del Potrero. I was inclined to think my guides took too much precaution with their pistols & sabres; but the first bit of news we heard on the road was, that the day before a traveller to M: Video had been found, with his throat cut, lying dead on the road. — it happened close to a cross, a record of a former murder. —

We dined at a Pulperia, where there were present many Gauchos (this name only means "countrymen" & those who dress in this manner & lead their life). — I here found out I possessed two or three things which created unbounded astonishment. — Principally a small pocket compass. — in *every* house, I|318| entered I was asked to show its powers, & by its aid told the direction of various places. — This excited the liveliest admiration, that I a perfect stranger should know the road (for direction & road is synonimous in this open country) to places where I had never been. — At one house, a young woman, who was ill sent to entreat me to come to her room & show her the compass. If their surprise was great, mine was much greater to find such ignorance; & this amongst people, who possess their thousands of cattle & "estancia's" of great extent. — It can only be accounted for by the circumstance that this retired part of the country has seldom been visited by foreigners. I was asked whether the earth or sun moved; whether it was hotter or colder to the North; where Spain was & many more such questions. — Most of the inhabitants have an indistinct idea, that

England, London, N. America are all the same place; the better
informed well know that England & N: America are separate countries
close together; but that England is a large town in London. — I had in
my pocket some promethians, which I ignited by biting them between
my teeth; to see this the whole family was collected; and I was once
offered a dollar for a single one. — My washing my face in the morning
caused at Las Minas much speculation; a superior tradesman closely
cross-questioned me about so singular a practice; & likewise why on
board we wore our beards, for he had heard from my guides that|319|
we did so. — He eyed me with much suspicion; perhaps he had heard
of ablutions in the Mahomedan religion; knowing me to be a Heretic;
probably he came to the conclusion that all Heretics are Turks. — It is the
universal custom to ask for a nights lodging at the first convenient
house. — The general astonishment at the compass and other things
was to a certain degree advantageous, as with that & the long stories my
guides told of my breaking stones, knowing venemous from harmless
snakes, collecting insects &c I paid them for their hospitality. — Being
able to talk very little Spanish, I was looked at with much pity, wonder
& a great deal of kindness. — Some few however, I think, gave me the
credit of having a good deal of the Dousterswivel about me. —[1] I am
writing as if I had been amongst the inhabitants of central Africa. Banda
Oriental would not be flattered by the comparison, but such was my
feeling when amongst them. — We slept at a friend of Gonzales; & in the
morning proceeded on to the town of Las Minas. —

[1] This sentence and the previous one are lightly crossed out in pencil. The description of
the character of Dousterswivel in Walter Scott's *Antiquary* is 'A tall, beetle-browed,
awkward-built man, who entered upon scientific subjects, as it appeared to my ignorance
at least, with more assurance than knowledge.'

10th   During this days ride, there was not much interest, excepting from
the novelty of this manner of travelling. — The country is much the
same; more uneven & hilly; a sort of miniature alpine|320| district; the
whole surface, however with the exception of the bare rock is covered
with a short green turf. — And this indeed is the picture of all which I
saw:—it sounds very delightful riding over so much turf; but positively
I at last became so tired of the endless green hills that I thought with
pleasure of iron-shod horses & dusty roads. — It is very rare to meet a
single individual, and we did not till close to Las Minas. — This night we
stopped at a Pulperia or drinking shop, which also sells a few other
things. — The evening was very tiresome as we were obliged to remain
the whole time amongst a set of drinking stranger before the counter &

with scarcely a place to sit down. — This was however the only night, in which we did not sleep at private houses. — During the evening a great number of young Gauchos came in to drink spirits & smoke cigars. — They are a singularly striking looking set of men. — generally tall, very handsome, but with a most proud, dissolute expression. — They wear their moustachios & long black hair curling down their necks. — With their bright coloured robes; great spurs clanking on their heels & a knife, stuck (& often used) as a dagger at their waist, they look a very different race of men from our|321| working countrymen. — Their politeness is excessive, they never drink their spirits, without expecting you to taste it; but as they make their exceedingly good bow, they seem quite ready, if occasion offered, to cut your throat at the same time. —

The town of Las Minas is considerably smaller than Maldonado, & of the usual symetrical figure. — it is seated in the plain of the Rio St. Francisco, & is surrounded on all sides by the low rocky mountains. — It has rather a pretty appearance, with its church in the middle. — the outskirting houses all arise out of the plain, like isolated beings, without the usual (to our eyes) accompaniment of a garden or court. — This is the case with all the houses in the country, & gives to them an unsociable appearance. —

11th In the morning we pursued rather a rambling course; as I was examining several beds of marble. — We crossed some fine plains abounding with cattle, here also were very many Ostriches. — I saw several flocks of between 20 & 30. — When seen on the brow of a hill against the clear sky they form a fine spectacle. — Some of them are very tame; if, after approaching close, you suddenly gallop in pursuit. — it is beautiful to see them, as a sailor would express it, "up with their helm" & make all sail, by expanding their wings right down the wind. — At night we came to the house of Don Juan Fuentes, a very rich|322| man, but a stranger to both my companions. — Upon arriving we entered the room where the Signora and Signoritas were sitting, & after talking on indifferent subjects (which I observe is always the formula) for a few minutes; permission was asked to pass the night there. — As a matter of course this is granted to all strangers; & a room allotted to us. — We then unsaddle our horses & bring the recon's[1] &c into the room. — this latter was not so good as cowshed, but it contained beds, & for bed-clothes the cloths belonging to the recon are used. — Shortly after our arrival one of the great herds of cattle was driven in. The cattle having so much space to wander over are very wild & it is necessary several times in the week to drive the herd into a Corral or enclosure of stakes, for the

night. —& thus accustom them to one central place. — About a dozen
Gauchos on horseback drove them in & near to the house separated a
few for the purpose of killing them. — This afforded a very animated
chace, for the cattle run nearly as fast as a horse, & the poor beasts know
full well the fatal Lasso. — After seeing such a herd & such a number of
horses the miserable house of Don Juan was curious. — The floor is
hardened mud; there are no glass windows, a few of the roughest chairs
& stools & two small tables was all the furniture|323| in the room. — For
supper there was a huge pile of roasted meat & another boiled with
some pumpkin. —in the centre was a mug of cold water. —there were
scarcely forks, plates or spoons sufficient, & every thing, table-cloth &c
filthily dirty. —there was no bread, salt, or vegetables, or anything
more than water to drink. —& this the house of a large landed prop-
rietor. — The evening was spent in smoking & with a little impromptu
singing accompanied by the Guitar. — All the women remained huddled up in one corner & did not sup with the men. — And such are the
luxuries which wealth here purchases! —

[1] The word 'recon' seems to be used as an alternative to 'recado', the type of saddle used
by the Gauchos.

12th   We crossed the Rio Marmaraga & proceeded to the Tapes; where a
widow woman, a friend of Gonzales gave us a most hospitable reception. — The above rivers, ultimately flow into the R. Grande & thus
belong to a different system from the others which we crossed. — On
the road Morante practised with success a method of catching partridges which I had often heard of but never seen. —it requires a long
stick, at the end of which there is a running noose, made of the stem of
an Ostriches feather. —as soon as a partridge is seen, & they are
wonderfully numerous, the man with the stick rides in a circle or spire
round & round the bird, gradually coming nearer & nearer; |324|the
partridge not knowing which way to run at last squats to conceal itself;
the noose is then quietly put over its head & the bird secured by a
jerk. —in this manner a boy sometimes catches 30 or 40 in one day. —

13th   In the evening arrived at a Pulperia North of the Rio Polanco. —it
was my furthest point: its distance in a straight line from Maldonado is
not much more than 70 miles; but this distance was much lengthened
by our route. — I here saw what I wanted in the geology & in the
morning returned to near our former sleeping place;

14th   the country continues very much the same; it was about the Polanco
more level & the hills less steep & there were a few trees about the

rivers, chiefly of the willow kind. — In the evening I saw rather a curious scene: an old Paraguay man, who had been our guide in the morning, got very drunk, & being offended at a man present was drawing his knife under his poncho; a Gaucho who sat by him knew what he was about & stopped him, & took his knife from him. — After this, to frighten the old gentleman, the others in jest pretended to stab him. — the method with which they dashed across the room, struck him upon the heart & then sprang out of the door, showed it to be the result of practice, at least in them. — The only manner of fighting amongst the Gauchos is thus stabbing each other; & this little scene showed me very plainly the way in which it|325| takes place. I wear a large clasp-knife, in the manner of sailors fastened by a string round my neck; I had often noticed that the Gauchos seemed to think this practice of confining the knife very strange. —

15th Bad weather; all the other days have been fine so that I have no cause to complain: we remained all day in the house & it was to me sufficiently tedious, as I had nothing to do but watch the rain the Gauchos smoking their cigars.

16th Returned by a route rather different, & slept at a house, 4 leagues from Las Minas. — Yesterdays rain had so filled the rivers, that they were difficult to be crossed. —this is a great disadvantage in a semi-civilized country that travelling is quite dependant on the weather. —

17th We again passed through Las Minas, & then proceeded across some low wild mountains to a very hospitable house. — The formation is all Slate & a few years ago a gold mine was discovered & here worked; but very small quantities having been procured, the works have ceased. — I believe this & its neighbourhead is the only place where gold has been found in the Banda Oriental. —

18th In the morning we rode to the house of Sebastian de Pimiento; a relation of Gonzales & a fine old Cavallero. — His house was better furnished than any I had seen—this probably was owing|326| to the presence of some pretty Signoritas, his daughters. — These same young ladies are universally quite out of character with the rest of the house. — They are dressed exceedingly well; & their whole appearance & manner is very lady-like. — Yet with all this, as in Pimiento's house, they superintend all the cooking & perform some of the lowest menial offices. — One of the greatest inconveniences in the manners of these people, is the quantity you are obliged to eat. —time after time they pile heaps of meat on your plate; having eat a great deal too much & having skilfully arranged what is left so as to make as little show as possible, a charming Signorita will perhaps present you with choice piece from her own plate with her own fork; this you must eat, let the consequence be what it may, for it is a high compliment. — Oh the difficulty of smiling sweet thanks, with the horrid & vast mouthful in view! —

Sunday 19ᵗʰ I got up early to ascend the Sierra de las Animas. — This & Pan de Azucar are well known land-marks in navigating the Plata; I should guess their height to be about 8 or 900 feet. — The scenery, by the aid of the rising sun almost looked pretty. — From the top there was a very extensive view. — to the West over a very flat country to the Mount at M. Video, & to the East over the mamillated plains of Maldonado. — |327| On the summit there were several small heaps of stones; which evidently had been there for many years. — my companion, an inhabitant of the place, declared it was work of the Indians in the old times. — They were like, although on a smaller scale, the heaps so common in the Welsh Mountains: How universal is the desire of Man to show he has ascended the highest points in every country. — In the evening I again partook & suffered from the overpowering hospitality in the house of Don Fran Pimiento: & the next morning started for Maldonado.

20ᵗʰ We arrived there in the afternoon. I am well satisfied with this little excursion, which besides an outline of the geology, has given me a very good opportunity of seeing both the country & its wild Gaucho inhabitants. — The Beagle on the 18ᵗʰ brought a party of working hands for the Schooner, but did not stay more than hour. — She left letters for me. — one from home, dated Jan. 13ᵗʰ —

21ˢᵗ–23ʳᵈ Em[p]loyed in arranging the fruits of my excursion, & in collecting in the neighbourhead of the Town.

24ᵗʰ The Beagle returned from M. Video. — Mʳ Hammond is discharged into the Pylades & ultimately intends leaving the service. —

25ᵗʰ & 26ᵗʰ Took a long walk to the Laguna del Potrero; my principal object at present is birds, of which there are a great number of very beautiful ones. — The weather is most delightful. Temp. in room about 60°. — |328|

28ᵗʰ & 29ᵗʰ Captain FitzRoy hired a small Schooner to go to the Rio Negro to bring Mʳ Wickham in order that he might take command of our Schooner. She arrived yesterday, & to day Mʳ King, who came with Mʳ Wickham paid me a visit. — They are heartily tired of their little vessels & are again as glad to see the Beagle as every one in her is to see them. —

The weather has generally been very fine; but the gale of the 12th of Jany reached them. — It appears however to have been miserable work & more than sufficiently dangerous: from the smallness of the vessels, it was scarcely possible to keep anything dry. —to possess a dry shirt or bed was an unusual luxury. — In addition to these discomforts, Mr Wickham & some of the others constantly suffered from sea-sickness. Mr Stokes & Mr Usborne (who has taken Mr Wickhams place) will continue to work in the neighbourhead of the Rio Negro. —

30th & 31st Usual quiet occupations; one days collecting & the next arranging. —

June 1st–7th The weather generally has been boisterous, so that very little work has been done with the Schooner as it is impossible to heave down to get to her bottom without quite calm weather. The delay is to me agreeable, although not serviceable, as there is not much more to be collected: — Birds insects & reptiles have been my chief game. —

8th Letters from home dated Feb 13th & March 3rd [1]

[1] That of 3 March was from his sister Susan. See *Correspondence* 1: 299–300.

Sunday 9th, 10th A heavy gale of wind; I think I may make my mind up for a fortnight more at this place. — |329|

11th–19th My time passes precisely in the same manner as the last 3 weeks. — My collection of the birds & quadrupeds of this place is becoming very perfect. — A few Reales has enlisted all the boys in the town in my service; & few days pass, in which they do not bring me some curious creature. — The progress with the Schooner has hitherto been very slow; but if the present fine weather lasts, another week will complete the coppering. — To day I returned from paying a visit to the vessel in order to see Mr Wickham after his return from the South. — The Beagle is in such a state of bustle, that I am sure I am for the present in the best quarters. —

20th–28th My only object is completing the collection of birds & animals; the regular routine is one day, shooting & picking up my mouse traps, the next preserving the animals which I take. — On Saturday I rode some leagues into the Camp & had some excellent rifle shooting at deer; I killed three bucks out of one herd. — My occupations are so very quiet, it gives me nothing more to say, than if I was living in an English village.

29th Arrived safely on board with all my Menagerie; am become such a complete landsman. —that I knock my head against the decks & feel the motion although in harbor. —

30ʰ, July 1ˢᵗ & 2ⁿᵈ   Have been employed in arranging & writing notes about all my treasures from Maldonado. — The Captain informs me that he hopes next summer to double the Horn. — My heart exults whenever I think of all the glorious prospects of the future. |330|

3ʳᵈ–7ᵗʰ   All hands of the Beagle continue to be employed in working at the Schooner (for the future the Unicorn). My occupations likewise are the same & I do not stir out of the Ship.

8ᵗʰ   It was discovered to day that one of the Mates, belonging to the Unicorn, had formerly been in the President, a vessel supposed to be piratical & which brought the English man of war, the Black Joke, to action.[1] It has, since the Trial, been suspected that this same ship took & murdered every soul on board the Packet Redpole. — Captain FitzRoy has determined to take the man a prisoner, to the Consul at M. Video. I have just been astonished to hear the order, "to reeve the running rigging, & bend sails". And we now a little before 12 at night have weighed anchor & are under sail. —

[1] HMS *Black Joke* was sent out by the Admiralty in 1829 to intercept slavers in West Africa.

9ᵗʰ   A fine breeze carried us into the harbor of M: Video by seven oclock. — The same wind brought a packet; with a letter for me dated May. —[1] After breakfast went on shore, to purchase numberless little &cˢ &cˢ. — M: Video has an air of great wealth & business, after the forlorn deserted streets of Maldonado. —

[1] Probably from his sister Caroline. See *Correspondence* 1: 309–10.

10ᵗʰ   So much wind & rain, could not go on shore. — The climate here is detestable; one feels it the more from the exposed anchorage where, we are pitching amongst the hillocks of muddy water. — I must say I like extremes in climate. For the last |331| year the summer of Tierra del F & the winter of the Rio Plata, the weather to us have been much the same. Constant gloomy sky, with much wind & rain, & the temperature raw & cold, but never sufficiently so to dry the atmosphere. My heart has revelled with delight to hear the orders for getting 12 months provisions ready for our next visit. —

11ᵗʰ & 12ᵗʰ   Spent the greater part of these two days in the city, transacting business. —

13ᵗʰ   After dinner sailed for Maldonado; arrived there at 11 at night, making a good passage. —

Sunday 14ᵗʰ   Enjoyed the rarity of clear, cloudless sky; the weather is cold; in the morning on shore there was some hoar frost. —

15ᵗʰ, 16ᵗʰ & 17ᵗʰ   The Schooners name has been changed into that of "Adventure", in commemoration of the Corvette employed in the former voyage with the Beagle, & likewise as being the name of one of Captain Cooks ships & therefore classical to all Surveying vessels. —[1]

All hands have been employed in getting her masts in & bringing on board her iron ballast. — There is a curious little history attached to this ballast. — the old Adventure having too much buried 30 tuns in the Island of Guritti. The Brazilians when they were in possession, had heard of it & made great efforts to find the spot. — The sepulcre was close to the well; so public a place having been chosen quite baffled the Brazilians; So that the ballast remained to be very serviceable to the young|332| Adventure.

[1] The ships engaged in surveying the southernmost part of the coast of South America in 1826–30 were HMS *Adventure* under the command of Capt P. P. King, and HMS *Beagle*, initially commanded by Capt Pringle Stokes (who was not related to John Lort Stokes, Mate and Assistant Surveyor on the *Beagle*, 1831–6), and after Capt Stokes's suicide at Port Famine in August 1828 by Capt FitzRoy. A tender, the *Adelaide*, was also taken on in Monte Video after Capt King, acting in a more circumspect manner than FitzRoy did later, had first secured the Admiralty's approval.

18ᵗʰ   At night the Packet fired guns to tell us she was on her way to Rio: This caused a scene of animation & bustle; for immediately orders were given "hands unmoor ship". The Captain. — having letters of importance, determined to stand out after her. — We were soon under weigh, & joining the Packet hailed her that we would keep company for a few days. —

19ᵗʰ   A calm & heavy fog, we were obliged occasionally to signalize by guns. —

20ᵗʰ   At noon a boat was lowered with the letters &c & my collections & taken on board the Packet; we then parted company; & are now sailing back for Maldonado. —

21ˢᵗ, 22ⁿᵈ   Gained the harbor of Maldonado. — the weather being very light & hazy detained us. — We had a strong instance of the dangerous navigation of the Plata; having good Latitude observations & having only left port for two days we were nearly 40 miles out of our reckoning. This was entirely owing to a strong current. — of which there was no means of previously ascertaining the existence.

23ʳᵈ, 24ᵗʰ   In the evening of the 24ᵗʰ, after it was dark, we got under weigh & started on our cruize to the Rio Negro. The whole sky was brilliant with lightning; it was a wild looking night to go to sea, but time is too precious to lose even a bad portion of it. —

25th–29th Our regular fortune followed us in the form of a sharp gale of wind. — It soon lulled but for two or three days a nasty head swell remained, which sadly hindered our progress. — The object of this cruize is to survey some|333| of the outer banks near the R. Negro & Bahia Blanca & likewise to pick up Mr Stokes & his party, who have been so laboriously employed with the little Schooners. —

30th–August 2nd Light contrary winds, interrupted by a few breezes: the whole passage a very tedious one; the ship being on a wind nearly all the time. —

3rd Arrived off the mouth of the Rio Negro, after firing several signal guns, the little Schooner La Lievre came out. In a short time I went on board her & we then returned within the mouth of the river. The Beagle stood out to sea to survey some of [the] outer banks which employment will occupy her a week. — We joined the other Schooner & I spent a very pleasant evening in hearing all their adventures. Every one in them may thank providence that he has returned in safety. To survey an unknown coast in a vessel of 11 tuns, & with one inch plank to live out in open sea the same gale in which we lost our whale-boat, was no ordinary service. — It seems wonderful that they could last one hour in a heavy gale, but it appears the very insignificance of small vessels is their protection, for the sea instead of striking them sends them before it. — I never could understand the success of the small craft of the early navigators.|334|

We then anchored near the Pilot's house & I went there to sleep.

4th Crossed the river and took a long walk to examine the South Barranca; the country is a level plain, which on the coast forms a perpendicular cliff about 120 feet high. having walked several miles along the coast, I with difficulty found a pass to ascend to the plain above. — This plain has a very sterile appearance it is covered with thorny bushes & a dry looking grass, & will for ever remain nearly useless to mankind. It is in this geological formation that the Salinas or natural salt-pans occur; excepting immediately after heavy rain no fresh water can be found. The sandstone so abounds with salt, that all springs are inevitably very brackish. — The vegetation from the same cause assumes a peculiar appearance; there are many sorts of bushes but all have formidable thorns which would seem to tell the stranger not to enter these inhospitable plains. —

5th Rode with Mr Stokes to the town of Patagones situated about 18 miles up the river; it was a pleasant ride, the road generally lying at the foot of the sloping cliff which forms Northern bank of the great valley of the R. Negro. —[1] We passed the ruins of some fine Estancias, which a few years since were destroyed by the Indians. They withstood several attacks; a man present at one gave me a very lively description|335| of what took place. — The Spaniards had sufficient notice to drive all the cattle & horses into the Corral which surrounded the house, & likewise to mount some small cannon. — The Indians were Araucanians from the South of Chili; several hundred in number & highly disciplined. — They first appeared in two bodies on a neighbouring hills; having there dismounted & taken off their fur mantles, they advanced naked[2] to the charge. — The only weapon of an Indian is a very long bamboo or Chusa ornamented with Ostrich feathers and pointed by a sharp spear head. My informer seemed to remember with the greatest horror, the quivering of these Chusas as they approached near. When close, the Cacique Pinchera, hailed the besieged to give up their arms or he would cut all their throats. — As this would probably have been the result of their entrance under any circumstances, the answer was given by a volley of musketry. The Indians with great steadiness came to the very fence of the Corral. — to their surprise they found the posts fastened together by iron nails instead of leathern thongs, & of course in vain attempted to cut them with their knives. This saved the lives of the Christians: many of the Indians were carried away by their companions, & at last one of the under Caciques being wounded the bugle sounded a retreat. They retired to their horses & seemed to hold|336| a council of war. — This was an awful pause for the Spaniards, as all their ammunition with the exception of a few cartridges was expended. — In an instant the Indians mounted their horses & galloped out of sight.

Another attack was still shorter; a cool Frenchman managed the gun, he stopped till the Indians had approached very close & then raked their line with grape shot. He thus laid thirty nine of them on the ground. Of course such a blow immediately routed the whole party. —

[1] Once again the pocketbook entries demonstrate CD's interest in the geographical ranges of allied ornithological species. On this day he wrote: 'Rode to the town, pleasant ride—banks of river, very unpicturesque country: Indians attacked a house: ornithology different: only small Icterus; not so very tame; some pidgeons; different parrots; different partridge; BBB birds common. Rose starling, Finch with black; sparrow. One days shooting, many new birds.' See Down House Notebook 1.14, and *CD and the Voyage* p. 186.

[2] A crossed-out note in the margin runs: 'with exception of Nutria skin round waist.'

6th & 7th   The town is built on the cliff which faces the river; many of the houses are actually excavated in the Sandstone. The river is here about four times as wide as the Severn at Shrewsbury & the stream rapid. — the many islands, with their willow trees & the headlands one seen behind the other, forming the Northern boundary of the flat valley, form by the help of the rising sun a view almost picturesque. — The number of inhabitants is not great, there are many Indians & Spaniards of pure blood & a far less mixture of the two races than is common in these countries. — The tribe of the Cacique Leucanee constantly have their Toldos outside the town. — Government supports them by giving them all the old horses to eat. —they also work in making Horse-rugs, boots of the horses legs &c. What their character may have gained by lessening their|337| ferocity, is lost by their entire immorality. Some of the younger men are however improving; they are willing to labour, & the other day a party agreed to go on a sealing voyage & behaved very well. — They were now enjoying the fruits of their Labour, by being dressed in very gay, clean clothes & being very idle. — The taste they show in their dress is admirable; if you could turn one of these young Indians into a statue of bronze, the drapery would be perfectly graceful.

8ᵗʰ   Rode to the great Salina, which is worked for the exportation of its salt, it is situated about 15 miles up & 3 from the river. —at this time it is nothing more than a large shallow lake of brine; but in summer it dries up & there is left a large field of snow white salt. — Both on the banks of the river & on the edge of the lake there were heaps of many hundred tuns ready for exportation. — The working time is as it were the harvest for Patagones, the whole population encamps on the bank of the river & every morning with the bullock waggons the men go to the lake to draw out the salt & form the Montes. There are other Salinas which are more distant & these are many leagues in circumference & the salt several feet thick, a quantity sufficient to supply the world. Yet at M: Video they use English salt to make salted|338| butter. (*Note in margin:* On account of the salt petre.) So little do the inhabitants profit by the natural advantages of their country.[1] In a like manner wheat in the province of B: Ayres produces an immense percentage yet a great deal of flour is imported from North America. (*Note in margin:* On account of the Pobrillo or red blight.) Killing an animal & flaying it does not give much trouble, & hides in consequence are nearly the only produce which these indolent people care about. —

Many of the geological facts connected with this Salina are curious & I returned highly satisfied with my ride. —[2]

[1] In the 2nd (1845) edition of the *Journal of Researches* pp. 65–6, CD was less hard on the inhabitants, and explained the inferiority of the local salt for the purpose of preserving

meat as actually arising from its excessive purity. He wrote: 'This salt is crystallized in great cubes, and is remarkably pure: Mr. Trenham Reeks has kindly analyzed some for me, and he finds in it only 0.26 of gypsum and 0.22 of earthy matter. It is a singular fact, that it does not serve so well for preserving meat as sea-salt from the Cape de Verd islands; and a merchant at Buenos Ayres told me that he considered it as fifty per cent. less valuable. The purity of the Patagonian salt, or absence from it of those other saline bodies found in all sea-water, is the only assignable cause for this inferiority: a conclusion which no one, I think, would have suspected, but which is supported by the fact lately ascertained, that those salts answer best for preserving cheese which contain most of the deliquescent chlorides.'

[2] CD wrote: 'On the road all rock sandstone which in places contains calcareous bed 4 or 5 inches thick. Vide specimen—light, porous; perhaps much of the calcareous formation also a Tosca bed; inferior (or mortar) owes its origin to this; but I do not conceive there is sufficient for this purpose:—

Much of the gravel is white-washed as at P Praya. It struck me that the cause of this & calcareous matter was owing to the rock being stretched at the base of the Andes.

Pumice stone conglomerate.

The gravel likewise in places contains concretions varying from size of fist to the head (not rotted of course) of small crystallised Gypsum (V Specimen); it is worked for burning to white wash the walls.—

The gravel bed must have been formed at bottom of sea (the interior shells show this); & tranquilly with some Chemical action; their half concretionary masses of Mortar & nodules of Gypsum show this.—'

See Down House Notebook 1.14, and *CD and the Voyage* p. 187.

9th Some months ago the government of B: Ayres sent out an army, under the command of General Rosas to exterminate the Indians.— They are now encamped on the Rio Colorado, in consequence the country is now very tolerably safe from Indians.—the only danger is meeting with a few stragglers; but a week since a man lost his whole troop of mares but it was on the Southern shore of the river.— As the Beagle intended to touch at Bahia Blanca, I determined to pass over land to that place.—

I made arrangements with a guide for a troop of horses, & M[r] Harris (of the little Schooner) who was going to take a passage to Buenos Ayres in the Beagle, agreed to accompany me.—

10th The weather was bad, so would not start: our party was increased by five more Gauchos who were going on business to the Encampment.—every body seemed glad of companions in this desolate passage.—|339|

11th We started early in the morning, but owing to some horses being stolen we were obliged to travel slow & accompany the Cargeroes or loaded horses.— The distance between Patagones & the pass of the Colorado is 85 miles, & in all this distance there are only two springs of fresh water.— They are called fresh, but even at this season were very brackish; in summer this must be a very disagreeable passage; from the heavy rain of yesterday we were well off, for there were several small puddles in the waggon ruts. We passed several small Salinas & in the distance there was one which was at least 3 or 4 leagues in length. — The country has one universal appearance, brown withered grass & spiny bushes; there are some depressions & valleys. —

Shortly after passing the first spring we came in sight of the famous tree, which the Indians reverence as a God itself, or as the altar of Walleechu. — It is situated on a high part of the plain & hence is a landmark visible at a great distance. — As soon as a tribe of Indians come in sight they offer their adorations by loud shouts. — The tree itself is low & much branched & thorny, just above the root its apparent diameter is 3 feet. It stands by itself without any neighbour, & was indeed the first tree we met with; afterwards there were others of the same sort, but not common. — |340| Being winter the tree had no leaves, but in their place were countless threads by which various offerings had been suspended. Cigars, bread, meat, pieces of cloth &c &c. — poor people only pulled a thread out of their ponchos. — The Indians both pour spirit & mattee into a hole & likewise smoke upwards, thinking thus to afford all possible gratification to Walleechu. — To complete the scene the tree was surrounded by the bleached bones of horses slaughtered as sacrifices. All Indians of every age & sex make their offerings, they then think that their horses will not tire & that they shall be prosperous. — In the time of peace the Gauchos who told me this had been witnesses of the scene; they used to wait till the Indians passed on & then steal from Walleechu their offerings. The Gauchos think that the Indians consider the tree itself as a God; but it seems far more probable that it is an altar. — The only cause which I can imagine for this choice, is its being a landmark in a dangerous passage. — The Sierra de la Ventana is visible at an immense distance & a Gaucho told me, that he was once riding with an Indian a few miles to the North of the R. Colorado when the latter began making the same noise which is usual at the first sight of the tree, & putting his hand to his head & then in the direction of the Sierra. Upon being asked the reason of this the Indian said|341| in broken Spanish "first see the Sierra". — This likewise would render it probable that the utility of a distant landmark is the first cause of its adoration. —

About two leagues beyond this very curious tree[1] we halted for the night: at this instant an unfortunate cow was spied by the lynx-eyed

Gauchos. Off we set in chase,[2] & in a few minutes she was dragged in by the lazo & slaughtered. — We here had the four necessaries for life "en el campo", — pasture for the horses, — water (only a muddy puddle) — meat — & fire wood. The Gauchos were in high spirits at finding all these luxuries, & we soon set to work at the poor cow. — This was the first night which I passed under the open sky with the gear of the Recado for a bed.[3] There is high enjoyment in the independence of the Gaucho life, (in margin) to be able at any moment to pull up your horse and say here we will pass the night. The death-like stillness of the plain, the dogs keeping watch, the gipsy-group of Gauchos making their beds around the fire, has left in my mind a strongly marked picture of this first night, which will not soon be forgotten. —

[1] Followed by the deleted words '& 11 from the town'.
[2] Spelling corrected from CD's usual 'chace' with a different pen.
[3] The words which follow are inserted from the margin, and the resulting sentence 'There is high . . . pass the night' replaces a deleted one.

12th The country continued the same. — it is inhabited by very few living beings: the most common is the hare or Agouti, there are likewise some Ostriches & Guanaco. — We passed the second|342| well, the water of which is brackish, but I think chiefly with saltpetre. — We found a good place for sleeping, the water was however so scanty that we could not take Mattee before starting the next morning. — The Gauchos when travelling only eat twice in the day, at night & before daylight in the morning; by this means one fire serves for 24 hours; an object of great consideration in many parts of this country.

13th Our distance was not more than 3 leagues from the R. Colorado; we soon left the desert sandstone plain & came to one of turf, with its flowers clover & *little owls*, the usual characteristic features of the Pampas. We passed a muddy swamp of considerable extent, which is occasionally overflowed by the Colorado. — It is a Salitràl that in summer is encrusted with saltpetre & hence is covered with the same species of plants which grow on the sea beach. —

We then arrived at the Colorado. The pass is about 9 leagues in a direct line from the mouth, but by water it is said to be not much less than 25. — Its width here is about 60 yards generally it must be once & half as wide as the Severn at Shrewsbury. The tortuous course of the river is marked by numerous willow trees & beds of reeds. — We were delayed crossing in the canoe by some immense troops of mares, which were swimming the river in order to follow the march of a division of

troops into the interior. Mares flesh is the only food of the soldiers|343|
when thus employed. This gives them a very great facility in movement;
for the distance & length of time horses can be driven over these plains
is quite surprising. — I have been assured an unloaded horse will travel
100 miles for *many* days successively. —

The encampment of General Rosas is close to the river; it is square of
3 or 400 yards, formed by waggons, artillery, straw huts &c. — The
soldiers are nearly all Cavalry. I believe such villainous Banditti-like
army was never before collected together: the greater number of men
are of a mixed race, between Negro, Indian & Spaniard: I know not the
reason, but men of such origin seldom have good expressions. — I
called on the Secretary to show my passport; he began to cross question
me in a most dignified & mysterious manner. — By good luck I had a
letter of recommendation from the Government of B. Ayres to the
Commandante of Patagones. (*Note in margin:* I am bound to express in
the strongest terms my obligation to the government of Buenos Ayres
for the most obliging manner in which passports to all parts of the
country was given me as Naturalist of the Beagle.) This was taken to
General Rosas, who sent me a very obliging message & the Secretary
returned all smiles & graciousness. — We took up our residence in the
Rancho or hovel of a curious old Spaniard, who had served with
Napoleon in the expedition against Russia. —

14th The weather was miserable & I had nothing to do: the surrounding
country is a swamp & in (December) summer overflowed by the
Colorado, which collects the snow water on the Cordilleras. — My chief
amusement was watching the Indian families as they came to buy little
articles at the Rancho where I staid. — It is said that General Rosas has
about six hundred Indian allies. — They certainly were very numerous.
The men are a tall exceedingly fine race; yet it is easy to see the same
countenance, rendered hideous by the cold, want of food & less
civilization, in the Fuegian savage. — Some authors in defining the
primary races of man have separated these two classes of Indians. — but
I cannot think this is correct. — Amongst the young women, or Chinas,
some deserved to be called even beautiful; their hair is coarse but
exceedingly bright & black; they wear it in two plaits hanging down to
the waist. — They have a high colour & eyes which glisten with
brilliancy. — Their legs, feet & arms are small & elegantly formed. —
Round the wrist & ancle they wear broard bracelets of blue beads.
Nothing could be more interesting than some of the family groups. —

Two or three Chinas (women) ride on one horse; a mother with her two daughters would thus often come to buy sugar & Yerba. —[1] They mount their horses with much delicacy; the horses have a broard band round their necks, which reaches just below the chest; this they use as a stirrup, but stand, when using it, even more in front of the horse than a man does. When on, they ride like a man, but with their knees tucked up much higher. — When travelling, the Chinas always ride the loaded horses, hence perhaps this habit. — Their duty is likewise to pack & unpack the horses & make the tents for the night;|345| they are in short, like the wives of all Savages useful slaves. — The men fight, hunt, take care of the horses & make the riding gear: One of their chief in doors occupation is the continual knocking of two stones together, till they are both round. — The bolas or balls are very important weapons with the Indian; he catches his game, & his wild horses with them; in fighting his first attempt is to throw his adversaries horse, & when entangled by the fall to kill him with the Chusa or long spear. — If the balls only catch the neck or body of an animal, they are often carried away & lost. — As the mere making of the stones round is the labour of two days, the manufacture of the balls is the most usual employment of the Indians. — Several of the men & women had their faces painted red, but I never saw the horizontal bands so common amongst the Fuegians. Once I saw a man with a little blue circle & straight line leading from it beneath each eye: Their chief pride is having all their things made of silver. I have seen a Cacique with silver stirrups, spurs, head-gear of silver chain, handle of knife &c. &c. —& occassionally some silver ornaments in the hair. —

From presents of General Rosas, their clothing was generally cloth, with some little fur. — They all had recado's, iron bits, & stirrups.

[1] The word 'Yerba' or 'hierba' is used in Spanish for every kind of herb.

15th   General Rosas[1] sent a message, that he should be glad to see me, before I started, by this means I lost a day, but subsequently his acquaintance was of the greatest utility.|346| General Rosas is a man of an extraordinary character; he has at present a most predominant influence in this country & probably may end by being its ruler. — He is said to be owner of 74 square leagues of country & has about three hundred thousand cattle. — His Estancias are admirably managed, & are far more productive of corn than any others in the country. He first gained his celebrity by his laws for his own Estancia & by disciplining several hundred workmen or Peons, so as to resist all the attacks of the

Indians. — He is moreover a perfect Gaucho: — his feats of horseman-
ship are very notorious; he will fall from a doorway upon an unbroken
colt as it rushes out of the Corral, & will defy the worst efforts of the
animal. He wears the Gaucho dress & is said to have called upon Lord
Ponsonby[2] in it; saying at the same time he thought the costume of the
country, the proper & therefore most respectful dress. — By these
means he has obtained an unbounded popularity in the Camp, and in
consequence despotic powers. — A man a short time since murdered
another; being arrested [&] questioned, he answered, "the man spoke
disrespectfully of General Rosas & I killed him"; in one weeks time the
murderer was at liberty. — In conversation he is enthusiastic, sensible
& very grave. — His gravity is carried to a high pitch. I heard one of his
mad buffoons (for he keeps two like the Barons of old) relate the
following anecdote. I wanted very much to hear a|347| piece of music,
so I went to the General two or three times to ask him, he said to me,
"go about your business for I am engaged". — "I went again"; "he said,
If you come again I will punish you". — A fifth time I asked him & he
laughed. — I rushed out of the tent, but it was too late; he ordered two
soldiers to catch & stake me. I begged by all the Saints in Heaven he
would let me off; but it would not do. — When the General laughs he
never spares mad man or sound man." — The poor flighty gentleman
looked quite dolorous at the very recollection of the Staking. — This is a
very severe punishment; four posts are driven into the ground, & the
man is extended by his arms & legs horizontally, & there left to stretch
for several hours. — the idea is evidently taken from the usual method
of drying hides. My interview passed away without a smile & I obtained
what I wanted, a passport and order for the government post horses, &
this he gave me in the most obliging and ready manner. — When
General Rosas, some months since, left B Ayres with his army, he
struck in a direct line across the unknown country, & in his march left
at wide intervals a posta of 5 men with a small troop of horses, so as to
be able to send expresses to the Capital. — By these I travelled to Bahia
Blanca & ultimately to Buenos Ayres. — I was altogether pleased with
my interview with the terrible General. He is worth seeing, as being
decidedly the most prominent character in S. America. — |348|

[1] General Juan Manuel de Rosas was a cattle rancher who served as Governor of Buenos
Aires in 1829–32 and 1835–52. From 1833 to 1835 he commanded a vigorous campaign of
extermination against the Indians in order to gain more territory. In 1852 he was deposed
from power as Dictator of Argentina, and retired to Swaythling in Hampshire.
[2] Lord Ponsonby was British Minister in Buenos Aires from 1826 to 1828.

16th Started early in the morning. Mr Harris did not accompany me as he was not quite well, & I was anxious to arrive at Bahia Blanca, not knowing when the ship would be there. We passed the Toldos of the Indians, which are without the regular encampment. — They are little round ovens covered with hides, with the tapering Chusa stuck in the ground by its entrance. — They were divided into separate groups, which belonged to the different Cacique's tribes, & each group of huts were divided into smaller ones, apparently according to the relationship of the owners. — The first Posta lay along the course of the Colorado. — the diluvial plains on the side appeared fertile & it is said are well adapted for the growth of corn: the advantage of having willows trees will be very great for the Estancias which General Rosas intends making here. — This war of extermination, although carried on with the most shocking barbarity, will certainly produce great benefits; it will at once throw open four or 500 miles in length of fine country for the produce of cattle. —

From the 2nd to 3rd Posta began the grand geological formation, which I believe continues the same to St Fe, a distance of at least 600 miles. — The country had a different appearance from that South of the Colorado: there were many different plants & grasses & not nearly so many spiny bushes, & these gradually became less frequent; untill a little to the North there is not a bush. — The plain is level & of a uniform brownish appearance; it is interrupted|349| by nothing, till about 25 miles North of the river, with a belt of red dunes stretching as far as the eye reaches to the East & West. — These are invaluable in the country, for resting on the clay the[y] cause small lakes in the hollows & thus supply that most rare article, fresh water. The extreme value of depressions & elevations in the land is not often reflected on. — the two miserable springs in the long passage between the Rios Nigro & Colorado are formed by two trifling inequalities in the plain, without which there absolutely would be none & of course boring would be quite unsuccessful. — The belt of sand hills is about eight miles wide, on the Northern edge the fourth Posta is situated; as it was evening & the fresh horses were distant we determined to pass the night here. —

The house is at the base of a ridge between one & two hundred feet high, a most remarkable phenomenon in this country — from this ridge there was an excellent view of the Sierra Ventana, stretching across the country & not appearing as at Bahia Blanca as a solitary mountain. — This posta was commanded by a Negro Lieutenant born in Africa & to his credit be it said there was not a Rancho between the Colorado & B.

Routes of CD's eight principal inland expeditions, as drawn for Nora Barlow's 1933 edition of The 'Beagle' Diary.

Ayres in nearly such neat order. He had a little room for strangers & a small Corral for the horses, all made of sticks & reeds. He had dug a ditch round the house, as a defence in case of being attacked; it would however be poor one if the Indians were to come. — His only comfort appeared to be that he would sell his life dear. |350| Some short time ago, a body of Indians had travelled past his house in the night.[1] If they had been aware of the Posta our black friend & his four soldiers would assuredly have been slaughtered. — I did not anywhere meet a more obliging man than this Negro; it was therefore the more painful to see that he would not sit down and eat with us. —

[1] Followed by the deleted words 'which is their usual time'.

17th In the morning he sent for the horses very early & we started for another exhilarating gallop. — We passed the Cabeza del Buey, an old name given to the head of a large marsh which extends from Bahia Blanca. Here we changed horses & passed through some leagues of swamps & saltpetre marshes; Changing horses for the last time, we again began wading through the mud. — My animal fell & I was well souzed in black mire, a very disagreeable accident, when one does not possess a change of clothes. — Some miles from the Fort we met a man who told us that a great gun had been fired, which is a signal that Indians are near. — We immediately left the road & followed the edge of a marsh, which when chaced, offers the best mode of escape; we were glad to arrive within the walls, when we found all this alarm was about nothing, for the Indians turned out to be friendly ones, who wished to join General Rosas. —

Sunday 18th The Beagle had not arrived. — I had nothing to do, no clean clothes, no books, nobody to talk with. — I envied the very kittens playing on the floor. — I was however lucky in a hospitable reception by Don Pablo, a friend of Harris. — |351|

19th I was anxious to see if the Beagle was in the mouth of the Bay. — The Commandante lent me a soldier as guide & two horses; on the road we picked up two more; yet they were all such miserable horses, that one was left behind & the three others could hardly reach Anchor Stock hill, a distance of about 25 miles, where the Ship waters. My guide two months ago had a wonderful escape, he was out hunting with two companions, only a few leagues from the fort. — when a party of Indians appeared, they balled the other two men & killed them. — They then balled his horse, he jumped off & with his knife liberated the horses legs. Whilst doing this he was obliged to dodge behind his horse & thus received two bad Chusa wounds. — Seizing an opportunity, he sprung on his horse, & could just manage to keep ahead of the Chusas till within sight of the Fort, when the Indians gave up the chace. — From that time, there was an order against any individual leaving the fort. — I did not know all this till near the coast, & had been surprised to see how earnestly my guide watched a deer which appeared to have been frightened from some other quarter. —

After two hours rest, & not seeing the Beagle, we made an attempt to return; but only could manage two or three leagues & even then left a horse behind. — In the morning we had caught an Armadillo, which was but a poor breakfast & dinner for two men. — Where we slept at night the whole ground was thickly|352| encrusted with saltpetre & of course no water. —

20th The next morning with nothing to eat or drink we started; the horses could hardly walk; at last that of the Gaucho was quite tired, & as a Gaucho cannot walk, I gave up my horse & took to my feet. — The sun was very hot & about noon the dogs killed a kid which we roasted & I eat some, which made me intolerably thirsty. The road was full of little puddles from some recent rain, yet every drop quite undrinkable. — At last I could walk no more, & was obliged to mount my horse, which was dreadful inhumanity as his back was quite raw. — I had scarcely been 20 hours without water & only part of the time with a hot sun; yet my thirst rendered me very weak. — How travellers manage to live in Africa I do not understand. — Although I must confess my guide did not suffer at all & was astonished that one days deprivation should be so trouble-some to me. — I do not know whether the poor horse or myself were most glad to arrive at the Fort. —

21st Bought a fine powerful young horse for 4$^{£}$..10^{s} & rode about the neighbouring plains. —

22nd So tired of doing nothing, I hired the same guide & started for Punta Alta, which is not so distant & commands a good view of the harbor. — I went this time better provided with bread & meat & horns with water & made up my mind to sleep there so as not to fatigue the horses. When not very far from our destination, the Gaucho spied 3 people on horse-back hunting. He immediately|353| dismounted & watched them intently. — He said they dont ride like Christians & nobody can leave the Fort. The three hunters joined company, & dismounted also from their horses, at last one mounted again & rode over the hill out of sight. — The Gaucho said, "We must now get on our horses, load your pistol" & he looked to his sword. — I asked are they Indians. — Quien Sabe? (who knows?), if they are no more than three it does not signify. — It then struck me that the one man had gone over the hill to fetch the rest of his tribe; I suggested this; but all the answer I could extort was, Quien sabe? — His head & eye never for a minute ceased scanning slowly the whole horizon. — I thought his uncommon coolness rather too good a joke; & asked him why he did not return home. I was startled when he answered: "We *are* returning, only near to a swamp, into which we can gallop the horses as far as they can go & then trust to our own legs. — So that there is no danger". — I did not feel quite so confident of this & wanted to increase our pace. — He said, no, not until they do. — When any little inequality concealed us, we galloped, but when in sight, continued walking. — At last we reached a valley, & turning to the left galloped quickly to the foot of a hill, he gave me his horse to hold, made the dogs lie down, & crawled on his hands & knees to reconnoitre. — He remained in this|354| position for some time & at last, bursting out in laughter, exclaimed: "Mugeres" (women). He knew them to be the wife & sister in law of the Majors son, hunting for Ostriches eggs. — I have described the mans conduct because he acted under the full impression they were Indians. As soon however as the absurd mistake was found out, he gave me a hundred reasons why they could not have been Indians; but all these were forgotten at the time. — After this we proceeded on to Punta Alta and ate our dinner in peace & quietness. — Punta alta is the place where I found so many bones last year. — I employed the evening in seeking for more & marking the places. — There was a beautiful sunset & everything was deliciously quiet & still. — But the appearances were false; an hour after being in bed, very heavy rain began, but I slept through it & was very little wet. —[1]

[1] A long and closely argued pocketbook entry for this day concludes as follows: 'My alteration in view of Geological nature of P. Alta is owing to more extended knowledge of country; it is principally instructive in showing that the bones necessarily were not coexistent with present shells, though old shells: they exist at M: Hermoso, pebbles from the beds of which occur in the gravel. Therefore such bones, if same as those at M. Hermoso must be anterior to present shells: How much so, Quien Sabe?' See Down House Notebook 1.14, and *CD and the Voyage* p. 194.

This was not CD's final conclusion on the geology of Patagonia and the fossils at Punta Alta, which was summarized in the *Journal of Researches* pp. 201–12, and had changed again in the 1845 edition (pp. 81–9) where he wrote that: 'gigantic quadrupeds, more different from those of the present day than the oldest of the tertiary quadrupeds of

Europe, lived whilst the sea was peopled with most of its present inhabitants; and we have confirmed that remarkable law so often insisted on by Mr. Lyell, namely, that the "longevity of the species in the mammalia is upon the whole inferior to that of the testacea."'

23rd In the morning the rain did not cease, so we started on our return. — In our path we saw a fresh track of a Lion & commenced an unsuccessful chace: the dogs seemed to know what we were about & were not eager to find the beast. —

In these plains a very curious animal, the Zorilla or Skunk, is sufficiently common. — Its habits resemble those of fitchet, but it is larger & the body much thicker in proportion. — Conscious of its power, it roams about the open camp by day & fears neither dogs or men. — if a dog is encouraged to attack one. — the fetid oil, which is ejected makes him instantly very|355| sick & run at the nose. — Clothes once touched are for ever useless. — Every other animal makes room for the Zorilla. —

On my return found my fellow traveller Harris arrived from the R. Colorado. A few days previously news had come that the Indians had murdered every soul in one of the Postas. — It was suspected that Bernantio's tribe, the same which the other day stopped here on the road to join General Rosas were the perpetrators. — Harris informed us, that a few miles from the Colorado he met these Indians, & that at the same instant an officer arrived bearing the following summary message, "that if Bernantio failed to bring the heads of the murderers, it should be his bitterest day for not one of his tribe should be left in the Pampas". —

24th The Ship was seen; its figure curiously altered by the refraction over the widely extended mud banks.

Sunday 25th Rode down to the creek: but there was too much wind for a boat to leave the ship. — In the evening Commandante Miranda arrived with 300 men; with orders to accompany Bernantios tribe & examine the "rastro" or track of the murderers. — If the latter was guilty, the whole tribe was to be massacred, if not to follow the rastro even if it led them to Chili. — Many of Mirandas troops were Indians; nothing could be more wild or savage than the scene of their bivouaccing. — Some of them drank the warm, steaming blood of the beasts which were slaughtered for supper. — |356|We subsequently heard that the *rastro* proved Bernantio to be guiltless. The Indians had escaped directly into the great plains or Pampas, & for some reason could not be pursued. — One glance at the Rastro tells to one of these people a whole history. —

Supposing they examine the track of about a thousand horses, they will at once know, by the canter, how many men were with them. —by the depth of the impression, how many loaded horses; by the regularity of the footstep how far tired; by the manner in which the food is cooked whether the Indians were travelling very fast; by the general appearance of the rastro how old it is.— They consider one of 10 days or a fortnight old quite recent enough to be hunted out.— We also heard that Miranda started from the West end of the Sierra Ventana in a direct line to the Island of Churichoel; situated 70 leagues up the Rio Negro. — This is a distance of 2 or 300 miles & through a country entirely unknown. What other troops in the world are so independent? With the sun for their guide, mares-flesh for food, & the Recado's for beds, as long as there is water, these men would penetrate to the worlds end. —

26th   A boat with Mr Chaffers arrived from the ship, we waited till the evening for a cow to be killed, to take fresh meat on board. We did not start till late, but the night was beautiful & calm. — The ship had moved her berth, & we had a long hunt after her, at last arrived on board at ½ after one oclock. —|357|

27th   Whole day consumed in telling my travellers tales.

28th   Actively employed in arranging things, in order to start to Buenos Ayres by land. —the feeling of excitement quite delightful after the indolence of the week spent at the fort of Bahia Blanca. —

29th   After dinner the Yawl started on a surveying cruize. I went in her. We slept at Punta Alta & I commenced a successful bone hunt; Leaving my servant & another man to continue their labours. —

30th   We the next morning set out for Fort; but did not arrive there till 9 oclock at night.

31st   My guide or Vacciano not having come, I rode to Punta Alta, in order to superintend the excavation of the bones. — It is a quiet retired spot & the weather beautiful; the very quietness is almost sublime, even in the midst of mud banks & gulls, sand hillocks & solitary Vultures.

September 1st   Returned in the evening. During the last week the weather has been very hot & dry; in consequence of this all the pools & shallow lakes, which before contained saline water, now presented a level plain of salt-petre, as white as snow. — This resemblance was the more complete from the edges of the pools appearing like drift heaps. —

2nd   Nothing to be done.

3rd Harris & Mr Rowlett went to the Creek, from thence in the Yawl on board. — in the road they would pick up my servant & the bones. — [1]|358|

[1]Syms Covington, originally 'Fiddler and boy to the Poop Cabin', had been appointed by FitzRoy as CD's servant on 22 May 1833 (see *Correspondence* 1: 311–15), though six months earlier CD appears already to have assumed some responsibility for clothing him, *vide* note 'Black duck—Covington Trousers' in Down House Notebook 1.14 (see *CD and the Voyage* p. 169). He served CD as a general amanuensis until 1839, when he emigrated to Australia. Of him CD wrote to Catherine Darwin: 'My servant is an odd sort of person; I do not very much like him; but he is, perhaps from his very oddity, very well adapted to all my purposes.' See *Correspondence* 1: 392.

4th–7th These four days were lost in miserable ennui. A man, whom I had engaged to be my Vacciano, disappointed me & ultimately at some risk & much trouble I hired another. — My only amusement was reading a Spanish edition published at Barcelona of the trial of Queen Caroline! — Moreover I heard many curious anecdotes respecting the Indians. — The whole place was under great excitement, there were continual reports of victories &c. — A prisoner Cacique had given information of some Indians at the small Salinas. — On the 5th a party of a hundred men were sent against them. — These Salinas only lie a few leagues out of the road between the Colorado & Bahia Blanca. The Chasca (or express) who brought this intelligence, was a very intelligent man & gave me an account of the last battle, at which he was present. — Some Indians, taken previously, gave information of a tribe North of the Colorado. Two hundred soldiers were sent. — They first discovered the Indians, by the dust of their horses, in a wild mountainous country. — My informer thought they were half as high as the Sierra Ventana, therefore between 1 & 2000 feet high. — The Andes were clearly in sight, so that it must have been very far in the interior. — The Indians were about 112, women & childer & men, in number. — They were nearly all taken or killed, very few escaped. (*Note in margin:* Only one Christian was wounded.)

The Indians are now so terrified that they offer no resistance in body; but each escapes as well as he|359| can, neglecting even his wife & children. — The soldiers pursue & sabre every man. — Like wild animals however they fight to the last instant. — One Indian nearly cut off with his teeth the thumb of a soldier, allowing his own eye to be nearly pushed out of the socket. — Another who was wounded, pretended death with a knife under his cloak, ready to strike the first who approached. My informer said, that when he was pursuing an Indian, the man cried out "Companèro (friend) do not kill me," at the same time was covertly loosening the balls from round his body, meaning to whirl them round his head & so strike his adversary. "I however struck him with my sabre to ground, then got off my horse & cut his throat." — This is a dark picture; but how much more shocking is the unquestionable fact, that all the women who appear above twenty years old, are massacred in cold blood. — I ventured to hint, that this appeared rather inhuman. He answered me, "Why what can be done, they breed so." — Every one here is fully convinced that this is the justest war, because it is against Barbarians. Who would believe in this age in a Christian, civilized country that such atrocities were committed? — The children of the Indians are saved, to be sold or given away as a kind of slave, for as long a time|360| as the owner can deceive them. — But I believe in this respect there is little to complain of. — In the battle four men ran away together, they were pursued, one was killed, the other three were caught. — They turned out to be Chascas (messengers or embassadors) of the Indians. — The Indians were on the point of holding a grand council, the feast of mares flesh was ready & the dance prepared. In the morning the Chascas were to return to the Cordilleras, where there is a great union of the Indians & from whence they were sent. — They were remarkably fine young men, very fair, & above 6 feet high, all of them under 30 years old. — The three surviving ones of course possessed very valuable information, to extort this they were placed in a line. — The two first being questioned; answered, "No sè", (I do not know), & were one after the other shot. — The third also said "No sè" adding, "fire, I am a *man* & can die". — What noble patriots, not a syllable would they breathe to injure the united cause of their country! The conduct of the Cacique has been very different; his life will perhaps be spared, & he has confessed all the plans; & betrayed the point of union in the Andes. It is said there are already six or seven hundred together & that there will be in Summer time twice that number. — Embassadors were to have been sent from this tribe to the Indians at the small Salinas near Bahia Blanca, whom I mentioned that this same|361| Cacique had betrayed. The communication therefore extends from the Cordillera to the East coast. — General Rosas's plan is to kill all stragglers & thus drive the rest to a common point. — In the summer, with the assistance of the Chilians, they are to be attacked in a body, and this operation is to be repeated for three successive years. — I imagine the summer is chosen as the time for the main attack, because the plains are then without water, & the Indians can only travel in particular directions. — The escape of the Indians to the South of the Rio Negro, where in such a vast unknown country they would be safe, is prevented by a treaty with the Tehuelches to this effect. — that Rosas pays them so much to slaughter every Indian who passes to the South of the river. — but if they fail in doing this, they themselves shall be exterminated. — The war is chiefly against the Indians near the Cordillera; for many of the tribes on this Eastern side are fighting with Rosas. The general however, like Lord Chesterfield, thinking that his friends may in a future day become his enemies, always places them in the front ranks, so that their numbers may be thinned. — If this warfare is successful, that is if all the Indians are butchered, a grand extent of country will be gained for the production of cattle: & the vallies of the R. Negro, Colorado, Sauce will be most productive in corn.[1] The country|362| will be in the hands of white Gaucho savages instead of copper-coloured Indians. The former being a little superior in civilization, as they are inferior in every moral virtue. —

By the above victory, a good many horses were recovered, which had been stolen from B. Blanca.[2] Amongst the captive girls, were two very pretty Spanish ones, who had been taken by the Indians very young & now could only speak the Indian language. From their account, they must have come from Salta, a distance in a straight line of nearly one thousand miles. This gives one a grand idea of the immense territory over which the Indians can roam. — Great as it is, in another half century I think there will not be a wild Indian in the Pampas North of the Rio Negro. — The warfare is too bloody to last; The Christians killing every Indian, & the Indians doing the same by the Christians.[3]

I also heard some account of an engagement which took place, a few weeks previously to the one mentioned, at Churichoel. — This is an island 70 leagues up the R. Negro & very important as being a pass for horses. — A division of the army has at present its head quarters there; when they first arrived, they found a tribe of Indians & killed between twenty & thirty men. The Cacique escaped in a manner which astonished every one. — The chief Indians always have one or two picked horses, which they keep ready for any urgent occasion. On one of these, an old white horse, the Cacique sprung taking with him his little son; the horse|363| had neither saddle or bridle; to avoid the shots, the Indian rode in the peculiar method of his nation namely an arm round the horses neck & one leg only on the back; thus hanging on one side, he was seen patting the horses head & talking to him. — The pursuers urged every effort in the chase; the Commandante three times changed his horse. — But all would not do. — The old Indian father with his son escaped & were free. — What a fine picture can one form in ones mind;. — the naked bronze like figure of the old Indian with his little boy, riding like a Mazeppa on the white horse, thus leaving far behind the host of his pursuers. —

I saw one day a soldier striking fire with a piece of flint; which I immediately recognized as having been a part of the head of an arrow. — He told me it was found near the island of Churichoel, & that they were frequently picked up there. — It was between two & three inches long, & therefore twice as large as those used in Tierra del Fuego; it was made of opake cream-coloured flint, but the point & barbs had been intentionally broken off. It is well known that no Pampas Indians now use bows & arrows; I believe a small tribe in Banda Oriental must be excepted, but they are widely separated from the Pampas Indians & border close to those tribes which inhabit the forest & live on foot. — It appears therefore to me that these heads of arrows are antiquarian relics of the Indians|364| before the great changes in habit consequent on the introduction of horses into South America. This & the invention of catching animals with the balls would certainly render the use of arrows in an open country quite superfluous. — In N: America bones of horses have been found in close proximity to those of the Mastodon; and I at St Fe Bajada found a horses tooth in the same bank with parts of a Megatherium;[4] if it had not been a *horses* tooth, I never should have for an instant doubted its being coeval with the Megatherium. — Yet the change of habits, proved by the frequency of the arrow heads, convinces me that the horse was not an original inhabitant. —

[1] This sentence has been marked in pencil to be deleted.

[2] There are marks in the margin suggesting that an insert stuck in at this point has been lost.

[3] In the *Journal of Researches* p. 121, CD adds at this point 'Since leaving South America we have heard that this war of extermination completely failed.'

[4] Since the horse's tooth was found by CD at St Fe Bajada only on 10 October (see Down House Notebook 1.14, and *CD and the Voyage* p. 210), this passage must have been written at least five weeks after the events described.

Sunday 8th   Having at last obtained a Vacciano & passport for government horses from General Rosas, I started for Buenos Ayres. — The distance is about 400 miles. — The weather was favourable, but remarkably hazy; I thought it the forerunner of a gale, but the Gauchos tell me it is the smoke from the camp at some great distance being on fire. — To the first Posta 4 leagues, the plain without any bushes but varied by vallies. — The 2nd Posta is on the R Sauce, a deep, rapid little river, not above 25 feet wide. It is quite impassable here & the whole distance to the sea, & forms by this means a useful barrier against the Indians.

Where the road crosses it, about a league further up, the water does not reach to the horses belly. The Jesuit Falkner, whose information, drawn from the Indians, is generally so very correct|365| in his map, makes it a great river arising in the Andes. — I think he is right. —for the soldiers say, that in the middle of summer, there are floods, at the same time with the Colorado; if so it is clear there must be a channel for the snow water, although it is probably dry during the greater part of the year. — The valley of the Sauce, appears very fertile, it is about a mile wide, there are large tracts of a wild Turnip much resembling the Europæan, they are good to eat but rather acrid. —

I arrived here in the afternoon, & getting fresh horses & a guide started for the Sierra de la Ventana. — The distance was about 6 leagues, & the ride interesting, as the mountain began to show its true form. — I do not think Nature ever made a more solitary desolate looking mountain; it well deserves the name of Hurtado or separated. —its height, calculated by angular measurement from the ship, is between 3 & 4000 feet. —it is very steep, rough & broken. — It is so completely destitute of all trees, that we were unable to find even a stick to stretch out the meat for roasting, our fire being made of dry thistle stalks. — The strangeness of its appearance chiefly is caused by its abrupt rise from the sea-like plain, which not only comes up to the foot of the mountain, but separates the parallel ridges or chains. — The uniformity|366| of the colouring gives extreme quietness to the view. — The whitish-grey of the quartz rock & the light brown colour of the withered grass of the plain is unbroken by the brighter tints of a single bush. — When we arrived at the foot of the main chain, we had much difficulty in finding water; & were afraid we should pass the night without any; it seems that all the streamlets, after flowing a few hundred yards in the plain bury themselves; at last we found some, it was then growing dark & we bivouacced for the night. —

The night was very clear & cold, the dew, which in the early part wetted the yergas[1] of the Recado, was in the morning ice. — The water in the kettle was also a solid block. — The place where we slept could not I think have been more than 700 feet above level of the sea, so that I suppose the neighbourhead of the mountain caused this unusual degree of cold. — The highest part of the Sierra is composed of four peaks in a gradually lowering order. — The two highest of these can alone be seen from Bahia Blanca. — To this part a ridge or saddle back appears to join. —our halting place was at the foot of this. —

[1] The yergas is the blanket of felt or coarse cloth placed under the gaucho's saddle.

9th In the morning the guide told me to ascend the ridge & that I could walk along its edge to the very summit. — The climbing up such very rough rocks was fatiguing; the sides are so indented that what is gained in one five minutes is often lost in the next. At last when I reached the summit of the ridge, my|367| disappointment was great to find a precipitous valley, as deep as the plain, separating me from the four peaks. — This valley is very narrow & the sides steep; it forms a fine horse pass, as the bottom is flat with turf, & connects the plains on each side of the mountain. — Whilst crossing it, I saw two horses grazing. I immediately hid myself in the long grass & began with my telescope to reconnoitre them, as I could see no sign of Indians, I proceeded cautiously on my second ascent. It was late in the day, & this part of the mountain, like the other was steep & very rugged. —[1] was on the top of the second peak by two oclock, but got there with extreme difficulty; every twenty yards I had the cramp in the upper parts of both thighs, so that I was afraid, I should not have been able to have descended; it was also necessary to find out a new road to the horses, as it was out of the question to return over the saddle-back. — I was thus obliged to give up the two higher peaks; their altitude was but little greater & every purpose of geology was answered; it was not therefore worth the hazard of any further exertion. — I presume the cause of the cramp was the great change in kind of muscular action from that of hard riding to still harder climbing. — It is a lesson worth remembering, as in some cases it might cause much difficulty. —

The ice which in many places coated the rocks was very refreshing & rendered superfluous the water, which I actually carried to the summit|368| in the corner of a cape of the Indian-rubber cloth. — Altogether I was much disappointed in this mountain; we had heard of caves, of forests, of beds of coal, of silver & gold &c &c, instead of all this, we have a desert mountain of pure quartz rock. — I had hoped the view would at least have been imposing; it was nothing; the plain was like the ocean without its beautiful colour or defined horizon. — The scene however was novel, & a little danger, like salt to meat, gave it a relish. — That the danger was very little was clear, by my two companions making a good fire, a thing never done when it is suspected Indians are near. — I returned by so easy a road, that if I had found it out in the morning I could have with ease reached the highest peak. — I reached the horses at sun-set, & drinking much mattee & smoking several little cigaritos, made up my bed for the night. — It blew furiously, but I never passed a more comfortable night. —

[1] The remainder of this paragraph is contained within square brackets, which evidently indicates that it is to be printed in *Journal of Researches*, since on pp. 126–7 it appears just as it does here.

10th In the morning we fairly scudded before the gale, & arrived by the middle of the day at the Sauce Posta. — On the road we saw very great numbers of deer & near the mountain a Guanaco. — I should think this latter animal was not to be found any further North on this side of S. America. — The plain which abuts against the Sierra is traversed by curious ravines, they are not above 20 feet wide & at least 30 deep; there are very few places where they are passable. — I staid the evening at the Posta, the conversation, |369| as is universally the case, being about the Indians. The Sierra de la Ventana, was formerly, a great place of resort for the Indians; three or four years ago there was much fighting there; my guide was present when many men were killed; the women escaped to the saddle back & fought most desperately with big stones; many of them thus saved themselves. —

11th Proceeded on to the 3rd Posta, in company with the Leutenant who commands it. — The distance is called fifteen leagues; but it is only guess-work & generally too much. — The road was uninteresting over a dry grassy plain, & on our left hand at a greater or less distance were low hills, a chain of which we crossed close to the Posta. — Before our arrival we met a large herd of cattle & horses, guarded by fifteen soldiers, but we were told that many had been lost. — It is very difficult to drive animals across these plains; if a lion or even a fox approaches the horses in the night, nothing can prevent their dispersing in every direction; and a storm will have the same effect. — A short time since, an officer left Buenos Ayres with 500 horses; when he arrived at the army he had under twenty.

Shortly afterwards we perceived by the cloud of dust that a party of horsemen were approaching; my companions perceived at a great distance, by the streaming hair, that they were Indians. — The Indians often have a narrow fillet round their heads, but never any covering; the long black hair |370| blowing across their faces heightens to an uncommon degree the wildness of their appearance. — They turned out to be a part of Bernantio's tribe going to a Salina for salt. The Indians eat much salt, the children sucking it like sugar; it is a curious contrast with the Gauchos, who living the same life, eat scarcely any. — My companions seemed to think there was not the slightest danger in meeting these gentlemen, & they know best. —but I heard the Commandante of

Bahia Blanca tell one of our officers, that he thought it unsafe for two or three to visit them, although they are professedly the most friendly Indians. —

12th   When at Bahia Blanca, General Rosas sent me a message to say that an officer with a party of men would in a day or two arrive there, & that they had orders to accompany me. As the Lieutenant of this Posta was a very hospitable person I determined to wait a couple of days for the soldiers. — In the morning I rode to examine the neighbouring hills; we were disappointed in not being able from the haziness to see the Ventana. — In coming to this Posta the day before, my guide showed what appeared to me a strong instance of the accuracy with which they know the bearings of different points. — When under a hill, & many leagues distant, I asked him where the Posta was. — After considering for some time, for he had nothing in front to guide him, he pointed out the direction; I marked it with a Katers Compass. Some|371| time afterwards we were on an eminence, from whence he knew the country certainly, again showing me the direction it was the same within 3 degrees that is the $\frac{1}{120}$^{th} part of the horizon. — After dinner the soldiers divided themselves into two parties for a trial of skill with the balls — two spears were stuck in the ground 35 yards apart, they were struck & entangled about once in four or five throws. The balls can be thrown between 50 & 60 yards. — but over 25 there is not much certainty. —

Our party had been increased by two men who brought a parcel from the next Posta to be forwarded to the General. — there were now besides myself & guide the Lieutenant & his four soldiers. — These latter were strange beings — the first a fine young Negro; the second half Indian & Negro; & the two others quite non descripts, one an old Chilian miner of the color of mahogany, & the other partly a mulatto; but two such mongrels, with such detestable expressions I never saw before. — At night, when they were sitting round the fire & playing at cards, I retired to view such a Salvator Rosa scene. — They were seated under a low cliff, so that I could look down upon them; around the party were lying dogs, arms, remnants of Deer & Ostriches, & their long spears were struck in the ground; further, in the dark background, were horses tied up, ready for any sudden danger. — If the stillness of the desolate plain was broken by one of the dogs barking, a soldier,|372| leaving the fire, would place his head close to the ground & thus slowly scan the horizon. Even if the noisy Teru-teru uttered its scream, there would be a pause in the conversation, & every head, for a moment, a little inclined. —[1] What a life of misery these men appear to us to lead. — They are at least ten leagues from the Sauce Posta, & since the murder committed by the Indians, twenty from another. The Indians are supposed to have made their attack in the middle of the night; for very early in the morning, after the murder, they were luckily seen approaching this Posta. — The whole party however escaped with the troop of horses, each one taking a line for himself, & driving with him as many horses as he was able. — The little hovel, built of thistle stalks, in which they slept neither keeps out the wind or rain, indeed in the latter case, the only effect the roof had was to condense it into larger drops. They have *nothing* to eat excepting what they can catch, such as Ostriches, Deer, Armadilloes &c & their only fuel is the dry stalks of a small plant somewhat resembling an Aloe. —[2] The sole luxury, which these men enjoyed was smoking the little paper cigars & sucking Mattee. — I used to think that the Carrion Vulture, the constant attendant on these|373| dreary plains, whilst seated on some little eminence, seemed by his very patience to say, "Ah when the Indians come, we shall have a feast".

[1] The next sentence has been deleted. It runs: 'There was too much appearance of danger, if a little fear is like salt, this assuredly was salted meat. —'
[2] Followed by deleted words: 'which grows in great abundance in all parts'.

13th   We all sallied forth to hunt; we had no success. —there were however some animated chaces & good attempts to ball various animals. The plain here abounds with three sorts of partridges; two, very large, like hen-pheasants. — Their destroyer, a small pretty Fox, is also singularly numerous; we could not in the course of the day have seen less [than] 40 or 50 of these animals. — They were generally near their holes; but the dogs killed one. — Two of our party had separated themselves from us; on our return we found they had been rather more successful, having killed a Lion & found an Ostriches nest with 16 eggs. — These latter afforded us an excellent supper. —

14th   As the men belonging to the next Posta meant to return, we should together make a party of five & all armed, I determined to start & not wait for the officer. — After galloping some leagues, we came to a low swampy country which extends for nearly 80 miles to the Sierra Tapalken. —in some parts there are fine damp grass plains; others black & rather peaty & very soft. —many extensive fresh water but shallow lakes, & large beds of reeds; it resembles the better part of the Cambridgeshire Fens. — This Posta, being a very long one, each of us had|374| two horses; Having passed many swamps, we found a dry spot & there passed the night. —

Sunday 15th   Rose very early in the morning; passed in the road the 4th Posta, where the men were murdered. — The Lieutenant, when found, had 18 Chusa wounds in his body. — Arrived in middle of the day at the 5th Posta. — Here are 21 men, as it is the central & most exposed part of the line of Postas. — The Rancho is built on the edge of a large lake, teeming with wild fowl. —amongst which the black necked swan was conspicuous. There was some difficulty about horses so I determined to sleep here. — In the evening the soldiers returned from hunting, bringing with them seven deer, 3 ostriches & 40 of their eggs. —many partridges & Armadilloes. — It is the constant habit of the soldiers wherever they go to fire the plain; we made several fires, which at night were seen burning with great brilliancy. —they do this to improve the pasture & perhaps also to puzzle any straggling Indians. — Slept in the open air, as the Rancho consisted only of an enclosure of reeds, without any roof. —

16th   To the 6th Posta; soil black & very soft, generally covered with long coarse herbage; —laborious travelling. Rancho here very neat; the posts & rafters were made by a dozen dry stalks bound together with thongs of hide. —by the aid of these Ionic looking columns|375| the sides & roof were thatched with reeds. To the 7th Posta, country improving, like Cottenham fen in Cambridgeshire. —a great abundance of beautiful wild fowl. — This posta is close to the Southern base of Sierra Tapalken; which Sierra is a low broken ridge of Quartz rock 2 or 300 feet high. —extending to the East to Cape Corrientes, but no great distance within the Interior. — I was here told a fact, which, if I had not partly ocular proof, I could not credit. That in the previous night there had been a hailstorm (I saw lightning to the North) & that the pieces of ice were as large as small apples & very hard. — They fell with such force as to kill almost all the small animals. — These men had already found twenty deers & I saw their fresh hides; one of the party a few minutes after my arrival, brought in seven of them; now I well know that one man without dogs would hardly kill 7 in a week; They thought they had seen about 15 dead ostriches; part of one I eat, likewise saw a large partridge with great black mark on its back, where it had been struck. — Many ducks & hawks were killed & ostriches were then running about, evidently blind in one eye. — My informer received a severe cut upon the head. — This extraordinary storm extended but for a short distance. —

To the 8th Posta; galloped very fast over an|376| extremely fine grass plain. — Arrived at the Posta on the R. Tapalguen after it was dark. At supper I was suddenly struck with horror that I was eating one of the very favourite dishes of the country, viz a half formed calf long before its time of birth. — It turned out to be the Lion or Puma; the flesh is very white & remarkably like Veal in its taste. — Dr Shaw was laughed at for stating that "the flesh of the Lion (of Africa) is in great esteem, having no small affinity with veal, both in colour, taste & flavour". — Yet the Puma & Lion are not, I believe, closer allied than any other two of the Cat genus. — The Gouchos differ much whether the Jaguar is good eating; but all agree that the Cat is excellent. —

17th To the 9th Posta, followed the course of the R. Tapalguen, very fertile country. — Tapalguen itself or the town of Tapalguen is a curious place. — It is a perfectly flat plain, studded as far as the eye reaches with the Toldos or oven like huts of Indians. — The greater part of the families of the men with Rosas live here. — There are immense herds of horses & some sheep. — We met & passed many young Indian women, riding by two's & three's on the same horse. These & many of the young men were strikingly handsome; their fine ruddy colour is the very picture of health. — Besides the Toldos there are three Ranchos, one with a Commandante, & two others Pulperia's or shops. —|377| We here bought some biscuit. — I had now been several days without tasting anything except meat & drinking mattee. — I found this new regimen agreed very well with me, but I at the same time felt hard exercise was necessary to make it do so. — I have no doubt that the Gauchos living so much on meat. — is the cause that they like other carnivorous animals can go a long time without food & can withstand much exposure. — I was told that some troops from Tandeel were in pursuit of some Indians, & that for three days they neither tasted water or food. — What other troops would not have killed their horses? —

To the 10th Posta; plain, partly swamp & partly good to the East of the R. Tapalguen. —

18th To the 11th & 12th Posta, a long ride, through a country similar to the last stage: We passed a small tribe of Indians going from Tapalguen to the Guardia del Monte for commerce. — The women rode the horses with goods. — these are of hides & articles woven by hand of wool, such as cloths or yergas & garters. — The patterns are very pretty & brilliantly coloured. — The workmanship is so good that an English merchant in Buenos Ayres declared that the ones, which I had, were of English manufacture. — He was not convinced to the contrary, untill he observed that the tassels were tied up with split sinew. —

12th to 13th to 14th Posta: we had to ride for a long distance in water above the horses knees. — By crossing the stirrups & riding Arab like with|378| the legs cocked up, we managed to keep pretty dry. — As it was growing dark we crossed the Salado; at this time it was about 40 yards wide, but very deep; in the summer it becomes nearly dry, the little water being as salt as the sea. — I ought to have mentioned that the 12th Posta, about 7 leagues to the South of the Salado, was the first Estancia where we saw cattle & a white woman. — Having crossed the Salado, we slept at the Posta, which was one of the great Estancias of General Rosas. It was fortified & of such extent that arriving in the dark I thought it was a Town & fortress. There were immense herds of cattle, as well there might be, the General here having 74 square leagues of land. — He used to have three or four hundred Peons working here & defied all the efforts of the Indians. — I was treated very hospitably, & [in] the morning started for Guardia del Monte.

19th This is a nice scattered little town, with many gardens full of peaches and quinces. — The camp here looked like that around B. Ayres. — the turf short & green (from the grazing & manuring by cattle?) with much clover, beds of thistles & Biscatcha holes. — I first noticed here two plants, which Botanists say have been introduced by the Spaniards. — Fennel which grows in the greatest abundance in all the hedge rows. —& a thistle looking plant which especially in Banda Oriental forms immense beds leagues in extent, & quite impenetrable by man or beast; it occurs in the most unfrequented places|379| near Maldonado. —in the vallies near Rozario, in Entre Rios, &c &c. The whole country between the Uruguay & M. Video is choked up with it; yet Botanists say it is the common artichoke, run wild. — An intelligent farmer on the R. Uruguay told me that in a deserted garden he had seen the planted Artichokes degenerating into this plant. — Of course this man had never heard of the theories of Botanists. — I certainly never saw it South of R. Salado. — The true thistle, (variegated green & white like the sort called sow-thistles,) & which chiefly abounds in the Pampas of Buenos Ayres, I noticed in the valley of the R. Sauce. — There is a very large fresh water Lake near the town, on the coast I found a perfect piece of the case of the Megatherium. — Whilst the postmaster sent for horses several people questioned me concerning the Army. — I never saw anything like the enthusiasm for Rosas & for the success of this "most just of all wars, because against Barbarians". — It is however natural enough, for even here neither man, woman, horse or cow was safe from the attacks of the Indians. The enthusiasm for Rosas was universal, & when some events which subsequently will be mentioned, happened, I was not at all surprised. —

To the 16th, 17th & 18th Posta. Country of one uniform appearance: rich green plain, abundance of cattle horses & sheep; here & there the solitary Estancia, with its Ombu tree. — In the evening torrents of rain, arrived after dark at the Posta; was told that if I travelled by the Post|380| I might sleep there; if not I must pass on, for there were so many robbers about, he could trust nobody. — Upon reading my passport, & finding that I was a Naturalista, his respect & civility were as strong as his suspicions had been before. — What a Naturalista is, neither he or his countrymen had any idea; but I am not sure that my title loses any of its value from this cause. —

20th   In two more Postas reached the city; was much delayed on the road from the rain of the day before. — Buenos Ayres looked quite pretty; with its Agave hedges, its groves of Olives, peaches & Willows, all just throwing out their fresh green leaves. — I rode to the house of Mr Lumb, an English merchant, who gave me a most hospitable reception; & I soon enjoyed all the comforts of an English house. —[1]

[1] Edward Lumb was most helpful to CD in procuring supplies and arranging for the shipment of some of his fossil specimens. See *Correspondence* 1: 355–6, 378–9, 386–8.

21st–26th   These few days of rest were very pleasant; I had plenty of business to transact; & was employed in obtaining letters of introduction, passport &c for St Fe. — My servant having arrived from M. Video, I despatched him to an English Estancia to shoot & skin birds. —

27th   At one oclock I managed to make a start. We rode for an hour in the dark & slept three leagues this side of the town of Luxan. —

28th   We passed it; the town is small & pretty looking, but all the Spanish towns are built on exactly the same model. — There is a fine wooden bridge over the R. Luxan, a most unusual luxury in this country. — We passed Areco, another small town: The country appears level, but it|381| is not so in fact; for in various places the horizon is extensive. — The Estancias are wide apart; for there is little good pasture, the plains being covered by thistles & an acrid clover. — The former was two thirds grown, reaching up to the horses back at this period; it grows in clumps & is of a brilliant green, resembling in miniature a fine forest. — In many parts where the ground was dry, the thistles had not even sprung from the surface, but all was bare & dusty like a turnpike road. — In summer, travelling is sufficiently dangerous for the thistles furnish an excellent retreat & home for numerous robbers, where they can live, rob & cut throats with perfect impunity. — There is little interest in passing over this country, few animals except the Biscatche, & fewer birds inhabit these great thistle beds. —

In the evening crossed the Arrecife, on a raft made of empty barrels lashed together. — We slept at the Post house on the further side. — I paid this day for 31 leagues, & with a burning sun, was but little fatigued. — When the days are longer, & riding a little faster, 50 leagues, as mentioned by Head, might be managed with no very great difficulty. — But then it must especially be remembered that a man, who pays for 50 leagues by the post, by no means rides 150 English miles. — the distance is so universally exaggerated. — My 31 leagues was only 76 English miles in a straight line; allowing 4 miles for|382| curvatures in the road will give 80 miles; Heads days journey reduced by the same proportion gives 129 miles; a much more credible distance than 150 geographical ones. —

Sunday 29th   Arrived in the evening at the town of St Nicholas; it is situated on one of the branches of the Parana. I here first saw this noble river. — There were some large vessels anchored at the foot of the cliff on which the town is built. —

30th   Crossed the Arrozo del Medio & entered the Province of St Fe. — I had been forewarned that nearly all the good people in this province are most dexterous thieves; they soon proved it, by stealing my pistol. — The road generally ran near the Parana, & we had some fine glimpses of it. — We crossed several streams; the water of the Pabon in a good body formed a cascade 20 feet high. — This must be a most unusual phenomenon in this country. — At the Saladillo I saw the curious occurrence of a rapidly running brook with water too salt to drink. — Entered Rozario, a large & striking looking town, built on a dead level plain which forms a cliff about 60 feet high over the Parana. — The river here is very broard with many islands which are low & wooded, as is also the coast of the opposite shore. — The view would resemble that of a great lake, if it were not for the linear shaped islands, which alone, give the idea of running water. The cliffs are the most picturesque part, sometimes absolutely perpendicular & of a red colour, at other times in large broken masses covered with Cacti & Mimosa trees. The real grandeur however of an immense river like this,|383| is derived from reflecting how important a means of communication & commerce it forms between one nation & another—to what a distance it travels—

from how vast a territory it drains the great body of fresh water which flows before your feet. — At Rozario, I had a letter of introduction to a most hospitable Spaniard, who was kind enough to lend me a Pistol. — Having obtained this most indispensable article; I galloped on as far as the Colegio de St Carlos. — A town known by the size of its church & it is said, the hospitality & virtue of the friars. — For many leagues to the North & South of St Nicholas & Rozario the country is really level; it deserves nearly all which travellers have written about these plains. — Yet I have never seen a spot where by slowly turning round, objects could not be seen at a greater distance in some points than in others; and this manifestly proves an inequality in the plain. — As at sea, the horizon is of course very limited; this entirely destroys a degree of grandeur which one would be apt to imagine a vast level plain would possess. — On the sea, your eye being 6 feet above the water, the horizon is distant $2\frac{4}{5}$ miles. —

October 1st Started by moonlight & arrived at the R. Carcavàna by sun rise. — This river is also called the Saladillo, & it deserves the name for the water is brackish. — I staid here the greater part of the day, searching for bones in the cliff. Old Falkner[1] mentions having seen great bones in this river;|384| I found a curious & large cutting tooth. Hearing also of some "giants" bones on the Parana, I hired a canoe; there were two groups of bones sticking out of a cliff which came perpendicular into the water. The bones were very large, I believe belonging to the Mastodon. — they were so completely decayed & soft, that I was unable to extract even a small bone. — In the evening rode on another stage on the road, crossing the Monge, another brackish stream. —

[1] See Thomas Falkner, *A description of Patagonia, and the adjoining parts of South America.* Hereford, 1774.

2nd Unwell & feverish, from having exerted myself too much in the sun. — The change in latitude between St Fe & Buenos Ayres is about 3 degrees; the change in climate is much greater. — everything shows it. — the dress & complexion of the inhabitants, the increased size of the Ombus, many new cacti, the greater beauty of the birds & flowers; all proves the greater influence of the sun. We passed Corunda, from the luxuriance of its gardens it is the prettiest village I have seen. — From this place to St Fe, the road is not very safe; it runs through one large wood of low prickly trees, apparently all Mimosa. — As there are no habitations to the West of this part of the Parana, the Indians sometime come down & kill passengers. — On the road there were some houses now deserted from having been plundered; there was also a spectacle, which my guide looked at with great satisfaction, viz the skeleton with the dried skin hanging to the bones, of an Indian suspended to a tree. The wood had a pretty appearance opening into glades like a lawn. |385| We changed our horses at a Posta where there are twenty soldiers: & by sun set arrived at St Fe. There was much delay on the road, on account of having to cross an arm of the Parana, St Fe being situated in a large island. — I was much exhausted & was very glad to procure an unfurnished room. —

3rd & 4th   Unwell in bed. — St Fe is nice, straggling town, with many gardens. —it is kept clean & in good order. — The governor of the province, Lopez, is a tyrant; which perhaps is the best form of government for the inhabitants. — He was a common soldier at the great revolution & has now been 17 years in power. — His chief occupation is killing Indians, a short time since he slaughtered 48 of them. — The children are sold for between 3 & 4 pound sterling.

5th   Crossed the Parana to St Fe Bajada, or as it is now called Parana, the capital of Entre Rios. — The passage took up four hours; winding about the different branches. —which are all deep & rapid; we crossed the main arm & arrived at the Port. — The town is more than a mile from the river; it was placed there *formerly* so as not to be exposed so much to the attacks of the Paraguay Indians. — I had a letter of introduction to an old Catalonian, who treated me with the most uncommon hospitality. — My original intention had been to cross the province of Entre Rios & return by the Banda Oriental to B. Ayres. Not being quite well and thinking that the Beagle would sail long before she eventually did, I gave up this plan, & determined to return immediately |386| to B. Ayres. I was unable to hire a boat so took a passage in a Balandra.

6th–11th   By the indolence of the master & from bad weather I was delayed five days. — The time passed pleasantly & was enabled to see the geology of the surrounding district. — And this possessed no common interest. — The Bajada itself is quiet town; about as large as St Fe or St Nicholas; it contained in 1825, 6000 inhabitants. — The whole province only contains 30,000. — Yet here there are representatives, ministers, standing army, governors &c &c. Few, as they are, none have suffered more from desperate & bloody revolutions. — In some future day however this will be one of the finest provinces. — As its name expresses, it is surrounded on every side by the magnificent rivers, the Parana & Uruguay. —the land is most fertile. — Here there is no fear of

the Indians; an immense advantage over their neighbours; to the North
of St Fe, there is not a single Estancia on the West of the Parana; & we
have seen that the road is not safe between the Capital & Corunda. —

My usual walk during these days was to the cliffs on the Parana to
admire the view of the river & pick up fossil shells. — Amongst the
fallen masses of rock, vegetation was very luxuriant; there were many
beautiful flowers, around which humming birds were hovering. — I
could almost fancy that I was transported to that earthly paradise,
Brazil. — |387|

12th Embarked on board the Balandra; a one masted vessel of a hundred
tuns; we made sail down the current. — The weather continuing bad,
we only went a few leagues & fastened the vessel to the trees on one of
the islands. — The Parana is full of islands; they are all of one character,
composed of muddy sand, at present about four feet above the level of
the water; in the floods they are covered. — An abundance of willows &
two or three other sorts of trees grow on them, & the whole is rendered
a complete jungle by the variety & profusion of creeping plants. —
These thickets afford a safe harbour for many capinchas & tigers. — The
fear of these latter animals quite destroyed all pleasure in scrambling in
the islands. — On this day I had not proceeded a hundred yards, before
finding the most indubitable & recent sign of the tiger. I was obliged to
retreat; on every islands there are tracks; as in a former excursion the
"rastro" of the Indians had been the constant subject of observation, so
in this was the "rastro del tigre".

The jaguar is a much more dangerous animal than is generally
supposed: they have killed several wood-cutters; occassionally they
enter vessels. There is a man now in the Bajada, who coming up from
below at night time was seized by a tiger, but he escaped with the loss
of the use of one arm. — When the floods drive the tigers out of the
islands; they are most dangerous. —a few years since a very large one
entered a church at St Fe. Two padres entering one after the other were
killed, a third|388| coming to see what was the cause of their delay,
escaped with difficulty. — The beast was killed by unroofing one corner
of the room & firing at it. — The tigers annually kill a considerable
number of young oxen & horses. — These islands undergo a constant
round of decay & renovation. —in the memory of the master several
large ones had disappeared, others again had been formed & protected
by vegetation. —

13th & 14th A constant gale & rain from the SE, remained at our moor-
ings. —the greater part of the time I passed in bed, as the cabin was too

low to sit up in. — there was also good sport in fishing, the river abounds in large & extraordinary looking fish, which are excellent food. —

15th We got under weigh; passed Punta Gorda, where there is colony of tame Indians from the province of Missiones. — We sailed rapidly down the current; before sunset from a silly fear of bad weather brought to in a narrow arm or "Riacho". — I took the boat & rowed some distance up the creek; it was very narrow, winding & deep; on each side there was a wall 30 or 40 feet high formed by the trees entwined with creepers, this gave to the canal a singularly gloomy appearance. — I here saw a very extraordinary bird, the scissor-beak. — the lower mandible is as flat & elastic as an ivory paper-cutter, it is an inch & a half longer than the upper. — With its mouth wide open, & the lower mandible immersed some depth in the water, it flies rapidly up & down the stream. Thus ploughing the surface, it occassionally seizes a small fish. — |389|

The evenings are quite tropical; the thermometer 79° — an abundance of fire flies, & the mosquitoes very troublesome. — I exposed my hand for five minutes, it was black with them: I do not think there could have been less than 50, all busy with sucking. — At night, I slept on deck, the greater coolness allowing the head & face to be covered up with comfort. —

16th Some leagues above Rozario we came to cliffs, which are absolutely perpendicular. — these form the West bank to below St Nicholas; & the whole coast more resembles that of a sea than a fresh-water river. — It is a great draw back to the scenery of the Parana, that from the soft nature of the banks, all the water is very muddy. — The Uruguay is much clearer, & I am told where the two waters flow in one channel, they may clearly be distinguished by their black & red colours. In the evening, the wind not being quite fair, the master was much too indolent to think of proceeding. — Moored 5 leagues above St Nicholas. —

17th Gale. — remained stationary.

18th & 19th Sailed quietly on with gentle winds, & anchored in the middle of the night near the mouth of the Parana, called Las Palmas. —

20th I was very anxious to reach B. Ayres, so that I determined to leave the vessel at Las Conchas & ride into town a distance about 20 miles. — After changing my vessel three times in order to pass the bar, I obtained a canoe, & we paddled quickly along to the Punta de St Fernando. — The channel |390| is narrow & several miles long. — On each side the islands were covered with peaches & Oranges. These have been planted by nature, & flourish so well, that the market of B. Ayres in the fruit season is supplied by them. — On one of the islands I saw a bevy of fine gallinaceous birds of a black colour & nearly the size of a Turkey. — Upon leaving the canoe, I found to my utter astonishment I was a sort of prisoner. — About a week before, a violent revolution had broken out; all the ports were under an embargo. — I could not return to my vessel, & as for going by land to the city it was out of the question. — After a long conversation with the Commandante I obtained permission to go the next day to General Rolor, who commanded a division of the rebels on this side of the Capital. —

21st Arrived early in the morning at Rolors encampment, the general, officers, & soldiers all appeared, & I believe really were, great villains. — The General told me, that the city was in a state of close blockade; that he could only give me a passport to the General in chief (of the rebels) at Quilmes. — I had therefore to take a great sweep round the city; & it was with very much difficulty that I procured horses. — When I arrived at the encampment, they were civil, but told me I could not be allowed to enter. This was General Rosases party; & his brother was there. — I soon began to talk about the Generals civility to me at the R. Colorado. — Magic could not have altered circumstances quicker than this|391| conversation did. At last they offered me the choice to enter the city on foot without my Peon horses &c &c & without a passport: I was too glad to accept it, & an officer was sent to give directions not to stop me at the bridge. The road, about a league in length, was quite deserted; I met one party of soldiers; but I satisfied them with an old passport. — I was exceedingly glad when I found myself safe on the stones of B. Ayres. —

This revolution is nothing more or less than a downright rebellion. — A party of men who are attached to General Rosas, were disgusted with the Governor; they left the city to the number of 70, & with the cry of Rosas, the whole country took arms. — The city was then closely blockaded: no provisions, cattle, or horses are allowed to enter; excepting this, there is only a little skirmishing, a few men daily killed. — The outside party well know that by stopping the supply of meat they will certainly be victorious. —

General Rosas could not have known of this rising; but I think it is quite consonant with his schemes. — A year ago he was elected Governor; he refused it, without the Sala would also give him extraordinary powers. — This they refused, & now Rosas means to show them that no other Governor can keep his place. — The warfare on both sides was avowedly protracted till it was possible to hear from Rosas. — A note arrived, a few days after my leaving B. Ayres, which stated that the|392| General disapproved of peace being broken, but that he thought the outside party had justice on their side. — Instantly, on the reception of this, the Governor & ministers resigned, & they with the military to the amount of some hundreds flew from the city. — The rebels entered, elected a new Governor, & were paid for their services to the number of 5500 men. — It is clear to me that Rosas ultimately must be absolute Dictator, (they object to the term king) of this country. —

22nd–November 1st These disturbances caused me much inconvenience; my servant was outside, I was obliged to bribe a man to smuggle him in through the belligerents. His clothese, my riding gear, collections from St Fe, were outside with no possibility of obtaining them. — I was, however, lucky in having them all sent to me at M: Video. — The residence in the town was disagreeable, it was difficult to transact any business, the shops being closed; & there were constant apprehensions of the town being ransacked. — The real danger lay with the lawless soldiery within; they robbed many people in the day time, & at night the very sentinels stopped people to demand money from them. — [1]

[1] In a letter to Caroline Darwin, CD wrote: 'Such a set of misfortunes I have had this month, never before happened to poor mortal. My servant (Covington by name & most invaluable I find him) was sent to the Estancia of the Merchants whose house I am staying in. — he the other day nearly lost his life in a quicksand & my gun completely. — . . . There literally is only one Gentleman in Buenos Ayres, the English Minister. — He has written to order the Beagle up. — But we sail under such particular instructions I know not whether the Captain will come. — If he does all will be right about Covington. — otherwise I shall be obliged to send some small vessel or boat to smuggle him off the coast. —' See *Correspondence* 1: 342–3.

2nd With sufficient trouble got on board the Packet; found it crowded with men, women & children, glad to escape from so miserable a town.

3rd & 4th After a long passage, arrived at M: Video; I went on board the Beagle: Was astonished to hear we were not to sail till the beginning of|393| December: the cause of this great delay was the necessity of finishing all charts, the materials for which had been collected by the Schooners. —

5th The poop-cabin being full of workers, I took up my residence on shore, so as to make the most of this additional month. —

6th Had a long gallop to the East end of the Barrancas de St Gregorio: was disappointed in the Geology, but had a pleasant gallop along the coast of the Plata. — It was necessary to cross the St Lucia near its mouth; we passed in a boat, the horses were obliged to swim at least 600 yards; I was surprised to see with what ease they performed it. — We did not return till so late, that I slept at a Rancho,

7*th* & returned home early in the morning. —

8*th*–13*th* I prepared for a ride to see the R. Uruguay & its tributary the R. Negro. — These days were lost by true Spanish delay in giving me my passport, letters &c &c. —

14*th* Started in the afternoon & slept in the house of my Vaqueano in Canelones.

15*th* In the morning we rose early in the hopes of being able to ride a good distance; it was a vain attempt, for all the rivers were flooded; we passed R. Canelones, St Lucia, San Josè in boats, & thus lost much time: at night we slept at the Post house of Cufrè. — In the course of the day, I was amused by seeing the dexterity with which some Peons crossed over the rivers. — As soon as the horse is out of its depth, the man slips backwards & seizing the tail is towed across; on the other side, he pulls|394| himself on again. — A naked man on a naked horse is a very fine spectacle; I had no idea how well the two animals suited each other: as the Peons were galloping about they reminded me of the Elgin marbles. —

16*th* Not being quite well, stayed the whole day at this house. In the evening the Post-man or letter carrier arrived; he was a day after his time, owing to the R. Rozario being flooded; it could not however be of much consequence, for although he passes through some of the principal towns in B. Oriental, his luggage consisted of two letters. — The view from the house was pleasing, an undulating green surface with distant glimpses of the Plata. — I find I look at this province with very different eyes from what I did upon first arrival. — I recollect I then thought it singularly level; but now after galloping over the Pampas, my only surprise is what could have induced me ever to have called it level; the country is a series of undulations; in themselves perhaps not absolutely great, but as compared to the plains of St Fe, real mountains. — From these inequalities there is an abundance of small rivulets, & the turf is green & luxuriant. —

17*th* We crossed the Rozario which was deep & rapid, & passing the village of Colla, arrived at mid-day at Colonia del Sacramiento. — The distance is twenty leagues, through a fine grass country, but which is very poorly stocked with cattle or inhabitants. I was invited to sleep at Colonia & to accompany on the following day a gentleman to his

Estancia, where there were some rocks of recent limestone. — The town is built on a stony promontory something in the same manner as M. Video: it is strongly fortified, but both fortifications & town suffered much from the Brazilian war. — It is very ancient, & from the irregularity of the streets & the surrounding groves of old Orange trees|395| & peaches had a pretty appearance. — The church is a curious ruin; it was used as a powder magazine & was struck by lightning in one of the ten thousand storms of the Rio Plata. — Two thirds of the building was blown away to the very foundation, & the rest stands a shattered & curious monument of the united powers of lightning and gunpowder. In the evening I wandered about the half demolished walls of the town. — It was the chief seat of the Brazilian war; a war most injurious to this country, not so much in its immediate effects, as in being the origin of a multitude of Generals, & all other grades of officers. More generals are numbered but not paid in the united provinces of La Plata than in Great Britain. — These gentlemen have learned to like power & do not object to a little skirmishing. Hence arises a constant temptation to fresh revolutions, which in proportion as they are easily effected, so are they easily overturned. — But I noticed here & in other places a very general interest in the ensuing election for the President; & this appears a good sign for the stability of this little country. — The inhabitants do not require much education in their representatives; I heard some men discussing the merits of those for Colonia; "that although they were not men of business, they could all sign their names". With this every reasonable man was satisfied.

18th   Rode with my host to his Estancia at the Arroyo de St Juan. — In the evening we took a circuit round the estate; it contained two square leagues & a half and was situated in what is called a rincon; that is one side is fronted by the Plata, & the two others are guarded by impassable brooks. There is an excellent port for little vessels, & an abundance of small wood, which is valuable as supplying fuel to Buenos Ayres. — I was curious to know the value of so complete an Estancia; — at present there are 3000 cattle &|396| it would well support three or four times the number. — there are 800 mares, 150 broken horses, 600 sheep; plenty of water & limestone; a rough house, excellent corrals, & a peach orchard. — For all this he has been offered 2000£ only wants 500£ additional, and probably would sell it for less. The chief trouble with an Estancia is driving all the cattle twice a week to a central spot, in order to make them tame & to count them. This latter would be thought a difficult operation, when there are ten or fifteen thousand head together; it is managed on the principle that the cattle invariably divide themselves into little troops from forty to an hundred. — Each troop is recognised by a few peculiarly marked animals, & its number is known: thus one being lost out of ten thousand is perceived by its absence from one of the tropillas. During a stormy night the cattle all mingle together; but the next morning all the tropillas separate as before. —

19th Passed the town of las Vacas; it is a straggling village, built on an arm of the Uruguay, & has a good deal of trade up the river. — Slept at a North Americans, who works a lime kiln on the Arroyo de las Vivoras. —

20th In the morning went out riding to Punta Gorda; on the road tried to find a Jaguar; saw very fresh tracks & the trees against which they are said to sharpen their claws: the bark was cut up & grooved by scratches a yard long: — we did not succeed in disturbing one. — The low, thick woods on the coast of the Uruguay afford an excellent harbour for such animals. — At Punta Gorda, the R Uruguay presented a noble body of water; its appearance is superior to that of the Parana from the clearness of the water & rapidity of the stream: on the opposite coast|397| there are several branches, which enter from the Parana, when the sun shines, the two colours of the water may be seen quite distinct. — The house & lime-kiln were for this country unusually old, being built 108 years since. — I was told a curious circumstance respecting the Lime-kiln. — At the instant of the revolution it was full of fresh burnt lime; from the state of [the] country it was left 18 years untouched: on the surface young trees were growing, whilst in the middle the lime was quick. — When they dug down to the place where the half-burnt wood is left, in a few minutes it kindled & burst out into flames. — This caused uncommon superstitious fears amongst the workmen; but the owner tells me this is always the case in a lime-kiln opened after a few months interval. —

At this Estancia many mares, mares are never ridden in this country, are killed weekly for their skins, which are worth 5 paper dollars each or about ½ a crown. — I heard of some feats in the lassoing line. — One individual will stand 12 yards from the gate of the Corral, & will bet that he will catch every horse by the legs as it rushes by him. — Another will enter on foot a Corral, catch a mare, fasten its front legs, drive it out, throw it down, kill, skin & stake the hide, (a tedious job) & this whole operation he will perform on 22 mares in one day; or he will skin 50 in the same time. — This is prodigious; for it is generally considered [a] good days work, solely to skin & stake 15 or 16 animals. —

In the evening started on the road to Mercèdes|398| or Capella Neuva on the R. Negro. — We passed through much Acacia wood, like that near Coronda & which invariably grows in the low bottoms near streams & rivers. — At night we asked permission to sleep at an Estancia at which we happened to arrive.[1] It was a very large estate, being ten leagues square, & the owner at Buenos Ayres is one of the greatest landowners in the country. — His nephew has charge of it & with him there was a Captain of the army, who the other day ran away from Buenos Ayres. — Considering their station their conversation was rather amusing. They expressed, as was usual, unbounded astonishment at the globe being round, & could scarcely credit that a hole would if deep enough come out on the other side. They had however heard of a country where there were six months light & six of darkness, & they said the inhabitants were very tall & thin. They were curious about the price & condition of horses & cattle; upon finding out we did not in England catch our animals with the Lazo, they added "Ah then, you use nothing but the bolas": The idea of an enclosed country was quite novel to them. — The Captain at last said, he had one question to ask me, & he should be very much obliged if I would answer him with all truth. — I trembled to think how deeply scientific it would be. — "it was whether the ladies of Buenos Ayres were not the handsomest in the world". I replied, "Charmingly so". — He added, I have one other question—"Do ladies in any other part of the world wear such large combs". I solemnly assured him they did not. — They were absolutely delighted. — The Captain exclaimed, "Look there, a man, who has seen half the world, says it is the|399| case; we always thought so, but now we know it". My excellent judgment in beauty procured me a most hospitable reception; the Captain forced me to take his bed, & he would sleep on his Recado. —

[1] Followed by a deleted sentence: 'I now travelled with hired horses, as there is no Post road.'

21st   Started at sunrise, & rode slowly during the whole day. — The geological nature of the country is here different from the rest of the province, & closely resembles that of the Pampas. — From this cause we here have immense beds of the thistle, as well as the cardoon:—the whole country indeed may be called one great bed. The two sorts grow separate, each plant in company with its own kind. — The cardoon is as high as a horses back, but the Pampas thistle often higher than the crown of the head of the rider. — To leave the road for a yard is out of the question, & the road itself is partly, & in some cases entirely, closed; pasture of course there is none; if cattle or horses once enter the bed they are for the time, completely lost. — For this reason, it is very hazardous to attempt to drive cattle at this season of the year, for when jaded enough to face the prickles, they rush amongst the thistles & are seen no more. — From the same cause there are but few Estancias, & these near damp vallies where the thistle will not grow. — As night came on before we could arrive at the house of an Englishman for whom I had a letter of introduction we slept at a Rancho. —

22nd Arrived at the Estancia of the Berquelo, near Mercedes, & found the owner not at home. — he returned in the evening & I spent the day in geologising the neighbouring country. — |400|

23rd Rode to the Capella Nueva; a straggling village: & saw the R. Negro; it is a fine river blue water & running stream; it is nearly as large as its namesake to the South. —

24th Went with my host to the Sierra del Pedro Flaco about 20 miles up the R. Negro: the greater part of the ride was through long grass up to the horses belly. — There are few Estancias & leagues of camp without a head of cattle. The country left to nature as it now is would *easily* produce 5 or 6 times the number of cattle. — Yet the annual exportation of hides from M. Video is 300 thousand; & the home consumption is something considerable. The view of the R. Negro from the Sierra is decidedly the most picturesque one I have seen in this country. The river is rapid & tortuous; it is about twice as large as the Severn (when banks full) at Shrewsbury; the cliffs are precipitous and rocky; & there is a belt of wood following the course of the river; beyond which an horizon of grass plain fills up the view. — The Peons horse was quite tired; so we rode to a Rancho; the master was not at home, but as a matter of course [we] entered the house, made a fire to cook some beef, & were quite at home in a strangers house. — We rode on but did not reach home till early in the night. —

25th We heard of some giants bones, which as usual turned out to be those of the Megatherium. — With much trouble extracted a few broken fragments. — In the evening a domidor or horse-breaker came to the house & I saw the operation of mounting a perfectly wild horse. — They were too fat to fight much: and there was little to see in the operation; the|401| horse is thrown down & the bridle is tied to the under jaw:—tying the hind legs together he is allowed to rise & is then saddled. — During these operations the horse throws himself down so repeatedly & is so beaten, that when his legs are loosed and the man mounts him, he is so terrified as hardly to be able to breathe, & is trickling down with sweat. — Generally however a horse fights for a few minutes desperately, then starts away at a gallop, which is continued till the animal is quite exhausted. — This is a very severe but short way of breaking in a colt. —

26th   Began my return in a direct line to M. Video; went by an Estancia where there was a part, very perfect, of the head of a Megatherium. I purchased it for a few shillings. — We had long gallop through a more rocky & hilly country than the coast road, to the R. Perdido, where we slept. — One of the Post-houses was kept by a man, apparently of pure Indian blood; he was half intoxicated. — My peon declares that he in my presence said I was a Gallego; an expression synonimous with saying he is worth murdering. — His companions laughed oddly:—& I believe what my Peon said was true; when I remonstrated with him on the absurdity, he only said, "you do not [know] the people of this country". — The motive must have been to sound my Peon, who perhaps luckily for me was a trust worthy man. — Your entire safety in this country depends upon your companion. —|402| At night there were torrents of rain; as the Rancho made but little pretensions to keep out water or wind, we were soon wet through. —

27th   In the morning had a long gallop: arrived at San Josè, from which point the road is the same by which I started. San Josè, Canelones, St Lucia are all rather nice little rectangular towns, & all just alike. — Slept one post beyond San Josè,

28th   & in the middle of the next day we arrived at Monte Video. The distance, paid by the Post, being about 70 leagues from Mercedes to the Capital. —

29th–December 4th   During these few days I resided on shore; the cause of the ships delay being the charts not being completed. —

During the last six months I have had some opportunity of seeing a little of the character of the inhabitants of these provinces. — The gauchos or country men are very superior to those who reside in the towns. — The gaucho is invariably most obliging, polite & hospitable. I have not met one instance of rudeness or inhospitality. He is modest both respecting himself & country, at the same time being a spirited bold fellow. — On the other hand there is much blood shed, & many robberies committed. — The constant presence of the knife is the chief cause of the former: it is lamentable to hear how many lives are lost in trifling quarrels; in fighting each party tries to mark the face of his adversary by slashing his nose or eye; deep & horrid looking scars often attest that one has been successful. — Robberies are a natural consequence|403| of universal gambling, much drinking & extreme indolence. — At Mercedes I asked two men why they did not work: — one said that the days were too long; the other that he was too poor. The number of horses & profusion of food is the destruction of all industry. — Moreover there are so many feast days; then again nothing can succeed without it is begun when the moon is on the increase; and from these two causes half the month must be lost. — Police & justice are quite inefficient, if a man commits a murder & should be taken, perhaps he may be imprisoned or even shot; but if he is rich & has friends he may rely on it nothing will happen. — It is curious that the most respectable people in the country will invariably assist a murderer to escape. — They seem to think that the individual sins against the government & not against the state. — A traveller has no other protection than his own arms; & the constant habit of carrying them chiefly prevents a more common occurrence of robberies. — The character of the higher & more educated classes who reside in the towns, is stained by many other crimes. — partaking in a lesser degree in the good parts of the Gaucho character; he is a profligate sensualist who laughs at all religion; he is open to the grossest corruption; his want of principle is entire. — An opportunity occurring not to cheat his friend would be an act of weakness; to tell the truth where a lie might be more serviceable, would be the|404| simplicity of a child. The term honor is not understood; neither it, nor any generous feeling, the remains of chivalry, have survived the long passage of the Atlantic. — If I had read these opinions a year ago, I should have accused myself of much illiberality: now I do not. — Every one, who has good opportunities of judging, thinks the same. In the Sala of B. Ayres I do not believe there are six men to whose honesty or principles you could trust. **Every** *public officer* is to be bribed; the head of the post office sells forged government francs: — the Governor and prime minister openly plunder the state. — Justice, where gold is in the case, is hardly expected. — I know a man (he had good cause) who went to the chief Justice & said "here are 200 dollars (sixpences) if you will arrest such a person *illegally*; my lawyer recommended me to take this step". The Chief Justice smiled acquisition[1] & thanked him; before night the man was in prison. — With this utter want of principle in the leading men; with the country full of ill-paid, turbulent officers; they yet hope that a Democratic form of government will last. In my opinion before many years, they will be trembling under the iron hand of some Dictator. — I wish the country well enough to hope the period is not far distant. —

On first seeing the common society of the people, two or three things strike one as remarkable: the excellent taste of all the women in dress: the general good manners in all grades of life: — but chiefly the remarkable equality of all ranks.|405| At the Colorado, men who keep the lowest little shops used to dine with General Rosas. — A son of a Major at B. Blanca gains a livelihood by making paper cigars; he wished to come as Vaqueano with me to B. Ayres; but his father was afraid. — Many in the army can neither read or write; yet all meet on perfect terms of equality. — In Entre Rios the Sala contains 6 members. — One of these was a sort of shopman in a store, & evidently by no means degraded by such an employment. — This is all what might be expected in a new country; nevertheless the abscence of Gentlemen par excellence strikes one as a novelty. —

My time at M. Video was spent in getting ready for our long cruize in Tierra del Fuego. — It was a pleasant employment preparing to leave for ever the uninteresting plains of the R. de La Plata. —

The Beagle & Adventure are both ready for sea, with a fine stock of provisions & excellent crews. — The other day, there was an instance of the unaccountable manner in which seamen sometimes run away from a ship. — Two men, petty officers in good favour & with 2 or 3 years pay owing them, ran away & the design must have been made sometime previously. — These men were allowed repeatedly to go on shore & held the first stations on board. — There is a degree of infatuation & childish want of steadiness in seamen, which to a landsman is quite incomprehensible & hardly to be credited. — |406|

I called one day on Mr Hood, the Consul General, in order to see his house which had been a short time previously struck by lightning. — The effects were curious: the bell wires were melted & the red hot globules dropping on the furniture drilled small holes in a line beneath them — when falling on glass vessels, they melted & adhered to them. — Yet the room was at least 15 feet high & the wire close to the ceiling. — In one of the walls the electric fluid exploded like gunpowder, & shot fragments of bricks with such force as to dent the wall on the opposite side. Where the bell wire ran, the paper was blackened by the oxide of the metal for nearly a foot on each side; in a like manner the frame of a looking glass was blackened; the gilding must have been volatilized, for a smelling bottle which stood near was firmly coated with some of it. — The windows were all broken & everything hanging up fell down by the Jar. — It happened very early in the morning. When I was at B: Ayres a short time previous to this, the church was much shattered & a vessel lost her main-mast. —

I ought not to conclude my few remarks on the Inhabitants of the Provinces of the R. de La Plata, without adding that a most perfect & spirited outline of their manners & customs will be found in "Heads rough notes". — I do not think that his picture is at all more exaggerated, than every good one must be—that is by taking strong examples & neglecting those of less interest. — I cannot however agree with him "in the ten thousand beauties of the Pampas". — But I grant that the rapid galloping & the feeding on "beef & water" is exhilarating to the highest pitch. —[2]|407|

[1] Corrected in pencil in the margin to 'acquiescence'.
[2] See Francis Bond Head, *Rough notes taken during some rapid journeys across the Pampas and among the Andes*. London, 1826.

5th   Took a farewell of the shore & went on board.

6th   The Beagle got under weigh at 4 oclock in the morning & ran up the river to take in fresh water. — We are now becalmed within sight of the Mount. — The Adventure is at anchor close to us. May kind fortune for once favor us with fine weather & prosperous breezes. —

7th   With a fair wind stood out of the river & by the evening were in clear water; never I trust again to enter the muddy water of the Plata. — The Adventure kept ahead of us, which rejoiced us all, as there were strong fears about her sailing. —it is a great amusement having a companion to gaze at. — The following changes have taken place amongst the officers. — M[r] Wickham commands the Adventure; he has with him M[rs] Johnstone & Forsyth & M[r] Usborne as under-surveyor. — M[r] Kent from the Pylades has joined us as surgeon. — M[r] Martens is on board the Beagle filling the place which M[r] Earle is obliged to vacate from ill health. —[1]

[1] Conrad Martens served as the *Beagle*'s official artist for the next 15 months. FitzRoy first described him to CD in the following terms: 'If Mr P. has written as he intended you have heard of Mr Martens— Earle's Successor, — a *stone pounding artist*—who exclaims *in his sleep* "think of me* standing upon a pinnacle of the Andes—or sketching a Fuegian Glacier!!!" By my faith in Bumpology, I am sure you will like him, and like him *much*—he is—or I am wofully mistaken—a "rara avis in navibus, — Carlo que Simillima Darwin". — Don't be jealous now for I only put in the last bit to make the line scan— you know very

well your degree is "rarissima" and that *your* line runs thus— Est avis in navibus Carlos rarissima Darwin. — but you will think I am cracked so seriatim he is a gentlemanlike, well informed man. — his landscapes are *really* good (compared with London men) though perhaps in *figures* he cannot equal Earle— He is very industrious— and gentlemanlike in his *habits*, — (not a *small* recommendation).' See *Correspondence* 1: 335. The first picture executed by Martens on board the *Beagle* was a double-page panorama in ink and sepia wash labelled 'Montevideo from the anchorage of H.M.S. Beagle. Decr 4 1833' (CM No. 72, *Beagle Record* pp. 168–9).

8th–23rd Arrived at Port Desire. — Our passage has been a very long one of seventeen days; the winds generally being light & foul. — with the exception of a fresh gale or two. —

The Adventure delayed us: she is found not to sail well on a wind; & at this place her sails will be altered. — The harbor of Port Desire, is a creek which runs up the country in the form of a river: the entrance is very narrow; but with a fine breeze, the 'Beagle entered in good style.[1]|408|

[1] See watercolour by Conrad Martens labelled 'Ruins, North side of the Harbour of Port Desire. Decr 23 1833' and initialled RF, with the *Adventure* at anchor in the foreground (CM No. 76, *Beagle Record* p. 124).

24th Took a long walk on the North side: after ascending some rocks there is a great *level* plain, which extends in every direction but is divided by vallies. — I thought I had seen some desart looking country near B. Blanca; but the land in this neighbourhead so far exceeds it in sterility, that this alone deserves the name of a desart. — The plain is composed of gravel with very little vegetation & not a drop of water. In the vallies there is some little, but it is very brackish. — It is remarkable that on the surface of this plain there are shells of the same sort which now exist. —& the muscles even with their usual blue colour. — It is therefore certain, that within no great number of centuries all this country has been beneath the sea. —[1] Wretched looking as the country is, it supports very many Guanacoes. — By great good luck I shot one; it weighed without its entrails &c 170 pounds: so that we shall have fresh meat for all hands on Christmas day. —

[1] This was a most important conclusion, the elevation of the coast at Port Desire by no less than 330 feet being thus established to have taken place in relatively recent times.

Christmas 25th After dining in the Gun-room, the officers & almost every man in the ship went on shore. — The Captain distributed prizes to the best runners, leapers, wrestlers. — These Olympic games were very amusing; it was quite delightful to see with what school-boy eagerness the seamen enjoyed them: old men with long beards & young men without any were playing like so many children. —certainly a much better way of passing Christmas day than the usual one, of every seaman getting as drunk as he possibly can. —[1]|409|

[1] See watercolour by Conrad Martens labelled 'Slinging the monkey, Port Desire, Decr 25 1833. Note Mainmast of Beagle a little farther aft, Miz. Mast to rake more' and initialled RF (CM No. 80, *Beagle Record* p. 173). Also watercolour development (CM No. 79, *Beagle Record* p. 172) of a drawing of Port Desire made the same day with the *Beagle* and *Adventure* at anchor. This was engraved by S. Bull as 'Anchorage, and Spanish Ruins. Port Desire; (*Narrative* **2:** facing p. 316).

26th The Beagle is anchored opposite to a fort erected by the old Spaniards. — It was formerly attempted to make a settlement here; but it quite failed from the want of water in the summer, & the Indians in the winter. — The buildings were begun in very good style, & remain a proof of the strong hand of old Spain. — Some of the enclosures & some cherry trees may yet be seen. — The fate of all the Spanish establishments on the coast of Patagonia, with the exception of the R. Negro, has been miserable. — Port Famine, as it is well known, expresses the sufferings of the settlers. — At St Josephs every man, excepting two, was massacred by the Indians on a Sunday when in church. — The two were prisoners some years with the Indians; one of them, now in extreme old age, I conversed with at R. Negro.

I walked this day to some fine cliffs, five miles to the South: here the usual geological story, of the same great oyster bed being upheaved in modern days was very evident. — In the evening weather very cold, — & a Tierra del Fuego gale of wind. —

28th The Yawl, under the command of M[r] Chaffers with three days provisions, was sent to survey the head of the creek. — In the morning we searched for some watering places mentioned in an old Chart of the Spaniards. — We found one creek, at the head of which there was a small rill of brackish water. — Here the tide compelled us to stay some hours. — I, in the interval, walked several miles into the interior.|410| The plain, as is universally the case, is formed of sandy chalk, & gravel; from the softness of these materials it is worn & cut up by very many vallies. — There is not a tree, &, excepting the Guanaco, who stands on some hill top a watchful sentinel over his herd, scarcely an animal or a bird. — All is stillness & desolation. One reflects how many centuries it has thus been & how many more it will thus remain. — Yet in this scence without one bright object, there is a high pleasure, which I can neither explain or comprehend. — In the evening, we sailed a few miles further & then pitched the tents for the night. —

29th By the middle of the day the Yawl could not get any higher, from the

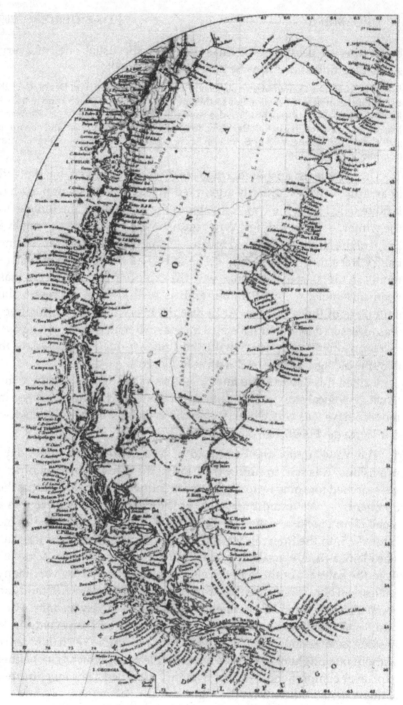

Patagonia and Tierra del Fuego (part of a map from *Narrative* 1).

shoalness of the water & the number of mud-banks. — One of the party happening to taste the water found it only brackish. — Mr Chaffers, directly after dinner started in the dingy, & after proceeding two or three miles found himself in a small fresh water river. — Small as it is, it appears to me probable, that it flows from the Cordilleras, the water is muddy as if flooded, & this is the time of year for the snow freshes of the Colorado, Sauce &c. — Mr Chaffers saw in a little valley a lame horse, with his back marked by the saddle; so that the Indians must have left him there or were then in the neighbourhead. The views here were very fine & rude; the red porphyry rock rises from the water in perpendicular cliffs, or forms spires & pinacles in|411| its very course. — Excepting in this respect the country is the same. — At night we were all well pleased at our discovery of the little river; which, however, was no discovery as a Sealer had said some years ago that he had been up it. —[1]

[1] See watercolour by Conrad Martens labelled 'The Bivouack. Decr 29' (CM No. 87, *Beagle Record* p. 176), which was engraved by S. Bull as 'Bivouac at the head of Port Desire inlet' (*Narrative* **2**: facing p. 316).

30th   We got under weigh at four oclock & reached Guanaco Island by midday. — as the weather was cold & wet, I determined to walk to the ship. — It turned out to be a very long one, from the number of inlets & creeks: The geology well repaid me for my trouble, & I found likewise a small pool of quite fresh water. —

January 1st   Walked to a distant hill; we found at the top an Indian grave. The Indians always bury their dead on the highest hill, or on some headland projecting into the sea. — I imagine it is for this reason they come here; that they do pay occasional visits is evident, from the remains of several small fires & horses bones near them. —

2nd   A party of officers accompanied me to ransack the Indian grave in hopes of finding some antiquarian remains. — The grave consisted of a heap of large stones placed with some care, it was on the summit of the hill, & at the foot of a ledge of rock about 6 feet high. — In front of this & about 3 yards from it they had placed two immense fragments, each weighing at least two tuns, & resting on each other. — These in all probability were originally in nearly the same position|412| & only just moved by the Indians to answer their purpose. — At the bottom of the grave on the hard rock, there was a layer of earth about a foot deep; this must have been brought from the plain below; the vegetable fibres, from the lodgement of water, were converted into a sort of Peat. — Above this a pavement of flat stones, & then a large heap of rude stones,

piled up so as to fill up the interval between the ledge & the two large
stones. — To complete the grave, the Indians had contrived to detach
from the ledge an immense block (probably there was a crack) & throw
it over the pile so as to rest on the two other great fragments. We
undermined the grave on both sides under the last block; but there were
no bones. —[1] I can only account for it, by giving great antiquity to the
grave & supposing water & changes in climate had utterly decomposed
every fragment. — We found on the neighbouring heights 3 other &
much smaller heaps of stones. —they had all been displaced; perhaps
by sealers or other Voyagers. — It is said, that where an Indian dies, he
is buried; but that subsequently his bones are taken up & carried to such
situations as have been mentioned. — I think this custom can easily be
accounted for by recollecting, that before the importation of horses,
these Indians must have led nearly the same life as the Fuegians, &
therefore in the neighbourhead of the sea. — The common|413| pre-
judice of lying where your ancestors have lain, would make the now
roaming Indians bring the less perishable part of their dead to the
ancient burial grounds. —

[1] A sketch in the margin appears to illustrate the arrangement of the stones.

3rd   During these days I have had some very long & pleasant walks. —
The Geology is interesting. I have obtained some new birds & animals. —[1] I also measured barometrically the height of the plain which
must so lately have been beneath the sea; it has an altitude of 247 feet. —
Yesterday I shot a large Guanaco, which must, when alive, have
weighed more than 200 pounds. — Two males were fighting furiously
& galloping like race horses with their ears down & necks low; they did
not see me & passed within 30 yards; & then I settled the contest by
shooting the Persecutor. —

[1] One of the new birds obtained on this occasion was the ostrich said by CD to be known
by the Gauchos as the 'Avestruz Petise', though he seems slightly to have misunderstood
the name in talking later about the 'Petises', because although 'avestruz' certainly means
an ostrich in Spanish, the word 'petiso' was used merely as a diminutive. The bird was
later named *Rhea darwinii* by John Gould. CD wrote: 'When at Port Desire in Patagonia
(Lat. 48°), Mr Martens shot an ostrich; I looked at it, and from most unfortunately
forgetting at the moment, the whole subject of the Petises, thought it was a two-third
grown one of the common sort. The bird was skinned and cooked before my memory
returned. But the head, neck, legs, wings, many of the larger feathers, and a large part of
the skin, had been preserved. From these a very nearly perfect specimen has been put
together, and is now exhibited in the museum of the Zoological Society.' See *Zoology* 2,
Part III: 123–5, and *Beagle Record* pp. 175–9. CD and Gould read a paper to the Zoological
Society on this subject on 14 March 1837. See *Correspondence* 2: 11.

Entrance to Port St Julian, by J. W. Cook after C. Martens (*Narrative* **2**: 248,
where it is labelled incorrectly as 'Entrance to Berkeley Sound').

4th   The Adventure not being ready for sea, the Captain determined to
run down to Port St Julian[1] about 110 miles to the South & to survey
some of the intermediate coast. — We floated with a strong tide out of
harbor; it is called backing & filling from a particular manner of sailing
the vessel & is a ticklish operation. — Having passed the narrows, made
sail: in a few minutes we struck rather heavily on a rock; — the tide was
ebbing, but with good fortune she struck only twice more & then went
over. — The Beagle, in her last voyage, struck in|414| the night & as is
now supposed, on the same rock. — the summit is so small that the next
day it could not be found by any efforts. — On both occasions the Beagle
has received no essential damage; for the which all in her ought to be
grateful. — At night we anchored off the coast. —[2]

[1] Puerto San Julián.
[2] FitzRoy wrote: 'In working out of Port Desire, the Beagle struck her fore-foot heavily
against a rock, so as to shake her fore and aft; but on she went with the tide, and as she
made no water, I did not think it worth while returning into port. I was instantly
convinced that we had hit the very rock on which the Beagle struck in 1829, in the
night—a danger we never again could find by daylight till this day, when I was, rather
imprudently, going out with the last quarter-ebb. At low-water there are but eight feet on
this rock, which is not far from mid-channel, just without the entrance.' See *Narrative* **2**:
317–18.

5th–9th   During these days we surveyed the coast & at night either
anchored or stood out to sea. There are many rocks & breakers lying
some way from the land & a ship ought not to come near them. The
table land of Port Desire, is continued to St Julian, but in many places
interrupted by great vallies; & large patches have been entirely

removed, so that the outline resembles fortifications. The Beagle anchored off the mouth of the harbor & the Captain went in to sound the bar.[1] He landed me & I found some most interesting geological facts. —[2] At sunset we went on board, & the Captain took the ship into the harbor. —

[1] See watercolour by Conrad Martens labelled 'Entrance to Port St Julian. Jany 9 1834.', initialled RF (CM No. 92, *Beagle Record* p. 181). This was engraved by S. Bull under the erroneous title 'Entrance to Berkeley Sound', in *Narrative* 2: facing p. 248. It was near here that in 1520 Luis de Mendoza and Gaspar Quesada led an unsuccessful revolt against Ferdinand Magellan, Captain General of the Spanish fleet, and that their dismembered bodies were impaled on stakes on the shore. Before sailing on to the south in search of a way to the Pacific, Magellan erected a cross at the summit of Monte Cristo, 4 miles north-west of the anchorage, which may be seen in the background on the right hand side of Martens's picture.

Fifty-eight years later, Francis Drake beheaded another mutineer, Thomas Doughty, beside the Spanish gibbet, carving his name in Latin on a rock 'that it might be better understood by all that should come after us.'

[2] It was at Port St Julian that CD found the remains of one of the most interesting of his fossil mammals, later named by Richard Owen *Macrauchenia patachonica*, a huge lipoptern the size of a camel with some of the features of a llama. See *Zoology* 1, Part I: 35–56.

10th   Went up to the head of the Harbor. —the boat being aground on a mud-bank, we were all obliged to lounch for a half mile through mud & water & did not reach the vessel till late at night & very cold we all were. — In the dark we were puzzled by seeing another ship. —it turned out to be a French whaler, which in the morning came over the bar neck or nothing. The French Government gives a great bounty to all Whalers, I suppose to encourage a breed of good seamen; but from what we have seen of them, it|415| will be a difficult task. —all the officers are brought up in the English trade & it is curious to hear every word of command in their boats given in English.

11th   Again I started with the Captain to the head of the harbor. —it suddenly came on to blow hard. —so the Captain ran the boat on shore & we & four of the boats crew all armed proceeded on foot. — It turned out to [be] a very long walk; in the evening two of the party could not walk any further & we were all excessively tired. — It was caused by a most painful degree of thirst; & as we were only 11 hours without water, I am convinced it must be from the extreme dryness of the atmosphere. Earlier in the day we experienced a great mortification; a fine lake was seen from a hill; I & one of the men volunteered to walk there, & not till quite close did we discover that it was a field solid of snow-white salt. — The whole party left their arms with the two who were knocked up & returned to the boat. Fresh men were then sent off with some water, & we made a signal fire, so that by 11 oclock we were all collected & returned to the Ship. — [1]

[1] FitzRoy wrote: 'One day Mr Darwin and I undertook an excursion in search of fresh-water, to the head of the inlet, and towards a place marked in an old Spanish plan, "pozos de agua dulce;" but after a very fatiguing walk not a drop of water could be found. I lay down on the top of a hill, too tired and thirsty to move farther, seeing two lakes of water, as we thought, about two miles off, but unable to reach them. Mr Darwin, more accustomed than the men, or myself, to long excursions on shore, thought he could get to the lakes, and went to try. We watched him anxiously from the top of the hill, named in the plan "Thirsty Hill", saw him stoop down at the lake, but immediately leave it and go on to another, that also he quitted without delay, and we knew by his slow returning pace that the apparent lakes were "salinas". We then had no alternative but to return, if we could, so descending to meet him at one side of the height, we all turned eastward and trudged along heavily enough. The day had been so hot that our little stock of water was soon exhausted, and we were all more or less laden with instruments, ammunition, or weapons. About dusk I could move no further, having foolishly carried a heavy double-barrelled gun all day besides instruments, so, choosing a place which could be found again, I sent a party on and lay down to sleep; one man, the most tired next to me, staying with me. A glass of water would have made me quite fresh, but it was not to be had. After some hours, two of my boat's crew returned with water, and we very soon revived. Towards morning we all got on board, and no one suffered afterwards from the over-fatigue, except Mr Darwin, who had had no rest during the whole of that thirsty day—now a matter of amusement, but at the time a very serious affair.' See *Narrative* 2: 319-20.

12th & 13th   I was not much tired although I reached the boat in the first division; but the two next days was very feverish in bed. — |416|

14th   Went out walking, & found some fine fossil shells. — The country precisely resembles that of Port Desire. — it is a little more uneven, & from the absence even of brackish water, there are fewer animals. The Guanacoe who drinks salt water is of course to be seen. — Two things have been found here for which we cannot account: on a low point there is a large Spanish oven built of bricks, & on the top of a hill a small wooden cross was found. — [1] Of what old navigators these are the relics it is hard to say. — Magellan was here & executed some mutineers; as also did Drake & called the Island "true justice". —

[1] It might conceivably have been Magellan's cross, if the wood had survived for over 300 years.

15th   A heavy gale of wind from the SW; several breezes from that quarter have reminded us of the neighbourhead of Tierra del Fuego. —

16th–18th   Bad weather preventing the completion of the survey has detained us these days.

19th   Made sail very early in the morning, & with a fair breeze ran up to Port Desire; next day anchored off the mouth & with the young flood entered the harbor. —

Patagonians at Gregory Bay, by T. Landseer after C. Martens (*Narrative* **2**: 136).

20th I landed directly the ship came to an anchor, & had some collecting. — On an headland projecting into the sea, I found a heap of stones similar to the ones already described.There was a tooth & head of thigh bone, all crumbling into earth. —in a few years no traces would be left: This explains the apparent absence of bones in the grave, made with so much|417| labor, on the top of the hill.

The Adventure is ready for sea & with her new square top-sail will doubtless sail well.

22nd The Adventure & Beagle stood out to sea. — At sunset the Adventure steered for West Falkland Island & we came to an anchor under Watchman Cape. —

23rd After Latitude observations at noon we made sail for the Straits of Magellan. —

26th With a fair wind, we passed the white cliffs of Cape Virgins & entered those famous Straits.

29th Came to an anchor in St Gregory Bay; these days we have beaten against strong Westerly gales. — the tide here rises between 40 & 50 feet & runs at the rate of between 5 & 6 miles per hour. Who can wonder at the dread of the early navigators of these Straits? On shore there were the Toldos of a large tribe of Patagonian Indians. — Went on shore with the Captain & met with a very kind reception.[1] These Indians have such constant communication with the Sealers, that they are half civilized. — they talk a good deal of Spanish & some English. Their appearance is however rather wild. —they are all clothed in large mantles of the Guanaco, & their long hair streams about their faces. — They resemble in their countenance the Indians with Rosas, but are much more painted; many with their whole faces red, & brought to a point on the chin, others black. — One man was ringed & dotted with white like a Fuegian. — The average height appeared to be more than six feet; the horses who carried these|418| large men were small & ill fitted for their work. When we returned to the boat, a great number of Indians got in; it was a very tedious & difficult operation to clear the boat; The Captain promised to take three on board, & every one seemed determined to be one of them. — At last we reached the ship with our three guests. — At tea they behaved quite like gentlemen, used a knife & fork & helped themselves with a spoon. — Nothing was so much relished as Sugar. They felt the motion & were therefore landed. —

[1]See watercolour of Patagonians at Gregory Bay by Conrad Martens (CM No. 96, *Beagle Record* p. 197), engraved by T. Landseer in *Narrative* **2**: facing p. 136.

30ᵗʰ   A large party went on shore to barter for mantles &c. The whole population of the Toldos were arranged on a bank, having brought with them Guanaco skins, ostrich feathers &c &c. The first demand was for fire-arms & of course not giving them these, tobacco was the next; indeed knives, axes &c were of no esteem in comparison to tobacco. — It was an amusing scene & it was impossible not to like these mis-named giants, they were so throughily good-humoured & unsuspecting. — An old woman, well known by the name of Santa Maria, recognized Mʳ Rowlett as belonging formerly to the Adventure & as having seen him a year & a half ago at the R. Negro, to which place a part of this tribe had then gone to barter their goods. Our semi-civilized friends expressed great anxiety for the ship to return & one old man wanted to accompany us. — Got under weigh & beat up to Elizabeth island & there|419| came to an anchor. Some Patagonians near Peckets harbor made three large fires, as did also the Fuegians on the more distant Southern shore. — Which signs of their proximity we are sorry to see. —

31ˢᵗ   The Ship came to an anchor in Shoal Harbor; but it was found inconvenient; she then doubled Cape Negro & again anchored in Lando Bay. — The boats were lowered & a party went on shore. — no good water could be found.

February 1ˢᵗ   So in the morning got under weigh to run to Port Famine; The wind fell light; so the Captain sent the ship back to her anchorage & proceeded in a boat to the head of Shoal Harbor. During the last voyage the Captain discovered a large inland sea (Skyring water), 50 miles long; From the end of Shoal harbor we walked 5 miles across the country in hopes of being able to see it; the distance turned out to be greater than was expected & we were disappointed, if it had been nearer, the Captain had intended to have put a whale-boat on wheels & dragged it across, which would have saved much time in the survey of this Water. As soon as we came on board, the anchor was weighed & with a light air stood down for Port Famine.

The country, in this neighbourhead, may be called an intermixture of Patagonia & Tierra del Fuego; here we have many plants of the two countries; the nature of the climate being intermediate: a few miles to the|420| South the rounded Slate hills & forests of evergreen beeches commence. — The country is however throughily uninteresting. —

February 2ⁿᵈ   We got into Port Famine in the middle of the night, after a calm delightful day. M. Sarmiento a mountain 6800 feet high, was visible although 90 miles distant. —

3rd, *4th*, *5th* We are now within a wet circle, in consequence every
morning there has been torrents of rain; in the evening I managed to
have some walks along the beach; which is the only place where it is
possible to proceed in any way but scrambling.[1]

[1]See double-page panorama by Conrad Martens (CM No. 99, *Beagle Record* pp. 216–17),
labelled 'Port Famine Feby 4 1834', indicating above on left 'Lomas Range the highest
2963' and on right 'Mount Tarn 2700'. Beneath the picture Martens has written: 'Port
Famine. So called by Cavendish 1594, who discovered only 3 survivors of the many
hundreds of Colonists who embarked under Philip II of Spain to found a Colony
there—about 1580. Capt Stokes, Second in command of the British Expedition, committed suicide here 1826 in consequence of having contended for 4 months against storm &
currents in the Straits of Magellan endeavouring to make passage through.' For another
watercolour labelled 'Port Famine and Mount Tarn. Feby 5.' (CM No. 101) see *Beagle
Record* p. 185. The modern name of Port Famine is Puerto Hambre.
   Martens's notes require some slight correction. Port Famine was founded by Don
Pedro de Sarmiento in 1584 as the Ciudad del Rey Felipe, a fortress intended by Philip II
to block the Straits of Magellan against the marauding British sailors. But few of the
settlers that he landed there survived even the first winter, and in 1587, westward bound
in his circumnavigation of the globe, Thomas Cavendish found the Ciudad deserted.
Sarmiento himself was captured by the British, released as an act of grace by Queen
Elizabeth, and recaptured by the French. In commemoration of his gallant failure, his
name was given to the highest mountain in Tierra del Fuego. See also *Narrative* 1: 29–34.

6th   I left the ship at four oclock in the morning to ascend Mount Tarn; this
is the highest land in this neighbourhead being 2600 feet above the sea.
For the two first hours I never expected to reach the summit.— It is
necessary always to have recourse to the compass: it is barely possible
to see the sky & every other landmark which might serve as a guide is
totally shut out.— In the deep ravines the death-like scene of desolation
exceeds all description. It was blowing a gale of wind, but not a breath
stirred the leaves of the highest trees; everything was dripping with
water; even the very Fungi could not flourish.— In the bottom of the
valleys it is impossible to travel, they are barricaded & crossed in every
direction by great mouldering trunks: when using one of these as a
bridge, your course will often be arrested by sinking fairly up to the
knee in the rotten wood; in the same manner it is startling to rest against
a thick tree & find a mass of decayed matter|421| is ready to fall with the
slightest blow.— I at last found myself amongst the stunted trees &
soon reached the bare ridge which conducted me to the summit.—
Here was a true Tierra del Fuego view; irregular chains of hills, mottled
with patches of snow; deep yellowish-green valleys; & arms of the sea
running in all directions; the atmosphere was not however clear, &
indeed the strong wind was so piercingly cold, that it would prevent
much enjoyment under any circumstances.— I had the good luck to

find some shells in the rocks near the summit. — Our return was much easier as the weight of the body will force a passage through the underwood; & all the slips & falls are in the right direction. —

7th The day has been splendidly clear; Sarmiento, appearing like a solid mass of snow, came quite close to us.[1] If Tierra del could boast one such day a week, she would not be so throughily detested, as she is by all who know her. — I made the most of it & enjoyed a pleasant stroll with Mr Rowlett & Martens. — There is little fear of Indians. —we found however a wigwam which was not very old. —& the marks of a horse; There can be little inducement for the Patagonians to come here, as they cannot leave the beach; it is one of the few spots where the Fuegian & Patagonian can meet. — Many of the trees are of a large size. I saw several near the Sedger river, 13 feet in circumference & there is one 18·9 inches. — I saw a Winters bark 4'.6" in circumference. — |422|

[1]See watercolour by Conrad Martens labelled 'Mnt Sarmiento 6800. Lomas Range, the highest 2963', with the *Beagle* at anchor (CM No. 107, *Beagle Record* p. 112). Pencil note in margin runs: 'description of snow mountain'.

10th As soon as observations were obtained, we made sail in order to leave the Straits & survey the East coast of Tierra del Fuego.

11th The next day we were almost becalmed. — It is a most extraordinary contrast with the last season. — A sealing Schooner in the course of the day sent a boat on board; which brought lamentable news from the Falkland Islands. —the Gauchos had risen & murdered poor Brisbane & Dixon & the head Gaucho Simon, & it is feared several others. — Some English sailors managed to escape & are now in the West Island. — Since this the Challenger has been there & left the Governor with six (!) marines. — A Governor with no subjects except some desperate gauchos who are living in the middle of the island. — Of course they have taken all the half wild cattle & horses: in my opinion the Falkland islands are ruined. —this second desperate murder will give the place so bad a name that no Spanish Gauchos will come there, & without them to catch the wild cattle, the island is worth nothing. —

This Sealer has been this summer at anchor for six weeks under the Diego Ramiroz islands; & without a gale of wind!— The very time during which last year we had a gale of a month. — He was last year at these same islands. —during the gale of the 13th his deck was fairly swept, he lost all his boats &c &c. — At this time two of his men were on one of the Diego rocks, where they were left miserably to perish, as|423| he was obliged to run for the Falkland I^{ds}.

12th   With very baffling winds we anchored late in the evening in
Gregory Bay, where our friends the Indians anxiously seemed to desire
our presence. During the day we passed close to Elizabeth Island, on
North end of which there was a party of Fuegians with their canoe
&c. — They were tall men & clothed in mantles; & belong probably to
the East Coast; the same set of men we saw in Good Success Bay; they
clearly are different from the Fuegians, & ought to be called foot
Patagonians. — Jemmy Button had a great horror of these men, under
the name of "Ohens men". — "When the leaf is red, he used to say,
Ohens men come over the hill & fight very much." —

13th   Early in the morning we paid the Indians a visit in hopes of being
able to obtain some Guanaco meat. — They were as usual very civil:
there is now married & living amongst them a native of M: Video (by
birth I should think 2/3 of Northern Indian blood) who has been four
years with them. — He tells us that they will remain here all the winter
& then proceed up the Cordilleras; hunting for ostrich eggs; but that
Guanaco meat never fails them in these parts. — The Captain is thinking
of exploring the R. Santa Cruz, & this man gave us some good news, viz
that there are very few Indians in that part & that the river is so deep,
that horses can no where ford it. — In the R. Chupat, much further
North, there are very many Indians; enemies to this|424| tribe. — But
that all the Southern Indians 900 in number are friends. — At this
present time there were two boat Indians paying the Patagonians a visit
(the men whom I have called foot Patagonians); they do not speak the
same language; but one of this tribe has learnt their dialect. — These
Indians appear to have a facility in learning languages: most of them
speak a little Spanish & English, which will greatly contribute to their
civilization or demorilization: as these two steps seem to go hand in
hand. —

   At mid-day we passed out of the first Narrows, & began to survey the
coast. — There are many & dangerous banks, on one of which we ran a
very good chance of sticking; to escape it was necessary to get in three
Fathom water. —

14th–21st   During this week a complete survey has been made of the East
coast of Tierra del Fuego. We landed only once, which was at the mouth
of what was formerly supposed to be St Sebastians Channel, it now
turns out only to be a large wild bay. — The country here is part of
Patagonia, open & without trees; further to the South, we have the
same sort of transition of the two countries which is to be observed in

the Straits of Magellan. The scenery has in consequence a pretty, broken & park-like appearance. — In St Sebastian bay, there was a curious spectacle of very many Spermaceti Whales, some of which were jumping straight up out of the water; every part of the body was visible excepting the fin of the tail. As they fell sideways into the|425| water, the noise was as loud as a distant great gun. — By the middle of the day we were, after very fortunate weather, at anchor in Thetis Bay, between C St Vincent & Diego.

Upon going on shore, we found a party of Fuegians; or the foot Patagonians, fine tall men with Guanaco mantle. — The wigwam was also covered with the skin of the same animal. — It is a complete puzzle to every-one, how these men with nothing more than their slight arrows, manage to kill such strong wary animals. —

22nd As soon as the Ship doubled C. St Diego she got into a very great & dangerous tide rip. The Ship pitched very heavily; in a weak vessel it would almost have been sufficient to have jerked out her Masts. We soon got out of these uncomfortable straits; where a strong tide, great swell, & a bottom so uneven as to vary from 16 to 60 fathoms & then to 5, almost always cause a great bubbling sea. — In the evening, it fell a complete calm, & the long Southerly swell set us far too close to the West end of Staten land.

23rd What a great useless animal a ship is, without wind; here the swell was setting us right on shore & in the morning we found ourselves at the East end of the island about 30 miles further from our destination, than on the day before. — Staten land is one of the most desolate places; it is the mere backbone of a mountain|426| forming a ridge in the ocean. Its outline is peaked, castellated & most rugged. —

24th Came to an anchor in the evening under Woollaston I^{sd}.

25th I walked or rather crawled to the tops of some of the hills; the rock is not slate, & in consequence there are but few trees; the hills are very much broken & of fantastic shapes. —

Whilst going on shore, we pulled alongside a canoe with 6 Fuegians. I never saw more miserable creatures; stunted in their growth, their hideous faces bedaubed with white paint & quite naked. — One full aged woman absolutely so, the rain & spray were dripping from her body; their red skins filthy & greasy, their hair entangled, their voices discordant, their gesticulation violent & without any dignity. (*Note in margin*: Woman with child.) Viewing such men, one can hardly make oneself believe that they are fellow creatures placed in the same world.

I can scarcely imagine that there is any spectacle more interesting &
worthy of reflection, tha[n] one of these unbroken savages. — It is a
common subject of conjecture; what pleasure in life some of the less
gifted animals can enjoy? How much more reasonably it may be asked
with respect to these men. — To look at the Wigwam; any little depression in the soil is chosen, over this a few rotten trunks of trees are placed
& to windward some tufts of grass. Here 5 or 6 human|427| beings,
naked & uncovered from the wind, rain & snow in this tempestuous
climate sleep on the wet ground, coiled up like animals. — In the
morning they rise to pick shell fish at low water; & the women winter &
summer dive to collect sea eggs; such miserable food is eked out by
tasteless berrys & Fungi. — (*Note in margin:* Jerk out little fish out of the
Beds of Kelp.) They are surrounded by hostile tribes speaking different
dialects; & the cause of their warfare would appear to be the means of
subsistence. —[1] Their country is a broken mass of wild rocks, lofty hills
& useless forests, & these are viewed through mists & endless storms.
In search of food they move from spot to spot, & so steep is the coast,
this must be done in wretched canoes. — They cannot know[2] the feeling
of having a home—& still less that of domestic affection; without,
indeed, that of a master to an abject laborious slave can be called so. —
How little can the higher powers of the mind come into play: what is
there for imagination to paint, for reason to compare, for judgement to
decide upon. —to knock a limpet from the rock does not even require
cunning, that lowest power of the mind. Their skill, like the instinct of
animals is not improved by experience; the canoe, their most ingenious
work, poor as it may be, we know has remained the same for the
last|428| 300 years. Although essentially the same creature, how little
must the mind of one of these beings resemble that of an educated man.
What a scale of improvement is comprehended between the faculties of
a Fuegian savage & a Sir Isaac Newton— Whence have these people
come? Have they remained in the same state since the creation of the
world? What could have tempted a tribe of men leaving the fine regions
of the North to travel down the Cordilleras the backbone of America, to
invent & build canoes, & then to enter upon one of the most inhospitable countries in the world. — Such & many other reflections, must
occupy the mind of every one who views one of these poor Savages. —
At the same time, however, he may be aware that some of them are
erroneous. — There can be no reason for supposing the race of Fuegians
are decreasing, we may therefore be sure that he enjoys a sufficient
share of happiness (whatever its kind may be) to render life worth

having. Nature, by making habit omnipotent, has fitted the Fuegian to the climate & productions of his country. —

[1] It was not until 28 September 1838 (see *Autobiography* p. 120) that CD 'happened to read for amusement Malthus on *Population*, and being well prepared to appreciate the struggle for existence which everywhere goes on from long-continued observation of the habits of animals and plants, it at once struck me that under these circumstances favourable variations would tend to be preserved, and unfavourable ones to be destroyed. The result of this would be the formation of new species.' This sentence comes close to anticipating that reading.

   It may be noted that nowhere in the Diary does CD mention the practice of cannibalism among the Fuegians, a possibly false accusation for which he was later criticized in some quarters. The story that under conditions of severe famine the oldest women of the tribe were sometimes eaten was first told by FitzRoy in *Narrative* 2: 183, and repeated by CD only in the 1845 edition of *Journal of Researches* p. 214. According to E. L. Bridges (*Uttermost part of the earth*, Hodder & Stoughton, London, 1948, pp. 33–4) it was based on misleading interviews with the unreliable Jemmy Button and a boy questioned by the trader William Low.

[2] Above 'They cannot know', the words 'The habitable land cannot support' are written in pencil.

26th   In the night it blew very hard & another anchor was let go. — The leaden sky, the water white with foam, brings one back to reason after all the fine weather. — Dear Tierra del has recollected her old winning ways. — The ship is now starting & surging with her gentle breath. — Oh the charming country.|429|

27th   The weather was very bad: we left Wollaston Island & ran through Goree roads & anchored at the NE end of Navarin Island.

28th   This not being found a good place, the ship was moved to within the East end of the Beagle Channel & was moored by a beautiful little cove, with her stern not 100 yards from the mountains side. We passed this way last year in the boats. —

March 1st   All hands employed in getting in a stock of wood & water. There were three canoes full of Fuegians in this bay, who were very quiet & civil & more amusing than any Monkeys. —[1] Their constant employment was begging for everything they saw; by the eternal word—yammer-scooner.— They understood that guns could kill Guanaco & pointed out in which direction to go. — They had a fair idea of barter & honesty.— I gave one man a large nail (a very valuable present) & without making signs for any return, he picked out two fish & handed them up on the point of his spear.— If any present was designed for one canoe & it fell near another, invariably it was restored to the right owner. — When they yammer-scooner for any article very eagerly; they by a simple artifice point to their young women or little children; as much as to say, "if you will not give it me, surely you will to them". —

[1] FitzRoy wrote: 'The 1st of March passed in replenishing our wood and water at a cove, where we had an opportunity of making acquaintance with some Yapoo Tekeenica natives, who seemed not to have met white men before' (*Narrative* 2: 323). See watercolour by Conrad Martens of a Fuegian and his canoe (CM No. 131, *Beagle Record* p. 97), engraved by T. Landseer as 'Fuegian (Yapoo Tekeenica) at Portrait Cove' in *Narrative* 2: frontispiece. Other watercolours and drawings of the Fuegians and their canoes (CM Nos. 121, 124, 125, 126 and 130) are reproduced in *Beagle Record* p. 187 etc. See also note 2 for 22 January 1833, p. 135.

2nd The Captain determined to make the bold attempt of beating against the Westerly winds & proceeding up the Beagle channel to Ponsonby Sound or Jemmy Buttons country.— The day was beautiful, but a calm. —|430|

4th Came to an anchor in the Northern part of Ponsonby sound. We here enjoyed three very interesting days: the weather has been fine & the views magnificent. The mountains, which we passed today, on the Northern shore of the Channel are about 3000 feet high.—they terminate in very broken & sharp peaks; & many of them rise in one abrupt rise from the waters edge to the above elevation. The lower 14 or 1500 feet is covered with a dense forest.— A mountain, which the Captain has done me the honour to call by my name, has been determined by angular measurement to be the highest in Tierra del Fuego, above 7000 feet & therefore higher than M. Sarmiento.—[1] It presented a very grand, appearance; there is such splendour in one of these snow-clad mountains, when illuminated by the rosy light of the sun; & then the outline is so distinct, yet from the distance so light & aerial, that one such view merely varied by the passing clouds affords a feast to the mind.— Till near Ponsonby Sound we saw very few Fuegians; yesterday we met with very many; they were the men Jemmy Button was so much afraid of last year, & said they were enemies to his tribe; the intervening & thinly inhabited space of ground, I suppose, is neutral between the belligerents.— We had at one time 10 or 12 canoes alongside; a rapid barter was established Fish & Crabs being exchanged for bits of cloth & rags.— It was very amusing to see with what unfeigned satisfaction one young & handsome woman with her face painted black,|431| tied with rushes, several bits of gay rags round her head.— Her husband, who enjoyed the very unusual priviledge in this country of possessing two wives, evidently became jealous of all the attention paid to his young wife, & after a consultation with his two naked beauties, was paddled away by them.— As soon as a breeze

sprung up, the Fuegians were much puzzled by our tacking; they had
no idea that it was to go to windward & in consequence all their
attempts to meet the ship were quite fruitless. — It was quite worth
being becalmed, to have so good an opportunity of looking & laughing
at these curious creatures; I find it makes a great difference being in a
ship instead of a boat. — Last year I got to detest the very sound of their
voices; so much trouble did it generally bring to us. — (*Note in margin:*
Yammer-schooner last & first word.) But now we are the stronger
party, the more Fuegians the merrier & very merry work it is. — Both
parties laughing, wondering & gaping at each other: we pitying them
for giving us good fish for rags &c; they grasping at the chance of
finding people who would exchange such valuable articles for a good
supper. —

[1] But see note 1 for 29 January 1833, p. 140.

5th   In the morning, after anchoring in Ponsonby Sound[1] we stood down
to Wullia or Jemmy Buttons country. This being a populous part of the
country, we were followed by seven canoes. — When we arrived at the
old spot; we could see no signs of our friends, & we were the more
alarmed, as the Fuegians made signs of|432| fighting with their bows
and arrows. — Shortly afterwards a canoe was seen coming with a flag
hanging up: untill she was close alongside, we could not recognise poor
Jemmy. It was quite painful to behold him; thin, pale, & without a
remnant of clothes, excepting a bit of blanket round his waist: his hair,
hanging over his shoulders; & so ashamed of himself, he turned his
back to the ship as the canoe approached. When he left us he was very
fat, & so particular about his clothese, that he was always afraid of even
dirtying his shoes; scarcely ever without gloves & his hair neatly cut. —
I never saw so complete & grievous a change. — When however he was
clothed & the first flurry over, things wore a very good appearance. —
He had plenty (or as he expressed himself too much) to eat. — was not
cold; his friends were very good people; could talk a little of his own
language! & lastly we found out in the evening (by her arrival) that he
had got a young & very nice looking squaw. This he would not at first
own to: & we were rather surprised to find he had not the least wish to
return to England. Poor Jemmy with his usual good feeling brought two
beautiful otter skins for two of his old friends & some spear heads &
arrows of his own making for the Captain. — He had also built a
canoe. —& is clearly now well established. The various things now
given to him he will doubtless be able to keep. — The strangest|433|

thing is Jemmys difficulty in regaining his own language. — He seems to have taught all his friends some English. — When his wife came, an old man announced her, "as Jemmy Buttons wife"! — York Minster returned to his own country several month ago, & took farewell by an act of consummate villainy: He persuaded Jemmy & his mother to come to his country, when he robbed them of every thing & left them. — He appears to have treated Fuegia very ill. —

[1] A watercolour showing the *Beagle* at Ponsonby Sound in the Beagle Channel was sold by Conrad Martens to CD for 3 guineas in Sydney on 17 January 1836 (CM No. 150, *Beagle Record* p. 116). It was developed from a pencil drawing (CM No. 148) dated 5 March in which Hoste Island is labelled as seen on the left of the picture, and the north side of the Beagle Channel behind the ship in the centre. The same scene was engraved by T. Landseer as 'Murray Narrow — Beagle Channel' in *Narrative* 2: facing p. 326.

6th   Jemmy went to sleep on shore but came in the morning for breakfast. — The Captain had some long conversations with him & extracted much curious information: they had left the old wigwams & crossed the water in order to be out of the reach of the Ohens men who came over the mountains to steal.[1] They clearly are the tall men, the foot Patagonians of the East coast. — Jemmy staid on board till the ship got under weigh, which frightened his wife so that she did not cease crying till he was safe out of the ship with all his valuable presents. — Every soul on board was as sorry to shake hands with poor Jemmy for the last time, as we were glad to have seen him. — I hope & have little doubt he will be as happy as if he had never left his country; which is much more than I formerly thought. — He lighted a farewell signal fire as the ship stood out of Ponsonby Sound, on her course to East Falkland Island. — [2] |434|

[1] These Fuegians, spelt 'Oens-men' by FitzRoy, were known as Onas by the Yàhgans.
[2] In a footnote on pp. 228–9 of the 1845 edition of *Journal of Researches*, CD wrote: 'Captain Sulivan, who, since his voyage in the Beagle, has been employed on the survey of the Falkland Islands, heard from a sealer in (1842?), that when in the western part of the Strait of Magellan, he was astonished by a native woman coming on board, who could talk some English. Without doubt this was Fuegia Basket. She lived (I fear the term probably bears a double interpretation) some days on board.'
What is known about the subsequent history of the three Fuegians repatriated by FitzRoy is related by E. L. Bridges in *Uttermost part of the earth*, Hodder & Stoughton, London, 1948, pp. 41–8 and 83–4. In November 1859 the Patagonian Missionary Society's ship *Allen Gardiner* attempted to establish a mission at Wulaia, but on Sunday 6 November the missionaries and all the crew except for the ship's cook were massacred by the Yahgans, led apparently by the traitorous Jemmy Button and his brother. At a subsequent enquiry, Jemmy tried to lay the blame on the Onas, but was not believed. Thomas Bridges first met Jemmy in 1863, and noted that he had three sons; but the date of his death is not recorded. Ten years later a party of Yahgans from the outer coast of Tierra del Fuego visited the Mission at Ushuaia, and turned out to include Fuegia Basket. After her marriage to York Minster in 1833 at Wulaia, she had had two children by him, but he had

Port Louis and Berkeley Sound,
by J. W. Cook after C. Martens (*Narrative* 2: 248).

subsequently been killed in retaliation for the murder of a man; in 1873 she was accompanied by a second husband much younger than herself. In February 1883, Bridges saw her for the last time among her own people, the Alacaloofs, at London Island in the extreme west of Tierra del Fuego. She was then 62 years of age, in weak condition and nearing her end.

10th Arrived in the middle of the day at Berkeley Sound, having made a short passage by scudding before a gale of wind. — Mr Smith, who is acting as Governor, came on board, & has related such complicated scenes of cold-blooded murder, robbery, plunder, suffering, such infamous conduct in almost every person who has breathed this atmosphere, as would take two or three sheets to describe. — With poor Brisbane, four others were butchered; the principal murderer, Antuco, has given himself up. —he says he knows he shall be hanged but he wishes some of the Englishmen, who were implicated, to suffer with him; pure thirst for blood seems to have incited him to this latter act. — Surrounded as Mr Smith [is] with such a set of villains, he appears to be getting on with all his schemes admirably well. —(?)

11th–14th The ship was moved to near the Town. —[1] The Adventure arrived, after an exceedingly prosperous voyage. They killed so many wild bulls, geese &c &c & caught so many fish, that they have not tasted salt meat; this with fine weather is the beau ideal of a sailors cruize. I went on shore, intending to start on a riding excursion round the island, but the weather was so bad I deferred it.

[1]See watercolour by Conrad Martens based on a drawing labelled 'Port Louis, East Faulklands. March 14 1834' (CM No. 164, *Beagle Record* p. 117). The scene was also engraved by J. W. Cook as 'Settlement at Port Louis' in *Narrative* 2: facing p. 248.

16th   Early in the morning I set out with 6 horses & two Gauchos. These were the only two Spaniards who were not directly concerned with the murder; but I am afraid my|435| friends had a very good idea of what was going to take place. — However they had no temptation to murder me & turned out to be most excellent Gauchos, that is they were dexterous hands in all the requisites of making the camp-life comfortable. — The weather was very boisterous & cold, with heavy hail storms. We got on however pretty well; excepting some little geology nothing could be less interesting. — The country is uniformly the same, an undulating moorland; the surface covered with light brown withered grass, & some few very low shrubs all growing out of an elastic peaty soil. — There is one main range of quartz rock hills, whose broken barren crests gave us some trouble to cross. Few sorts of birds inhabit this miserable looking country: there are many small flocks of geese feeding in the valleys, & solitary snipes are common in all parts. — On the South side of the range of hills we came into the best country for the wild cattle; we did not however see very many, because the Murderers had by hunting them so much, driven them amongst the mountains. These men only killed the cows, & then took out the tongue & piece of meat from the breast, when this was finished they killed another. By their own account they must have killed more than 200 head. — We saw plenty of the half|436| decayed carcases. — In the evening we came across a nice little herd. St Jago soon separated a fat cow, he threw his balls, they hit her legs, but did not entangle her: he dropped his hat to mark the place where the balls fell, uncoiled his lazo & again we commenced the chace; at last he caught her round the horns. — The other Gaucho had gone on with the horses, so that St Jago had some difficulty in killing the furious beast. The horses generally soon learn for their own safety to keep the lazo tight when their rider dismounts, when this is the case the man can easily hamstring & thus secure the beast. Here the horse would not stand still, & it was admirable to see with what dexterity St Jago dogged about the cow till he contrived to give the fatal touch to the main tendon of the hind leg. After which, driving his knife into the head of the spinal marrow the animal dropped as if struck by lightning. — St Jago cut off enough flesh with the skin, & without any bones, to last for our expedition. We then rode on to our sleeping place. Meat roasted with its skin (carne con cuero) is known over all these parts of S. America for its excellence. — it bears the same relation to common beef, which venison does to mutton. — I am sure if any worthy alderman was once to taste it; carne|437| con cuero would soon be celebrated in London. —

17th   During the night it rained, & the next day was very stormy with
much hail & snow. From the number of cows which have been killed
there is a much larger proportion of bulls. — These wander about by two
& threes or by themselves & are very savage.— I never saw such
magnificent beasts; they truly resemble the ancient sculptures, in
which the vast neck & head is but seldom seen amongst tame animals.
The young bulls run away for a short distance, but the old ones will not
stir a step excepting to rush at man & horse;— & many horses have thus
been killed. — One old bull crossed a boggy stream & took up his stand
on the side opposite to us. We in vain tried to drive him away & failing
were obliged to make a large circuit. — The Gauchos in revenge were
determined to render him for the future innocuous; it was very interesting to see how art completely mastered huge force. One lazo was
thrown over his horns as he rushed at the horse, & another round his
hind legs;— in a minute the monster was stretched harmless on the
ground. —

During the whole time we only saw one troop of wild horses & this
was to the North of the hills—it is [a] curious thing that these horses
although very numerous always remain in the East end of the island. —
The Gauchos cannot account for it. —|438| We slept in a valley in the
neck of land which joins the rincon del toro, the great peninsula to the
SW point of the island.[1] The valley was pretty well sheltered from the
cold wind; but there was very little brushwood for making a fire; the
Gauchos soon found what to my surprise made nearly as hot a fire as
coals, it was the bones of a bullock, lately killed but all the flesh picked
off by the Vultures. They told me that in winter time they have often
killed an animal, cleaned the flesh from the bones with their knives, &
then with these very bones roasted the meat for their dinner. What
curious resources will necessity put men to discover! —

[1] Port Darwin, later named after CD, is close to the point where they crossed the isthmus
on this occasion.

18th   It rained during nearly the whole day; so that at night it began to be
very miserable work. We managed however with our Recado's to keep
pretty warm & dry; but the ground on which we slept was every night
more or less a bog & there was not a dry spot to sit on after our days
work. — The best wood in the island for burning is about the size of
large heath it has however the good property of burning when green. —
It was very surprising to see the Gauchos in the midst of rain, &
everything soaking wet, with nothing more than a tinder box & piece of
rag immediately make a fire. — They seek beneath the bushes for some
dry twigs or grass & this they rub into fibres & then (somewhat like a

birds nest) surround|439| it with coarser twigs; they put the rag with its spark of fire in the centre & then covering it up with the fibrous matter, hold it up to the wind. When by degrees it smokes more & more & at last burst out into flames. — I am sure no other method would have any chance of succeeding with such damp materials. —

19th The weather continued so bad I was determined to make a push & try to reach the Ship before dark, which I succeeded in doing. From the great quantity of rain this boggy country was in a very bad state. — I suppose my horse fell at least a dozen times & sometimes the whole six were floundering in the mud together: All the little streams have their sides soft, so that it is a great exertion for the horses to jump over them, & from the same cause they repeatedly fall. — To finish our misery, we crossed an arm of the sea, which was up to the top of the horses backs, & the little waves from the violent winds broke over us. — So that even the Gauchos were not sorry to reach the houses. —

20th–30th The Adventure sailed to continue her survey. — We are detained owing to some prisoners who are in irons on board: we are waiting till a Cutter returns which will be chartered to take them to Rio. — My time passes very evenly. — one day hammering the rocks; another pulling up the roots of the Kelp for the curious little Corallines which are attached to them.|440|

April 7th Finally weighed our anchor on our passage to the coast of Patagonia. Several causes have delayed us. — The Cutter has not returned & in consequence to our great sorrow we are obliged to keep the two prisoners & the "Kings evidence". — the weather has been very bad: & lastly a French Whaler came in with her rudder injured, a bad leak, & mutinous crew. The latter wished to desert the vessel & live on shore; in the present state of affairs this of course could not be allowed, & we were obliged to bully them, & finally to see the vessel on her way to Rio de Janeiro. — Having thus removed two of the worst prisoners, there are little fears for Mr Smith's safety. — Two Gauchos yet remain free, & they are to be trusted: with their assistance sufficient wild cattle can be caught for the subsistence of the Colony. — Perhaps this may be the first start in that prosperity which these islands must ultimately obtain. —

13th Dropped our anchor within the mouth of the river of Santa Cruz: our passage has been a fortunate one; only six days, & this against the constant Westerly breezes. — It blew very strong in the morning, & we could only just manage to fetch in. — I have never seen His Majesty's vessel under a greater press of sail or much closer to a lee-shore. —

Tomorrow a place will be sought out to lay her aground to look at her bottom. Her top-masts|441| & everything excepting main masts will be on deck & her guns, anchors &c on shore. —

14th & 15th Took two very long walks. The country is, as at Port Desire, an elevated but perfectly level plain: it is dry & sterile in the extreme. — Its natural productions, plants, birds & animals, are the same as in other parts of the Coast. The air is dry & sky clear; & this at least makes exercise very pleasant. —

16th The Ship was laid on shore;[1] it was found that several feet of her false keel were knocked off, but this is no essential damage; one tide was sufficient to repair her & after noon she floated off & was again moored in safety. Nothing could be more favourable than both the weather & place for this rather ticklish operation. —

[1] See engraving (reproduced on title page) by T. Landseer after Conrad Martens entitled 'Beagle laid ashore, River Santa Cruz' in Narrative 2: facing p. 336. This was the only occasion on which Martens made a drawing of the Beagle at close quarters.

18th In the morning three whale-boats started under the command of the Captain to explore as far as time would allow the Santa Cruz river: During the last voyage, Capt. Stokes procceeded 30 miles, but his provisions failing, he was obliged to return. — Excepting, what was then found, even the existence of this large river was hardly known: We carried three weeks provisions & our party consisted of 25 souls; we were all well armed & could defy a host of Indians. With a strong flood tide & a fine day we made a good run, soon drank some of the fresh water, & at night were nearly above the tidal influence. The river here assumed a size & appearance, which, even at|442| the highest point we ultimately reached, was scarcely diminished. It is generally from three to four hundred yards broard, & in the centre about seventeen feet deep; & perhaps its most remarkable feature is the constant rapidity of the current, which in its whole course runs at the rate of from four to six knots an hour. The water is of a fine blue color with a slight milky tinge, but is not so transparent as would be expected; it flows over a bed of pebbles, such as forms the beach & surrounding plain. —[1] The valley is in a very direct line to the westward, in which the river has [a] winding course, but it varies from five to 10 miles in width, being bounded by perfectly horizontal plains of 3 to 500 feet elevation. —

[1] Followed by the deleted words: 'excepting where there are cliffs of a sandy clay'.

19th In so strong a current it was of course quite impossible either to pull or sail so that the three boats were fastened astern of each other, two hands left in each, & the rest all on shore to track, (we brought with us collars all ready fitted to a whale line). —[1] As the general arrangements were very good for facilitating the work, I will describe them; the party which included *every one*, was divided into two spells, (at first into three) & each of these pulled alternately for an hour & a half. — The officers of each boat lived with, eat the same food, & slept in the same tent with their crew; so that each boat was quite independent of the others; After sunset, the first level place where there were any bushes was chosen for our nights lodging. The boats-|443| crew took it in turns to be cook; immediately the boat was hauled up, the cook made his fire, two others of the men pitched the tent, the coxswain handed the things out of the boat, & the rest, carried them up to the tents & collected fire wood. — By this means in half an hour, every thing was ready for the night. A watch of two men & an officer was always kept, whose duty it was to look after the boats, keep up the fires & look out for Indians; each in the party had his one hour every night. —

During this day we tracked but a short distance, for there are in this part many islands, which are covered with thorny bushes, & the channels between them are shallow, these two causes hindered us much.

[1] The words in brackets have been marked for deletion in pencil.

20th We passed the islands & set to work; our regular days work, although it was hard enough, carried us, on an average, only ten miles, in a straight line, & perhaps 15 or 20 as we were obliged to go. — A large smoke was seen at some distance, & a skeleton & other signs of horses; by which we knew that Indians were in the country. Beyond the place, where we slept was completely terra incognita, for there Capt Stokes turned back. — in the course of the day an old boat-hook was picked up (with the Kings mark). One of the boats crew, who had been up the river on the former voyage, remembered that it was then lost. So that the boat-hook after lying 6 or 7 years in Patagonia, returned to its|444| proper home, the Beagle. — Both this & the last night was a severe frost & some of the party felt the cold. —[1]

[1] FitzRoy wrote: 'Mr Darwin tried to catch fish with a casting net, but without success; so strong a stream being much against successful fishing. A very sharp frost again this night. The net and other things, which had occupied but little room in the boat, were frozen so hard as to become unmanageable and very difficult to stow.' See *Narrative* **2**: 343–4.

21st In the morning, tracks of a party of horses & the long spear or Chusa which trails on the ground, were found; they were so fresh that it was generally thought they must have reconnoitred us during the night. — Shortly afterwards we came to a place where there were fresh footsteps

Basalt Glen on the Rio Santa Cruz, by T. Landseer after C. Martens (*Narrative* **2**: 348).

of men, dogs, children & horses at the edge of the river & beneath the
water; on the other side of the river there were also recent tracks & the
remains of a fire: it is very clear that this is the place where the Indians
cross, it must be both a difficult & dangerous passage. The Spaniard
who lives with the Gregory Bay indians told me that they crossed in the
manner which the Gauchos call "a pilota"; that is the corners of a hide
are tied up & thus a sort of canoe is made which generally is pulled over
by catching hold of the horses tail. — After a mile or two beyond this
there were for many days no signs of men or horses. — We saw however
fresh smoke of the party whom we left behind, from which I think they
never saw us, but that we accidentally passed within a day or two's
march of each other. — The Spaniard told me he believed there were no,
or very few Indians at S. Cruz; perhaps they are the same small tribe
which occassionally frequent Port Desire, & whose lame horse was
seen up the river. — |445| A Guanaco was found dead under water, but
in a shallow place; the meat was quite fresh: upon skinning its head, a
bruise was found, we imagine that the Indians must have struck it with
their balls & that going to the water to drink, it died. — Whatever its end
might have been, after a few doubtful looks it was voted by the greater
number better than salt meat, & was soon cut up & in the evening eat. —

22nd The country remains the same, & terribly uninteresting. the great
similarity in productions is a very striking feature in all Patagonia, the
level plains of arid shingle support the same stunted & dwarf plants; in
the valleys the same thorn-bearing bushes grow, & everywhere we see
the same birds & insects. Ostriches are not uncommon, but wild in the
extreme. The Guanaco, however, is in his proper district, the country
swarms with them; there were many herds of 50 to 100, & I saw one,
with, I should think 500. — The Puma or Lion & the Condor follow &
prey upon these animals; The footsteps of the former might almost
every-where be seen on the banks of the river. The remains of several
Guanaco with their necks dislocated & bones broken & gnawed,
showed how they met their deaths. Even the very banks of the river &
of the clear little streamlets which enter it, are scarcely enlivened by a
brighter tint of green. The curse of sterility is on the land. — The very
waters, running over the bed|446| of pebbles, are stocked with no fish:
Hence there are no water-fowl, with the exception of some few geese &
ducks. —

23rd Rested till noon, to clean arms, mend clothes & shoes, the latters
already began to show symptoms of hard work. —

24th Like the old navigators approaching an unknown land, we examined & watched for the most trivial signs of a change; the drifted trunk of a tree, a boulder of primitive rock were hailed with joy, as if we had seen a forest growing on the stony ridges of the Andes. — But the most promising, & which eventually turned out true sign, was the tops of a heavy bank of clouds which constantly remained in nearly the same place. — These at first were taken for mountains themselves, instead of the clouds condensed by their icy summits. A Guanaco was shot, which much rejoiced those who could not compel their stomachs to relish Carrion. —

25th & 26th This day I found, for the first time, some interesting work; the plains are here capped by a field of Lava, which at some remote period when these plains formed the bottom of an ocean, was poured forth from the Andes.[1] This field of Lava is on a grand scale; further up the river it is more than 300 feet thick, & the distance from its source is great. — The most Southern Volcanic rocks in the Andes hitherto known are many hundred miles to the North, not far|447| from the island of Chiloe. — The Lava caused many small springs,[2] the valleys here were greener & I recognised many plants of Tierra del Fuego. — The Guanaco was in his element amongst the rugged low prœcipices. It is curious how in many cases the scenery is totally dependent on the geology; some of the valleys so precisely resembled those at St Jago, that if I could have added the warmth of a Tropical day I should have looked about me to recognize old-frequented spots. —[3]

[1] Note in margin: 'Action of current. Origin of Valley'.
[2] Note in margin: 'Cause of springs'.
[3] See pencil drawing by Conrad Martens labelled 'Valley with a small stream running into Santa Cruz River, the hills crowned with Volcanic Rock, the most southern yet discovered. April 26' (CM No. 173, *Beagle Record* p. 200), and watercolour developed from it (CM No. 174, *Beagle Record* p. 205). FitzRoy wrote of it: 'The glen above mentioned is a wild looking ravine, bounded by black lava cliffs. A stream of excellent water winds through it amongst the long grass, and a kind of jungle at the bottom. Lions or rather pumas shelter in it, as the recently torn remains of guanacoes showed us. Condors inhabit the basaltic cliffs. Near the river some imperfect columns of basalt give to a remarkable rocky height, the semblance of an old castle. Altogether it is a scene of wild loneliness quite fit to be the breeding place of lions.' See *Narrative* 2: 348. When the scene was engraved by T. Landseer as 'Basalt Glen—River Santa Cruz' in *Narrative* 2: facing p. 348, the lions were duly added.

27th The bed of the river is rather narrower hence the stream more rapid; it generally runs nearly 6 knots an hour. — in the channel there are great blocks of Lava—which together make the tracking both laborious & very dangerous. — Yesterday two holes were knocked through the sides of one of the boats, but she was got on shore & repaired, without any further damage. —[1] I shot a condor, it measured from tip to tip of wing 8 & ½ feet; — from beak to tail 4 feet. — They are magnificent birds; when seated on a pinnacle over some steep precipice, sultan-like they view the plains beneath them. I believe these birds are never found excepting where there are perpendicular cliffs: Further up the river where the lava is 8 & 900 feet above the bed of the river, I found a regular breeding place; it was a fine sight to see between|448| ten & twenty of these Condors start heavily from their resting spot & then wheel away in majestic circles. —[2]

[1] See engraving by T. Landseer after C. Martens entitled 'Repairing boat', in *Narrative* 2: facing p. 336.

[2] See watercolour by Conrad Martens depicting condors preying on a dead guanaco (CM No. 197, *Beagle Record* p. 208). In *Journal of Researches* pp. 219–24, CD proceeds to summarize his observations on the habits of the condor.

28th Found a tripod of wood, fastened together by hide; it had floated down the river; the first sign of the reappearance of man. —

29th From the high land, we hailed with joy the snowy summits of the *Cordilleras*, as they were seen occasionally peeping through their dusky envelope of Clouds. —

30th & May 1st We continued to get on but slowly. The Captains servant shot two Guanaco:[1] Before the men could arrive to carry them to the boats the Condors & some small carrion Vultures had picked even the bones of one clean & white, & this in about four hours. — The Guanaco probably weighed 170 or 180 pounds. — When the men arrived, only two Condors were there & some small Vultures within the ribs were picking the bones. —

[1] FitzRoy wrote: 'The order of our march was usually one or two riflemen in advance, as scouts — Mr Darwin, and occasionally Mr Stokes, or Mr Bynoe, upon the heights — a party walking along the banks, near the boats, ready to relieve or assist in tracking, and the eight or ten men who were dragging the three boats along at the rate of about two miles an hour over the ground, though full eight knots through the water.' See *Narrative* 2: 347, and watercolours by Conrad Martens (CM Nos. 185 and 190, *Beagle Record* pp. 212–13).

2nd & 3rd The river was here very tortuous, & in many parts there were great blocks of Slate & Granite, which in former periods of commotion have come from the Andes:[1] Both these causes sadly interfered with our progress. — We had however the satisfaction of seeing in full view the long North & South range of the Cordilleras. —[2] They form a lofty & imposing barrier to this flat country; many of the mountains were steep & pointed cones, & these were clothed with snow. — We looked at them with regret, for it was evident we had not time to reach them; We were obliged to imagine|449| their nature & grandeur, instead of standing as we had hoped, on one of their pinnacles & looking down on the plains below. During these two days we saw signs of horses & several little articles belonging to the Indians, such as a bunch of Ostrich feathers, part of a mantle, a pointed stick. From a thong of cows hide being found; it is certain that these Indians must come from the North. — They probably have no connection with those whose smoke we saw nearer to the Coast; but that during the Summer they travel along the foot Andes, in order to hunt in fresh country. — The Guanaco being so excessively abundant I was at first much surprised that Indians did not constantly reside on the banks of this river; the cause of their not frequenting these plains must be their stony nature (the whole country is a shingle bed) which no unshod horse could withstand. — Yet in two places, in this very central part, I found small piles of stones which I think could not have been accidentally grouped together. — They were placed on projecting points, over the highest lava cliffs; & resembled those at Port Desire, but were on a smaller scale; They would not have been sufficient to have covered more than the bones of a man. —

[1] Pencil note in margin: 'Character of upper plain altered'.
[2] See watercolour by Conrad Martens sold to CD as 'River Santa Cruz' for 3 guineas in Sydney on 21 January 1836 (CM No. 193, *Beagle Record* p. 201), which shows the line of men hauling the boats, and the Cordillera of the Andes in the distance.

4th The Captain determined to take the boats no further; the mountain were between 20 & 30 miles distant & the river very serpentine. — Its apparent dimensions &|450| depth nearly the same; its current equally strong. — The country & its productions remained equally uninteresting. — In addition to all this our provisions were running short; we had been for some days on half allowance of biscuit. — This same half allowance, although really sufficient, was very unpleasant after our hard work; & those who have not tried it will alone exclaim about the comfort of a light stomach & an easy digestion. It was very ridiculous how invariably the conversation in the evening turned upon all sorts, qualities & kinds of food. —

The Captain & a large party set off to walk a few miles to the Westward. — We crossed a desert plain which forms the head of the valley of S. Cruz, but could not see the base of the mountains. — On the North side, there is a great break in the elevated lava plain, as if of the valley of a river. — It is thought probable that the main branch of the S Cruz bends up in that direction & perhaps drains many miles of the Eastern slope of the chain. — We took a farewell look at the Cordilleras which probably in this part had never been viewed by other Europæan eyes, & then returned to the tents. — At the furthest point we were about 140 miles from the Atlantic, & 60 from the nearest inlet of the Pacific.[1]

[1] At their furthest west, the party were probably within a few miles of Lago Argentino, which connects with Lago Viedma and Lago San Martin, and which was first described by J. H. Gardiner in 1867. FitzRoy read a paper about the expedition to the Royal Geographical Society on 8 May 1837 (*Journal of the Royal Geographical Society of London* 7: 114–26).

5th   Before sun-rise, we began our descent.|451| We shot down the stream with great rapidity; generally at the rate of 10 miles an hour; what a contrast to the laborious tracking. — We effected in this day. what had cost us five days & a half; from passing over so much country, we as it were condensed all the birds & animals together & they appeared much more numerous. —

6th   We again equalled five & half days tracking: the climate is certainly very different near to the mountains; it is there much colder, more windy & cloudy. —

7th   Slept at the place where the water nearly ceases to be fresh. — A tent & party was left to try to shoot some Guanaco. —

Almost every one is discontented with this expedition; much hard work, & much time lost & scarcely any thing seen or gained. — We have however to thank our good fortune, in enjoying constant fine dry weather & blue skys. To me the cruize has been most satisfactory, from affording so excellent a section of the great modern formation of Patagonia. —

9th–11th   I took some long walks; collecting for the last time on the sterile plains of this Eastern side of S. America. — The Sportsman have altogether been very lucky; Ten guanaco have been killed & eaten; several Condors & a large wild Cat have been killed, & M[r] Stuart shot a very large Puma. — |452|

12th   We put to sea; & steered in search of an alleged rock (the L'aigle) between the Falklands & mouth of Straits of Magellan; after an unsuccessful hunt, we anchored on the 16th off C. Virgins. —

16th   The weather has been bad, cold, & boisterous (& I proportionally sick & miserable). — It never ceases to be in my eyes most marvellous that on the coast of Patagonia there is constant dry weather & a clear sky, & at 120 miles to the South, there should be as constant clouds rain, hail, snow & wind. —

21ˢᵗ During these days we have been beating about the entrance of the Straits, obtaining soundings & searching for some banks (a dangerous one was found); at night we came to an anchor;

22ⁿᵈ & before daylight the Adventure was seen on her passage from the Falklands. Shortly after we left Berkeley Sound, a man of war came in; she has taken away all the prisoners, & now the island is quite quiet. — We received our letters; mine were dated October & November. —[1] We shall now in a few days make the best of our way to Port Famine; the days are of course very short for surveying; the weather however, gracias a dios, is pretty fine for these Southern latitudes. — It is a very curious fact; that it now being only one month from the shortest day & in such a latitude, that the temperature is scarcely perceptibly colder, than during the summer; we|453| all wear the same clothes as during last years visit. —

[1] These were from his sisters and several friends (see *Correspondence* 1: 336–42, 345–8, 350–1 and 353–8).

29ᵗʰ We anchored in Gregory Bay & took in six days water; our old friends the Indians were not there. — The weather has lately been very bad, & is now very cold. — The Thermometer has been all day below the freezing point & much snow has fallen: This is rather miserable work in a ship, where you have no roaring fire; & where the upper deck, covered with thawing snow is as it were, the hall in your house. —

June 1ˢᵗ Arrived at Port Famine. I never saw a more cheer-less prospect; the dusky woods, pie-bald with snow, were only indistinctly to be seen through an atmosphere composed of two thirds rain & one of fog; the rest, as an Irishman would say, was very cold unpleasant air. —
Yesterday, when passing to the S. of C. Negro, two men hailed us & ran after the ship; a boat was lowered & picked them up. — They turned out to be two seamen who had run away from a Sealer & had joined the Patagonians. They had been treated by these Indians with their usual disinterested noble hospitality. — They parted company from them by accident & were walking down the coast to this place to look out for some vessel. I dare say they were worthless vagabonds, but I never saw more miserable ones; they had for some days been living on Muscles &c & berrys & had been exposed night & day to all the late constant rain & snow. — What will not man endure! —|454|

2ⁿᵈ–8ᵗʰ The Adventure rejoined us, after having examined the East side of this part of the Straits. —

The weather has during the greater part of the time been very foggy & cold; but we were in high luck in having two clear days for observations. On one of these the view of Sarmiento was most imposing: I have not ceased to wonder, in the scenery of Tierra del Fuego, at the apparent little elevation of mountains really very high. — I believe it is owing to a cause which one would be last to suspect, it is the sea washing their base & the whole mountain being in view. I recollect in Ponsonby Sound, after having seen a mountain down the Beagle Channel, I had another view of it across many ridges, one behind the other. — This immediately made one aware of its distance, & with its distance it was curious how its apparent height rose.

The Fuegians twice came & plagued us. — As there were many instruments, clothese &c & men on shore, the Captain thought it necessary to frighten them away. — The one time, we fired a great gun, when they were a long way off; it was very amusing to see through a glass their bold defiance, for as the shot splashed up the water, they picked up stones in return & threw them towards the ship which was then about a mile & a half off. — This not being sufficient, a boat was sent with orders to fire musket balls wide of them. — The Fuegians hid themselves behind the trees, but for every discharge of a musket they fired an arrow.|455| These fell short of the boat; & the officer pointing to them & laughing made the Fuegians frantic with rage (as they well might be at so unprovoked an attack); they shook their very mantles with passion. — At last seeing the balls strike & cut the trees, they ran away; their final decampement was effected by the boat pretending to go in chase of their canoes & women. Another party having entered the bay was easily driven to a little creek to the north of it: the next day two boats were sent to drive them still further;[1] it was admirable to see the determination with which four or five men came forward to defend themselves against three times that number. — As soon as they saw the boats they advanced a 100 yards towards us, prepared a barricade of rotten trees & busily picked up piles of stones for their slings. — Every time a musket was pointed towards them, they in return pointed an arrow. — I feel sure they would not have moved till more than one had been wounded. This being the case we retreated.

We filled up our wood & water; the latter is here excellent. The water we have lately been drinking contained so much salt that brackish is almost too mild a term to call it. — Amongst trifling discomforts there is none so bad as water with salts in it: when you drink a glass of water, like Physic, & then it|456| does not satisfy the thirst. Mere impure,

Mount Sarmiento as seen from Warp Bay, by T. Landseer after C. Martens (*Narrative* **2**: 359).

stinking water is of little consequence: especially as boiling it & making tea generally renders it scarcely perceptible. —

[1] Note in margin: 'Rockets, noise, dead silence'.

8th   We weighed very early in the morning: The Captain intended to leave the St.[s] of Magellan by the Magdalen channel, which has only lately been discovered & very seldom travelled by Ships. The wind was fair, but the atmosphere very thick, so that we missed much very curious scenery. The dark ragged clouds were rapidly driven over the mountains, nearly to their base; the glimpses which we had caught through the dusky mass were highly interesting; jagged points, cones of snow, blue glaciers, strong outlines marked on a lurid sky, were seen at different distances & heights. — In the midst of such scenery, we anchored at C. Turn, close to Mount Sarmiento, which was then hidden in the clouds. At the base of the lofty & almost perpendicular sides of our little cove there was one deserted wigwam, and it alone reminded us that man sometimes wandered amongst these desolate regions; imagination could scarcely paint a scene where he seemed to have less claims or less authority; the inanimate works of nature here alone reign with overpowering force. —

9th   We were delighted in the morning by seeing the veil of mist gradually rise from & display Sarmiento. — I cannot describe the pleasure of viewing these enormous, still, &|457| hence sublime masses of snow which never melt & seem doomed to last as long as this world holds together.[1] —

The field of snow extended from the very summit to within $\frac{1}{8}$[th] of the total height, to the base, this part was dusky wood. — Every outline of snow was most admirably clear & defined; or rather I suppose the truth is, that from the abscence of shadow, no outlines, but those against the sky, are perceptible & hence such stand out so strongly marked. — Several glaciers descended in a winding course from the pile of snow to the sea, they may be likened to great frozen Niagaras, & perhaps these cataracts of ice are as fully beautiful as the moving ones of water. — By night we reached the Western parts of the Channel; in vain we tried to find anchoring ground, these islands are so truly only the summits of steep submarine mountains. — We had in consequence to stand off & on during a long, pitch-dark night of 14 hours, & this in a narrow channel. — Once we got very near the rocks; The night was sufficiently anxious to the Captain & officers. —

[1] See watercolour by Conrad Martens (CM No. 208, *Beagle Record* p. 113) based on a

drawing (CM No. 207) initialled RF and labelled as follows: 'The grand glacier, Mount Sarmiento. The mountain rises to about 3 times the height here seen, but all is here hidden by dark misty clouds—a faint sunny gleam lights the upper part of the glacier, giving its snowy surface a tinge which appears almost of a rose colour by being contrasted with the blue of its icy crags—a faint rainbow was likewise visible to the right of the glacier, but the whole was otherwise very grey & gloomy. June 9 1834'. See *Beagle Record* p. 398, and also CM Nos. 210 and 213, *Beagle Record* pp. 220–1.

10ʰ   In the morning, in company with the Adventure, we made the best of our way into the open ocean. —[1] The Western coast generally consists of low, rounded, quite barren|458| hills of Granite. Sir J. Narborough called one part of it South Desolation. — "because it is so desolate a land to behold", well indeed might he say so. — Outside the main islands, there are numberless rocks & breakers on which the long swell of the open Pacific incessantly rages. — We passed out between the "East & West Furies"; a little further to the North, the Captain from the number of breakers called the sea the "Milky way". — The sight of such a coast is enough to make a landsman dream for a week about death, peril, & shipwreck.

[1]See watercolour by Conrad Martens based on drawing labelled 'Lowe Cockburn Channel—Still morning—the vessel working out of Warp Cove from whence the sketch was taken. The ship's track was from behind this nearest point, which is the termination of Magdalen Channel. June 10 1834' (CM No. 210, *Beagle Record* p. 221).

28ᵗʰ   Early in the night we came to an anchor in the port of S. Carlos[1] in the island of Chiloe. It had been the Captains original intention to have gone direct to Coquimbo. —but a constant succession of Northerly gales compelled him first to think of Concepcion & ultimately to come in here. — Never has the Beagle had such ill luck; night after night, furious gales from the North put us under our close-reefed main top-sail, fore try-sail & stay-sail; when the wind ceased, the great sea prevented us making any way. — Such weather utterly destroys for every good end the precious time during which it lasts. — On leaving Tierra del Fuego, we congratulated ourselves too soon, in having escaped the usual course of its storms. —

On the 27ᵗʰ the purser of the Beagle, Mʳ Rowlett expired;|459| he had been for some time gradually sinking under a complication of diseases; the fatal termination of which were only a little hastened by the bad weather of the Southern countries. Mʳ Rowlett was in his 38ᵗʰ year; the oldest officer on board; he had been on the former voyage in the Adventure; & was in consequence an old friend to many in this ship; by whom & everyone else he was warmly respected. — On the following day the funeral service was read on the quarter-deck, & his body

lowered into the sea; it is an aweful & solemn sound, that splash of the
waters over the body of an old ship-mate. —

[1] Now Ancud.

29th In the morning, many of the poor people who had houses on the
point, rowed off to us in their little boats; it was quite pleasing to see the
unaffected joy with which they welcomed the Ship & those who were
formerly in her. They told us that the money they gained from cutting
wood &c &c had enabled them to buy sheep & that they had ever since
been much better off. They all appear to have a great mixture of Indian
blood & widely differ from almost every other set of Spaniards in not
being Gauchos. The country is so thickly wooded that neither horses or
cattle seem to increase much. Potatoes & pigs & fish are the main
articles of food; the obtaining these|460| requires labor, & has con-
sequently induced a different set of manners from what is found in
other parts of S. America. — In the middle of the day I took a short walk,
following up one of the winding creeks: Seen from a considerable
distance the country bears a very close resemblance to T. del Fuego; the
country is hilly & entirely clothed in thick wood, excepting a few
scattered green patches which have been cleared near to the Cottages. — The woods are incomparably more beautiful than those of T.
del Fuego, instead of the dusky uniformity of that country we have the
variety of Tropical scenery; excepting in Brazil I have never seen such
an abundance of elegant forms. — Chiloe, situated on the West coast,
enjoys a very uniform temperature, & an atmosphere saturated with
moisture; the soil resulting from Volcanic ashes appears very fertile;
hence arises the teeming luxuriance of the forests. — The high thatched
roofs of the cottages with the little railed paddocks of grass surrounded
by lofty evergreens, reminded me of some drawings of the houses in
the S. Sea Islands. — This resemblance to Tropical scenery is chiefly to
be attributed to a sort of arborescent grass or Bamboo, which twines
amongst the trees to the height of 30 or 40 feet & renders the woods
quite impervious. — to this may be added some large ferns, the trees
also are all evergreens, & the stems are variously coloured white, & red
&c. — This walk called to my mind all the delights of the sublime
scenery of Brazil. — |461|

30th–July 8th I staid in the town of S. Carlos three days, during the greater
part of this time the weather was very fine; the inhabitants themselves
wondering at such an event. — I do not suppose any part of the world
is so rainy as the Island of Chiloe. — The Cordilleras are very rarely in

San Carlos de Chiloe, by S. Bull after C. Martens (*Narrative* 1: 275).

sight; one morning before sun-rise we had a very fine view of the Volcano of Osorno; it stood out in dark relief; it was curious to see as the sun rose, the outline gradually lost in the glare of the Eastern sky. —[1] During the fine weather I enjoyed some very pleasant walks about the town & examined the structure of the rocks. — This island like the plains of Patagonia is only an appendage to the Andes; it is formed of the debris of its rocks & of streams of Lava. — These submarine beds have been elevated into dry land only in a very recent period. — The soil resulting from the decomposition of these rocks is very fertile; but agriculture is as yet in its rudest forms; to this the structure of the mills & boats & their method of spinning quite correspond. — The inhabitants, judging from their complexions & low stature, have ¾[s] of Fuegian or Boat Indian blood in their veins; they are all dressed in coarse strong woollen garments, which each family makes for themselves & dyes with Indigo of a dark blue color. —

Although with plenty to eat, they are excessively|462| poor; there is little demand for labor, & from the scarcity of money nearly all payments are made with goods. — Men carry on their backs from long distances, bags of charcoal, (the only fuel used in the town) to obtain the most trifling luxuries: the joy which the sight of a few Reals gave to these poor men was quite surprising; after making them a present, they always insisted on having your hand to shake it as a sign of their gratitude. — One day I walked a few miles on the road to Castro. This place was the former Capital & is now the second town in the island. The road is the only one which goes directly through the interior of the country. About two miles from S. Carlos it enters the forest, which covers the whole country & has only been rendered passable by the aid of the axe.[2] For its whole length there are not more than two or three houses; the road itself was made in the time of the old Spaniards & is entirely formed of trunks of trees squared & placed side by side. From the gloomy damp nature of the climate, the wood had a dreary aspect; in the Tropics such a scene is delightful from the contrast it affords with the brilliancy & glare of every open spot. The country generally is only inhabited round the shores of the creeks & Bays, & in this respect it resembles T. del Fuego;|463| the road by the coast is in some places so bad that many houses have scarcely any communication with others excepting by boats. —

The capital itself is worthy of the island, it is a small straggling dirty village; the houses are singular from being entirely built sides, roofs, partitions &c of plank.[3] The Alerce or cedar from which these planks are

made grows on the sides of the Andes; they possess the curious
property of splitting so evenly that by planing the planks are nearly as
well-formed as if sawed. — These planks are the staple export of the
Islands, to which may be added potatoes & hams. —

[1] See pencil drawing by Conrad Martens labelled 'Volcano of Osorno, from Chiloe. Bright
morning sky — sun just rising' (CM No. 226, *Beagle Record* p. 251).
[2] See pencil drawings by Conrad Martens of a forest scene in Chiloe (CM Nos. 234 and 235,
Beagle Record pp. 245 and 249).
[3] See watercolours by Conrad Martens showing wooden houses in San Carlos (CM Nos.
228 and 230, *Beagle Record* pp. 224–5).

13th   Got under weigh, we only managed to reach an outer harbor when
the wind failing, obliged us to anchor for the night: on the following day
we with difficulty got an offing by beating against the swell of anything
but "Pacific" Ocean. — We were all glad to leave Chiloe; at this time of
year nothing but an amphibious animal could tolerate the climate. Even
the inhabitants have not a word to say in its favor; very commonly I was
asked what I thought of the Island; ¿no es muy mala? is it not a
miserable place? I could not muster civility enough to contradict
them. — In summer, when we return, I dare say Chiloe will wear a|464|
more cheerful look. I hear of swarms of insects at that season; this
plainly tells me there must be a wide difference between this country &
Tierra del Fuego (although at present appearing nearly the same). In
the latter place in the midst of summer, the air can boast of few
inhabitants; the insect world requires a more genial climate. Besides the
Climate, it is disagreeable to see so much poverty & discontent. Poverty
is a rare sight in S. America; even here it is not the poverty of Europe;
there is an abundance of plain food, coarse clothes, & fire-wood; but the
poverty lies in the difficulty of gaining sufficient to buy even the
smallest luxuries. The greater part of the inhabitants are strongly
inclined to the old Spanish cause (it is well known with what difficulty
they were conquered by the Patriot forces), this feeling is kept up by
their having reaped no advantages by the revolution. The grand
advantage in other parts is the cheapness of Europæan articles of
luxury, of these the inhabitants of Chiloe can afford to enjoy but very
few. — Many of the old men whom I talked with, had good cause to
regret former times; they had been veterans in the Spanish armies, &
with the fall of the Spanish flag of course they have lost the half-pay to
which during their whole lives they had been looking forward to. —
Seventeen of the inhabitants were|465| executed, when the first Gover-
nor arrived from the Patriots, for having faithfully served their kings:

these things must rankle long in the minds of men who live the uniform
& retired life such as the inhabitants of Chiloe. —

22nd We were becalmed off Valparaiso; we made but an indifferent
passage; we enjoyed however the very unusual novelty to us of seeing
several vessels & speaking two of them; it is always interesting to see
ships, like great animals of the sea, come up & reconnoitre each other. —

23rd–31st Late in the night the Beagle & Adventure came to an anchor. —
When morning came everything appeared delightful; after Chiloe & T.
del Fuego we felt the climate quite delicious; the sky so clear & blue, the
air so dry & the sun so bright, that all nature seemed sparkling with
life. — The view from the Anchorage is very pretty; the town is built on
the very foot of [a] range of hills, which are 1600 feet high, & tolerably
steep; the surface is worn into numberless little ravines, which exposes
a singularly bright red soil between patches of light green grass & low
shrubs. — It is perhaps for this reason & the low white-washed houses
with tile roofs, that the view reminded *me* of Teneriffe & others of
Madeira. — The harbor is not large & the shipping is crowded together.[1]
In a NE direction there are some fine glimpses of the Andes. — These
however appear much grander when viewed from the neigh-
bouring|466| hills; we then better perceive how far distant they are
situated. The Volcano of Aconcagua is especially beautiful. — The
Cordilleras, however, viewed from this point owe the greater part of
their charms to the atmosphere through which they are seen; when the
sun sets in the Pacific it is admirable to watch how clearly the rugged
outline of their peaks can be seen, yet how varied & how delicate is the
tint of their colours. — When in T. del Fuego, I began to think the
superiority of Welsh mountain scenery only existed in my imagination.
Now that I have again seen in the Andes a grand edition of such
beauties, I feel sure of their existence. —

I have taken several long walks, but I have not ceased to be surprised
to find one day after another as fine as the foregoing. — what a difference
does climate make in the enjoyment of life. — How opposite are the
sensations, when viewing black mountains half enveloped in clouds, &
seeing another range through the light blue haze of a fine day: the one
for a time may be very sublime, the other is all gayety & happy life. —

The town of Valparaiso is from its local situation a long straggling
place; wherever a little valley comes down to the beach the houses are
piled up on each other, otherwise it consists of one street running
parallel to the coast. We all, on board, have been much struck by the

great superiority in the English residents over other towns in S. America.|467| Already I have met with several people who have read works on geology & other branches of science, & actually take interest in subjects no way connected with bales of goods & pounds shillings & pence. — It was as surprising as pleasant to be asked, what I thought of Lyells Geology. — Moreover every one seems inclined to be very friendly to us, & all hands expect to spend the two ensuing months very pleasantly. —

[1] See drawing by Conrad Martens of shipping and buildings at the quayside in Valparaiso (CM No. 256, *Beagle Record* p. 223).

August 2nd Took up my residence with Mr Corfield, who has taken the most obliging pains to render me all assistance in my pursuits. — His house is situated in the Almendral, which is an extensive suburb built on [a] small sand-plain, which very recently has been a sea-beach. — The house is a very pleasant one; one story high, with all the rooms opening into a quadrangle, there is a small garden attached to it, which receives a small stream of water 6 hours in the week. — Another gentleman lives with Mr Corfield; the expences of the house, table, wine, 2 men servants, 3 or 4 horses, is about 400 pounds sterling per-annum. — I should think this same establishment in England would at least cost double this sum. —

5th I have taken several long walks in the country. The vegetation here has a peculiar aspect; this is owing to the number & variety of bushes which seem to supply the place of plants; many of them bear very pretty flowers & very commonly the whole shrub has a strong resinous|468| or aromatic smell. In climbing amongst the hills ones hands & even clothes become strongly scented. — With this sort of vegetation I am surprised to find that insects are far from common; indeed this scarcity holds good to some of the higher orders of animals; there are very few quadrupeds, & birds are not very plentiful. I have already found beds of recent shells, yet retaining their colors at an elevation of 1300 feet; & beneath this level the country is strewed with them. It seems not a very improbable conjecture that the want of animals may be owing to none having been created since this country was raised from the sea. — [1]

[1] In both the 1839 and 1845 editions of the *Journal of Researches* the evidence for the relatively recent elevation of the coast is similarly described, but this final sentence is omitted.

14th I managed to set out on a geological excursion to the base of the Andes. Our first day's ride was along the Northern shore; we passed over a pleasant undulating country & after dark arrived at the Hacienda of Quintero; the estate which formerly belonged to Lord Cochrane. My object in coming here was to see the great beds of recent shells which are dug out of the ground to make lime.

15th On the next day I returned to wards the valley of Quillota. The country was exceedingly pleasant, just what I fancy Poets mean by Pastoral, green open lawns separated by small valleys & rivulets; the cottages of the sheepherds being scattered on the hill sides. At the base of the Sierra de Chilicauquen, which we were obliged to pass,|469| there were many fine evergreen forest trees, which however only flourish in the ravines where there is running water. A person who had only seen the country near Valparaiso would never dream there were such picturesque spots in Chili.

As soon as we reached the brow of the Sierra, the valley of Quillota was immediately under our feet. The prospect was one of remarkable artificial luxuriance. The valley is very broard & quite flat, & is thus easily irrigated in all parts; the little square gardens are crowded with orange & olive trees & every sort of vegetable. On each side huge bare mountains arise & this contrast renders the patch-work valley the more pleasing. — Whoever called Valparaiso the "valley of Paradise" must have been thinking of Quillota. — We crossed over to the Hacienda de San Isidro, situated at the very foot of the Bell mountain. —

Chili as may be seen in the maps is a narrow strip between the Cordilleras & the Pacific; & this strip is itself traversed by several lines of high hills parallel to the great range. — At the foot of the Andes there is a succession of level basins, generally connected together & extending chiefly to the South; in these the principal towns are situated, S. Felipe, St Jago, S. Fernando &c. — These basins or plains, together with the flat valleys which connect them with the coast, are the bottoms of ancient inlets & great bays such as the present|470| intersect every part of Tierra del Fuego & the West coast of Patagonia &c. Chili, at one time, must have in the configuration of its land & water exactly resembled these latter countries. This resemblance was occasionally seen with great force when a fog bank extended over the whole of the lower parts; the white vapor curling into all the ravines, beautifully represented the little coves & bays. Here & there a solitary hillock peeping up through the mist showed that it formerly had stood as an islet. — The scenery from the above causes must be I should think nearly unique; anyhow to me it was quite new & very interesting. From the natural slope to seaward of these plains they are, as I have said, very easily irrigated &

singularly fertile; without this process, the ground will produce scarcely anything, as the sky during the whole summer is cloudless. — The mountains & hills are dotted over with bushes & low trees, with the exception of this the vegetation is very scanty. Nevertheless many half wild cattle find sufficient pasture. The owners of lands in the plains possess each so much hill country where their cattle feed, & once a year there is a grand "Rodeo" when the cattle are all driven down, marked & counted & a certain number separated for fattening in the artificial fields in the valleys. —

Wheat is extensively cultivated & a good deal of Indian corn; a sort of bean is however the main article of food for the Common Labourers. — The orchards produce an enormous abundance of Peaches, Figs & Grapes. With all these advantages|471| the inhabitants of the country ought to be much more prosperous than they are.

16th   The Major Domo of the Hacienda was good enough to give me a guide & fresh horses; in the morning we set out to ascend the Campana or Bell, a mountain which is 6400 feet high.[1] The paths were very bad, but both the geology & scenery amply rapaid the trouble. We reached, by the evening, a spring called the Agua del Guanaco, which is situated at a great height. This must be an old name, for it is very many years since a guanaco has drunk its waters. During the ascent I noticed that on the Northern slope nothing but bushes grew, whilst on the Southern a sort of bamboo about fifteen feet high. In a few places there were palms, & I was surprised to see one at an elevation of at least 4500 feet. This palm is for its family an ugly tree: its stem is very large & of a curious form, being thicker in the middle than at the base or top. They are excessively numerous in some parts of Chili & valuable on account of a sort of honey made from the sap. — On one estate near Petorca they counted many hundred thousand trees; each year in August or September many are cut down, when lying on the ground (& it is necessary, I am told, that the trees should fall up the hill) the crown of leaves is cut off, & the sap begins to flow this continues for many months, but it is necessary every morning that a thin slice should be cut off, so as to expose a fresh surface. A good tree will give 90 gallons of Sap, which must all have been contained in the|472| apparently dry trunk; it is said to flow much more quickly on those days when the sun is powerful. — The sap is concentrated by boiling, & is called honey, which in its taste it resembles. — We unsaddled our horses near the spring & prepared to pass the night. — The setting of the sun was glorious, the valleys being black whilst the snowy peaks of the Andes yet retained a ruby tint. —

When it was dark, we made a fire beneath a little arbor of bamboos, fried our charqui (or dried strips of beef), took our matte & were quite comfortable. There is an inexpressible charm in thus living in the open air. — The evening was so calm & still; the shrill noise of the mountain bizcacha & the faint cry of the goatsucker were only occasionally to be heard. Besides these, few birds or even insects frequent these dry parched up mountains. —

[1] A plaque on the summit now commemorates this ascent.

17th   We climbed up to the highest ridge of the rough mass of greenstone. The rock as is so generally the case was much shattered & broken into angular fragments. I observed, however, here one remarkable difference, that the surfaces of many enormous fragments presented every degree of freshness, from what appeared quite fresh, to the state when Lichens can adhere. I felt so forcibly that this was owing to the constant earthquakes that I was inclined to hurry from beneath every pile of the loose masses. —

We spent the whole day on the summit, & I never enjoyed one more throughily. Chili & its boundaries the Andes & the Pacifick were seen as in a Map. The pleasure from the scenery, |473| in itself beautiful was heightened by the many reflections which arose from the mere view of the grand range, its lesser parallel ones and of the broard valley of Quillota which directly cuts these in two. Who can avoid admiring the wonderful force which has upheaved these mountains, & even more so the countless ages which it must have required to have broken through, removed & levelled whole masses of them?

The appearance of the Andes was different from what I expected; the lower line of the snow was of course horizontal, & to this line the even summits of the range appeared quite parallel. At long intervals, a mass of points or a single cone showed where a Volcano had or does now exist. — It hence looked more like a wall, than a range of separate mountains, & made a most complete barrier to the country.[1]

Almost every part of this mountain has been drilled by attempts to open Gold mines. I was surprised to see on the actual summit, a small pit where some yellow crystals had induced some people thus to throw away their labor; & this on a point which can only be reached by climbing. The rage for mining has left scarcely a spot in Chili unexamined, even to the regions of eternal snow. —

I spent the evening, as before, talking round our fire with my two companions. — The Guassos of Chili, which correspond to the|474| Gauchos of the Pampas, are however a very different set of beings. Chili

is the more civilized of the two countries; & the inhabitants in consequence have lost much individual character. Gradations in rank are
much more strongly marked; the Huasso does not by any means
consider every man his equal; I was quite surprised to find my companions did not like to eat at the same time with myself. This is a necessary
consequence of the existence of an aristocracy of wealth; it is said that
some few of the greater land owners possess from five to ten thousand
pounds sterling per annum. — This is an inequality of riches which I
believe is not met with in any of the cattle-breeding countries to the
eastward of the Andes. — A traveller by no means here meets that
unbounded hospitality which refuses all payment, but yet is so kindly
offered, that no scruples can be raised in accepting it. Almost every
house in Chili will receive you for the night, but then a trifle is expected
to be given in the morning: even a rich man will accept of two or three
shillings. — The Gaucho, although he may be a cut-throat, is a gentleman; the Huasso is in few respects better, but at the same time is a
vulgar, ordinary fellow. — The two men although employed much in
the same manner are different in their habits & clothes; and the
peculiarities of each are universal in their respective countries. The
Gaucho seems part of his horse & scorns to exert himself excepting
when on its back.|475| The Huasso can be hired to work as a labourer in
the fields. — The former lives entirely on animal food, the latter nearly
as much on vegetable. — We do not here see the white boots, the broard
drawers & scarlet Chilipa, the picturesque costume of the Pampas; here
common trowsers are protected by black & green worsted leggings: —
the poncho however is common to both. — The chief pride of the
Huasso lies in his spurs, these are absurdly large; — I measured one that
was six inches in the *diameter* of the rowel, & the rowel itself contained
upwards of thirty points: the stirrups are on the same scale, each one
consisting of a square carved block of wood, hollowed out, yet weighing
three or four pounds. — The huasso is perhaps more expert with the
lazo than the gaucho, but from the nature of the country, does not know
the use of the bolas. —

[1] A long pencil note in the margin here is not readily decipherable.

18ᵗʰ   Descended the mountain by rather a better track & passed some
beautiful little spots, with rivulets & fine trees; & arrived early at the
same Hacienda. —

19ᵗʰ & 20ᵗʰ   Passed the town of Quillota, which is more like a collection of
nursery gardens than a town, & followed up the valley. — The orchards

were beautiful, presenting one mass of peach-blossoms. I saw in one
or two places the Date-Palm; it is a most stately tree. I should think a
group of them in their native deserts must be superb. — We also
passed|476| S. Felipe, a large pretty straggling town like Quillota. —
The valley has here expanded into one of the basins or plains already
mentioned as so curious a part of the scenery of Chili. — We crossed
it & proceeded to the mines of Jajuel, situated in a ravine in the very
Andes.

21st These copper mines are superintended by a shrewd but ignorant
Cornish miner; he has married a Spanish woman & does not mean to
return, yet his admiration for Cornwall was unbounded; he never
ceased to descant upon the wonders of the mines. Amongst other
questions, he asked me, now that George Rex was dead, how many of
the family of Rex's were yet alive. This Rex certainly is a relation of Finis
who wrote all the books. —

The copper ore is shipped to Swansea to be smelted, hence the mines
have a singularly quiet aspect to those in England, here there is no
smoke or furnaces or great steam-engines to disturb the quiet of the
surrounding mountains. — The government encourages the searching
for mines by every method, the discoverer may work a mine in any
ground by paying 5 shillings, & before paying this he may try for 20
days, although it might be in the very garden of another man. It is now
well known that the Chilian method is the cheapest of working the
mines; my host here says that only two great improvements have been
introduced by the foreigners; the|477| one is reducing, by previous
roasting, the white copper ores, (*Note in margin:* Copper Pyrites.) which
being some of the best in Cornwall, the miners were here astonished to
find thrown away; the other is stamping the scoriæ which comes from
the furnaces, by which process small particles of copper are recovered
in abundance. Improvements likewise have been introduced into some
of the simple machinery; but even to this day, in some mines the water
is removed by men carrying it up the shaft in skins on their backs! — The
labouring men work very hard; they have little time allowed for their
meals, & during summer & winter they begin when it is light & leave off
at dark. — They are paid one pound sterling a month & their food given
them: consists for breakfast of sixteen Figs & two small loaves of bread;
for dinner boiled beans, for supper broken roasted wheat grains. They
scarcely ever taste meat; as with the twelve pound per annum they have
to clothe themselves & support their families. — The miners who work

in the mine itself have twenty-five shillings per month, & are allowed a little Charqui. — But these men only come down from their bleak habitation once every fortnight or three weeks.

I staid here 5 days & throughily enjoyed scrambling in all parts of these huge mountains; the geology was, as might be expected, very interesting; the shattered & baked rocks traversed by dykes of formerly melted greenstone showed what commotion has taken place during|478| the formation of these mountains. — The appearance of the mountains is the same as has been described; dry, barren mountains dotted over with bushes. — The Cacti were very numerous; I measured one of a depressed globular figure; including the dense spines, it was 6 feet 4 inches in circumference, height 1^{ft}..9 ins. The height of the ordinary cylindrical branching kind is from 12 to 15 feet, circumference of a limb (with spines) 3^{ft}: 7 inches. —

Two days before I left there was a heavy fall of snow in the mountains, which prevented me from taking some interesting excursions. — I attempted to reach a lake, which the inhabitants for some unaccountable reason believe to be an arm of the sea. — During a very dry season, it was proposed to attempt cutting a channel for the sake of the water, but the Padre after consultation declared it was too dangerous as all Chili would be inundated, if as generally supposed, the lake really was connected with the Pacific. — We ascended to a great height, but becoming involved in the snow drifts, failed in reaching this wonderful lake & had some difficulty in returning. I thought we should have lost the horses; for there was no means of guessing how deep the drifts were & the animals when led could only move by jumping. — The black sky showed that a fresh snow storm was gathering, & we therefore were not a little glad when we escaped. — By the time we reached the base, the storm commenced; & it was lucky for us that this did not happen three hours earlier in the day.

26th We left Jajuel & again crossed the basin of S. Felipe. |479|The day was truly Chilian, glaringly bright & the atmosphere quite clear. The thick & uniform covering of newly fallen snow rendered the view of the Volcano of Aconcagua & the main chain quite glorious. — We were now on the road to St Jago, the capital of Chili. — We crossed the Cerro del Talguen, & slept at a little Rancho. The host, talking about the state of Chili as compared to other countries, was very humble; "Some see with two eyes & some with one, but for his part he did not think that Chili saw with any". —

27th After crossing many low hills we descended into the small land-
locked plain of Guitròn. In these basins which are elevated from 1000 to
2000 feet above the sea, two species of Acacia, which are stunted in their
forms, and stand wide apart from each other, grow in great numbers. I
do not know the cause, but they never seem to live near the sea, & this
gives another characteristic mark to the scenery of these basins. —[1] We
crossed a low ridge which separates Guitron from the plain on which St
Jago stands: the view was here preeminently striking, the dead level
surface, covered in parts by woods of Acacia, & with the city in the
distance, abutted horizontally against the base of the Andes, their
snowy peaks bright with the evening sun. — This was one of those
views, where immediate inspection convinced me that a plain now
represents the extent of a former inland sea. There was equally little
doubt, how much more beautiful a foreground a plain makes, where
|480| distances can be measured, than an expanse of water. — We
pushed our horses into a gallop & reached the city before it was dark. —

[1] A crossed-out note in the margin runs: 'they grow in the Casa Blanca basin 8–11 ft only
above the sea'.

28th I staid a week in St Jago & enjoyed myself very much: in the
mornings I rode to various places in the plain, & in the evenings dined
with different merchants. — A never failing source of delight was to
mount the little pap of rock (Fort of St Lucia) which stands in the middle
of the city; the scenery certainly is very striking, & as I have said, very
peculiar. — I am informed that this same character is common to some
of the Mexican cities. — Of the town itself there is nothing to be said;
generally it is not so fine or so large as B. Ayres, but built on same model.

September 5th I had arrived here by a circuit to the North, & I determined
to return to Valparaiso by a longer circuit to the South. By the middle of
the day we crossed one of the famous suspension bridges of Hide. —
They are miserable affairs & much out of order. —the road is not level as
at the Menai, but follows the curvature of the suspending ropes. —the
road part is made of bundles of sticks & full of holes; the bridge
oscillates rather fearfully with the weight of a man leading a horse. — In
the evening we reached a very nice Hacienda; where there were several
very pretty Signoritas; they turned up their charming eyes in pious
horror at my having entered a Church to look about me; they asked|481|
me, why I did not become a Christian, "for our religion is certain"; I
assured them I was a sort of Christian; they would not hear of it,
appealing to my own words, "Do not your padres, your very bishops,
marry"— The absurdity of a Bishop having a wife particularly struck

them, they scarcely knew whether to be most amused or horrified at such an atrocity. —

September 6th   Rode on to Rancagua, never leaving the level plain. —the country here is divided by mud walls & hedges, like England & of course well irrigated.

7th   Left the great road to the South, turned up the valley of the R. Cachapol to the hot-baths of Cauquenes, long celebrated for medicinal properties. We were obliged to cross the above river; it is very disagreeable crossing these torrents; the bed is composed of very large stones, they are shallow & broard, but foaming with the rapidity with which they run. When in the middle it is almost difficult to tell whether your horse is moving or standing still; the water rushes by so quick that it quite confuses the head. — In summer these torrents are of course quite impassable, the scene of violence which their beds show at this time of year may give one some idea of their strength & fury. Generally the Suspension bridges which are necessary for the Summer, are taken down during the winter & this was the case in the present instance. —

The buildings attached to these Hot Springs consist of a square of hovels, each with a|482| table & stool. The situation is in a narrow deep valley not far from the Andes, there are only one or two houses higher up. — It is a solitary quiet spot with a good deal of beauty. — I staid here five days, being detained a prisoner during the last two by heavy rain; & this has been the last rain which has fallen this summer in Chili.

I rode one day to the last house in the valley; shortly above, the Cachapual divides into 2 deep tremendous ravines which penetrate right into the great range. I scrambled up one very high peaked mountain, the height of which could not be much less than 6000 feet; (*Note in margin:* Probably more.) here, as indeed everywhere else scenes presented themselves of the highest interest. — It was by one of these ravines (valle del Yeso) that Pinchero entered & ravaged Chili. — This is the same man whose attack on an Estancia at R. Negro I have described. — He was a Renegade Spaniard, who collected a great body of Indians together, & established himself by a stream in the Pampas, which none of the forces sent after him could ever find. — (*Note in margin:* A prisoner of his told me all this.) From this point he sallied forth, & crossing the Cordilleras by unknown passes, ravaged Chili & drove the flocks of cattle to his own recret rendezvous. — This man was a capital horseman, & he made all round him equally good, for he invariably shot any person who even hesitated to follow him. It was against this man & other wandering tribes of true Indians, that Rosas waged the war of exterminations.— I have since heard|483| from B. Ayres, that this was not so completely effected as it was supposed. The Indians had decamped 8 or 10 hundred miles & were hovering in great numbers about the borders of Cordova.

During my stay at this place, I had observed that there were very few Condors to be seen; yet one morning there were at least twenty wheeling at a great height over a particular spot: I asked a Guasso what was the cause, he said that probably a Lion had killed a cow or that one was dying; if the Condors alight & then suddenly all fly up; the cry is then "a Lion" & all hurry to the chace.— Capt Head states that a Gaucho exclaimed "a Lion" upon merely seeing one wheeling in the air.— I do not see how this is possible. The Lion after killing an animal & eating of it, covers the carcase up with large bushes & lies down at a few yards distance to watch it. If the Condors alight, he springs out & drives them away, & by this means commonly discovers himself. There is a reward of Colts & Cows. — I am assured that if a Lion has once been hunted, he never again watches the carcase, but eating his fill, wanders far away. They describe the Lion in these hunts as very crafty; he will run in a straight line & then suddenly return close to his former track & thus allow the dogs to pass by & completely puzzle them. The Guasso's possess|484| a particular breed of small dogs, which by instinct (like pointers set) know how to spring at the Lions throat & will very commonly kill him single-handed. The man at the baths had one. I never saw a more miserable creature to attempt fighting with so large an animal as the Puma. — From the uneven nature of the country nearly all these animals must be killed with dogs. — It is rather singular that the Lions on this side [of the] Cordilleras, appear to be much more dangerous than on the other. At Jajuel I heard of a man being killed & here of a woman & child; now this never happens in the Pampas. There being no deer or ostriches in Chili obliges them to kill a far greater number of Cattle; by this means perhaps they learn to attack a man. — It would also appear that the Lion is here more noisy, roaring when hungry & when breeding. —

13th We escaped from our foodless prison, & rejoining the main road slept at the village of Rio Claro.

14th From this place we rode on to the town of S. Fernando.— Before arriving there, the inland basin expands into a great plain, which to the South is so extensive that the snowy summits of the distant Andes were seen as over the horizon of the sea.— S. Fernando was my furthest point to the South, it is 40 leagues from St Jago. From this point I turned at right angles to seaward. — We slept at the gold mines of Yaquil near Rancagua, in the possession of Mr Nixon, an American gentleman. — |485| I staid at this place four days, during two of which I was unwell. — Where Mr Nixon lives the Trapiche or grinding mill is erected; the mine itself is at the distance of some leagues & nearly at the summit of [a] high hill. On the road we passed through some large woods of the Roble or Chilian oak; this tree from its ruggedness & shape of leaf & manner of growth deserves its name. (*Note in margin:* The Roble of Chili is different from the Roble of Chiloe.) This is its furthest limit to the North. I was glad to see anything which so strongly reminded me of England. — To the South there was a fine view of the Andes including the Descabezado described by Molina. — To the North I saw part of the lake of Taguatagua, with its floating islands: these islands (described by M. Gay)[1] are composed of various dead plants; with living vegetation on the surface, they float about 4 feet above the surface: as the wind blows they pass over the lake, carrying with them cattle & horses. —

When we arrived at the mine, I was struck by the pale appearance of many of the men, & enquired from Mr Nixon respecting their state. The mine is altogether 450 feet deep, each man brings up on his back a quintal or 104 lbs[2] weight of stone. With this load they have to climb up the alternate notches cut in a trunks of trees placed obliquely in the shaft. Even beardless young men of 18 & 20 years with little muscular development of their bodies (they are quite naked excepting drawers) carry this great load from nearly the same depth. — A strong man, who is not accustomed to|486| this sort of exercise perspires most profusely with merely carrying their own bodies up. — With this very severe labor they are allowed only beans & bread; they would prefer living entirely upon the latter; but with this they cannot work so hard, so that their masters, treating them like horses, make them eat the beans. — Their pay is 25 or 30^s per month. — They only leave the mine once in three weeks, when they remain with their families two days. — This treatment, bad as it sounds, is gladly accepted; the state of the labouring Agriculturist is much worse, many of them eat nothing but beans & have still less money. — This must be chiefly owing to the miserable feudal-like system by which the land is tilled. The land-owner gives so much land to a man, which he may cultivate & build on, & in return has his services (or a proxy) for every day for his life gratis. Till a father has a grown up son to pay his rent by his labor, of course there is no one to take care of the patch of ground. Hence poverty is very common with all the labouring classes. —

One of the rules of this mine sounds very harsh, but answers pretty well. —the method of stealing gold is to secrete pieces of the metal & take them out as occasion may offer. Whenever the Major-domo finds a lump of ore thus hidden, its full value is stopped out of the wages of all the men, who thus are obliged to keep watch on each other. — The ore is sent down to the mills on mules.|487| I was curious to enquire about the load which each mule carries: on a *level road* the regular cargo weighs 416 pounds.[3] In a troop there is a muleteer to every six mules. — Yet to carry this enormous weight, what delicate slim limbs they have; the bulk of muscle seems to bear no proportion to its power. The mule always strikes me as a most surprising animal: that a Hybrid should possess far more reason, memory, obstinacy, powers of digestion & muscular endurance, than either of its parents. —One fancys art has here out-mastered Nature. —

When the ore is brought to the Mill it is ground into an impalpable powder; the process of washing takes away the lighter particles & amalgamation at last secures all the gold dust. The washing when described sounds a very simple process: but it is at the same time beautiful to see how the exact adaptation of the current of water to the Specific Gravity of the gold so easily separates it from its matrix. It is curious how the minute particles of gold become scattered about, & not corroding, at last accumulate even in the least likely spots. Some men asked permission to sweep the ground round the house & mill; they washed the earth & obtained 30 dollars worth of gold. —

In M[r] Nixons house a German collector Renous was staying. I was amused by a conversation which ensued between Renous |488|(who is taken for a Chilian) & an old Spanish lawyer. Renous asked him what he thought of the King of England sending out me to their country to collect Lizards & beetles & to break rocks. The old Gentleman thought for some time & said, "it is not well, hay un gato encerrado aqui" "there is a cat shut up here"; no man is so rich as to send persons to pick up such rubbish; I do not like it; if one of us was to go & do such things in England, the King would very soon send us out of the country". And this old gentleman, from his profession is of course one of the more intelligent classes!—Renous himself, two or three years ago, left some Caterpillars in a house in S. Fernando under charge of a girl to turn into Butterflies. This was talked about in the town, at last the Padres & the Governor consulted together & agreed it must be some Heresy, & accordingly Renous when he returned was arrested.

[1]See Claude Gay, Aperçu sur les recherches d'histoire naturelle faites dans l'Amérique

du Sud, et principalement dans le Chili, pendant les années 1830 et 1831. *Annales des Sciences Naturelles* 28: 369–93, 1833.

[2] Pencil note added in margin runs: 'I believe this is a mistake, the weight is greater', and the figure is corrected to 'about 200 pounds weight of stone' in *Journal of Researches* p. 323, with a footnote 'In another mine, as will hereafter be mentioned, I picked out a load by hazard, and weighed it: it was 197 pounds.'

[3] A note in the margin runs: '"Mr Miers" states that in the mining (i.e. mountainous) district each mule carries 312£b. —' It is followed by a calculation dividing 14 into 416 by long division to yield 29.

19th   We took leave of Yaquil & followed the flat valley, formed like Quillota, in which the R. Tinderidica flows. — The climate even this little way South of St Jago is much damper: in consequence there were fine tracks of pasture ground which were not irrigated.

20th   We followed this vally till it expanded into a great plain which reaches from the sea to the mountains West of Rancagua. We shortly lost all trees & even|489| bushes; the inhabitants are nearly as badly off for fire-wood as in the Pampas. Never having heard of these plains, I was quite astonished to meet with such a country in Chili. — These plains are traversed by numerous great valleys, & there is more than one set of plains, all of which plainly bespeaks the residence & retreat of the ocean. In the steep sides of these valleys, there are some large caves; one of which is celebrated as having been consecrated: Formerly the Indians must have buried their dead in it, as various remains have been found. — I felt during the day very unwell, & from this time to the end of October did not recover.

21st   Rode but a short distance & obliged to rest. —

22nd   Continued crossing green plains without a tree, which almost resembled the Pampas, till we arrived at the village of Navedad, South of the mouth of the R. Rapel. — We passed during the day immense flocks of sheep, which appear to thrive better than the cattle. — We found a rich Haciendero, who received us in his house close to the sea. —

23rd   I staid here the whole ensuing day, & although very unwell managed to collect many marine remains from beds of the tertiary formation of which these plains consist. —[1]

[1] The presence here of beds of early Tertiary rocks 800 feet thick containing fossil shells that could only have lived in shallow water provided useful proof for CD's theory that at such places there must have been a slow subsidence of the sea-bottom in a previous era.

24th   Our course now lay directly to Valparaiso, still passing over the same plains. — At night I was exceedingly exhausted; but had the uncommon luck of obtaining some clean straw for my bed.|490| I was amused afterwards by reflecting how truly comparative all comfort is. If I had been in England & very unwell, clean straw & stinking horse cloths would have been thought a very miserable bed. —

25th Necessity made me push on & I contrived to reach Casa Blanca. —it was wretched work. —to be ill in a bed is almost a pleasure compared to it. —

26th I sent to Valparaiso for a carriage & so reached the next day Mr Corfields house.

27th Here I remained in bed till the end of October.[1] It was a grievous loss of time, as I had hoped to have collected many animals. — Capt FitzRoy very kindly delayed the sailing of the Ship till the 10th of November, by which time I was quite well again. — During my absence, some great changes took place in the affairs of the expedition.[2] The Adventure was sold; in consequence Mr Wickham has returned as 1st Lieutenant: Every one feels the want of room occassioned by this change; it is indeed in every point of view a great but unavoidable evil. — Only one good has resulted, that necessarily the perfecting of the former survey in Tierra del Fuego is given up & the Voyage has become more definite in its length. —Mr Martens, the artist has been obliged from want of room to leave the Beagle. —

[1]In a letter to Caroline Darwin CD wrote: 'I have been unwell & in bed for the last fortnight, & am now only able to sit up for a short time. As I want occupation I will try & fill this letter. — Returning from my excursion into the country I staid a few days at some Goldmines & whilst there I drank some Chichi a very weak, sour new made wine, this half poisoned me, I staid till I thought I was well; but my first days ride, which was a long one again disordered my stomach, & afterwards I could not get well; I quite lost my appetite & became very weak. I had a long distance to travel & suffered very much; at last I arrived here quite exhausted. But Bynoe with a good deal of Calomel & rest has nearly put me right again & I am now only a little feeble. —' See *Correspondence* 1: 410.

Although it was not until six months later (see entry for 25 March 1835) that CD recorded actually having been bitten by the Benchuca bug (*Triatoma infestans*), which is the vector for trypanosomiasis, Chagas's disease is endemic in Chile and other parts of South America, and it is possible that this otherwise unexplained illness from 20 September to the end of October resulted from his initial infection with *Trypanosoma cruzi*. For discussions of the pros and cons of the theory that CD was a victim of Chagas's disease, see articles by S. W. Adler (Darwin's Illness. *Nature, London* 184: 1102–3, 1959) and A. W. Woodruff (Darwin's Health in Relation to his Voyage to South America. *British Medical Journal* I: 745–50, 1965). The symptoms of the disease are not at all well defined, and it remains possible that it might have served as a trigger for the psychosomatic condition discussed by Colp (see R. Colp, *To be an invalid: the illness of Charles Darwin*. University of Chicago Press, 1977), from which CD seems undoubtedly to have suffered in later life.

[2]In a letter to Catherine Darwin dated 8 November, CD wrote: 'Capt FitzRoy has for the last two months, been working **extremely** hard & at same time constantly annoyed by

interruptions from officers of other ships: the selling the Schooner & its consequences were very vexatious: the cold manner the Admiralty (solely I believe because he is a Tory) have treated him, & a thousand other &c &c has made him very thin & unwell. This was accompanied by a morbid depression of spirits, & a loss of all decision & resolution. The Captain was afraid that his mind was becoming deranged (being aware of his heredetary predisposition). all that Bynoe could say, that it was merely the effect of bodily health & exhaustion after such application, would not do; he invalided & Wickham was appointed to the command. By the instructions Wickham could only finish the survey of the Southern part & would then have been obliged to return direct to England. — The grief on board the Beagle about the Captains decision was universal & deeply felt. — One great source of his annoyment, was the feeling it impossible to fulfil the whole instructions; from his state of mind, it never occurred to him, that the very instructions order him to do as much of West coast, as *he has time* for & then proceed across the Pacific. Wickham (very disinterestedly, giving up his own promotion) urged this most strongly, stating that when he took the command, nothing should induce him to go to T. del Fuego again; & then asked the Captain, what would be gained by his resignation. Why not do the more useful part & return as commanded by the Pacific. The Captain, at last, to every ones joy consented & the resignation was withdrawn. —' See *Correspondence* 1: 418.

FitzRoy wrote: 'At this time I was made to feel and endure a bitter disappointment; the mortification it caused preyed deeply, and the regret is still vivid. I found that it would be impossible for me to maintain the Adventure much longer: my own means had been taxed, even to involving myself in difficulties, and as the Lords Commissioners of the Admiralty did not think it proper to give me any assistance, I saw that all my cherished hopes of examining many groups of islands in the Pacific, besides making a complete survey of the Chilian and Peruvian shores, must utterly fail. I had asked to be allowed to bear twenty additional seamen on the Beagle's books, whose pay and provisions would then be provided by Government, being willing to defray every other expense myself; but even this was refused. As soon as my mind was made up, after a most painful struggle, I discharged the Adventure's crew, took the officers back to the Beagle, and sold the vessel. Though her sale was very ill-managed, partly owing to my being dispirited and careless, she brought 7,500 dollars, nearly £1,400, and is now (1838) trading on that coast, in sound condition.' See *Narrative* 2: 361–2.

November 10th   The Beagle made sail for Chiloe.

21st   Arrived in the harbor of S. Carlos. Considering the time of year, with almost constant Southerly winds, |491| our passage was a pretty good one. — The island wore quite a pleasing aspect, with the sun shining brightly on the patches of cleared ground & dusky green woods. At night however we were convinced that it was Chiloe, by torrents of rain & a gale of wind. —

24th   The Yawl & whale-boat under the command of Mr Sullivan proceeded to examine the correctness of the charts of the East Coast of Chiloe, & to meet the Beagle at the Southern extremity at the Is^{d} of S. Pedro. — I accompanied the expedition: instead of going in the boats, the first day I hired horses to take me to Chacao. The road followed the coast, every now & then crossing promontories covered with fine forests. — In these shaded paths, it is absolutely necessary to make the whole road of logs of trees, such as described on the main road to Castro. —otherwise the ground is so damp from the suns rays never penetrating the evergreen foliage, that neither man nor horse would be able to pass along. — I arrived at the village of Chacao shortly after the tents belonging to the boats had been pitched. — The land in this neighbourhead is extensively cleared & there are many quiet & most picturesque nooks in the forest. Chacao formerly was the principal port; but many vessels have been lost owing to the dangerous currents & rocks in the Straits; the Spanish government burnt the Church & thus arbitrarily compelled the greater number of inhabitants to migrate to S. Carlos. —

In a short time the bare-footed Governor's son came down to reconnoitre us; seeing the English flag hoisted to the yawls mast head|492| he asked with the utmost indifference whether it was always to fly at Chacao. In several places, the inhabitants were much astonished at the appearance of Men of wars boats, & hoped & believed it was the forerunners of a Spanish fleet coming to recover the Island from the patriot Government of Chili. — All the men in power however had been informed of our intended visit & were exceedingly civil. — Whilst eating our supper, the Governor paid us a visit; he had been a Lieut. Colonel in the Spanish service, but was now miserably poor. — He gave us two sheep & accepted in return two cotton handkerchiefs, some brass trinkets & a little tobacco. —

25th Torrents of Rain: we managed however to run down the coast as far as Huapilenou. The whole of this Eastern side of Chiloe has one aspect; it is a plain broken by vallies or divided into little islands, the whole of which are thickly covered with an impervious blackish-green forest. On the margins there are some cleared spaces surrounding high-roofed cottages. The plain in this part is only 100 to 200 ft high, further Southward it is double of this.

26th The day rose splendidly clear: The Volcano of Osorno was spouting out volumes of smokes; this most beautiful mountain, formed like a perfect cone & white with snow, stands out in front of the Cordillera. Another great Volcano, with a saddle shaped summit, also emitted from its immense crater little jets of steam or white smoke. Subsequently we saw the lofty peaked Corcobado, well deserving the name of "el famoso|493| Corcovado". Thus we saw at one point of view three great active Volcanoes, each of which had an elevation of about seven thousand feet. —[1] In addition to this, far to the South, there were other very lofty cones of snow, which although not known to be active must be in their origin volcanic. — The line of the Andes is not, however, in this neighbourhead nearly so elevated as in Chili, neither does it appear to form so perfect a barrier between the regions of the earth. — This great range though running in a North & South line, from an occular deception always appeared more or less semicircular; because the extreme peaks being seen standing above the same horizon with the nearer ones, their much greater distance was not so easily recognized. —

When landing on a point to take observations, we saw a family of pure Indian extraction; the father was singularly like to York Minster; some of the younger boys, with their ruddy complexions, might be mistaken for Pampas Indians. Everything I have seen convinces me of the close connection of the different tribes, who yet speak quite distinct languages. — This party could muster but little Spanish & talked to each other in their own dialect. — It is a pleasant thing in any case to see the aboriginal inhabitants, advanced to the same degree of civilization, however low that may be, which their white conquerors have attained. — More to the South we saw many pure Indians, indeed some of the Islands as |494|Chauques &c &c have no other inhabitants but those retaining the Indian surname. — In the census of 1832 there were in Chiloe & its dependencies 42 thousand souls, the greater number of these appear to be little copper-colored men of mixed blood. — Eleven thousand actually retain their Indian surname, but probably not nearly all their pure blood & they are all Christians; dress & manners of living like the rest of the poor inhabitants; cultivating potatoes & picking up, like their brethren in T. del Fuego, shellfish at low water. — They however to this day hold superstitious communication in caves with the devil; the particulars of the ceremony are not known; because formerly every one convicted of this offence was sent to the inquisition at Lima. — Many of the people who are not included in the eleven thousand cannot be told by their appearance from Indians. Gomez, the governor of Lemuy, is descended from noblemen of Spain on both sides, but by constant intermarriages with natives, the present man is an Indian. — On the other hand, the Governor of Quinchao boasts much of his pure Spanish blood. — The Indians belong to the tribes of the Chawes (or Chahues) and Ragunos, who both speak dialects of the Beliche language. — They are not however believed to be the original inhabitants of Chiloe; but rather the Bybenies, who speak quite a distinct language. This nation, when they found so many intruders, migrated, no one knows exactly where; Mr Lowe on a sealing voyage met a large party of Indians|495| in the channels South of C. Tres Montes; they had canoes built of plank like the Periaguas & pulled by oars; & in the head of each canoe there was a cross. — Were not these men descended from the ancient inhabitants of Chiloe? The Chawes & Ragunos are believed to be descended from Indians sent from the North to the first Spanish settlers "en nomiendas commendo", that is to be taught the Christian religion & in return to work,[2] in short be slaves to their Christian teachers; & likewise from a large tribe, who remained faithful at the surprisal of Osorno & other Spanish towns: they were given at first the territory of Cabluco, from whence they have spread over other Islands. — Of the original Bybenies only a few families remain, chiefly in Caylen, & these have lost their own dialect. The Indians yet retain their Caciques, but they scarcely have any power;[3] when the land-surveyor or other government officer visits their village the Cacique appears with a silver-headed cane. — (all the above particulars I heard from Mr Douglas who is employed in the boats as a pilot, & has been long resident in the island). —

We reached at night a beautiful little cove North of the Isd of Caucahue; the people here were all related one to the other & complained of the want of land. — This is partly owing to their own negligence in not clearing the woods & partly to restrictions of the Government, which makes it necessary before buying ever so small a piece to pay two shillings to the Surveyor for measuring each quadra|496| (150 yards square) & whatever price he fixes for the value of the land. — After his valuation, the land must be put up three times to auction & if no one bids more, the purchaser can have it at that rate. — All these enactions must be a serious check to clearing the ground, where the inhabitants are so extremely poor. — In most countries, forests are removed without much difficulty by the aid of fire, in Chiloe however from the damp nature of the climate & sort of trees, it is necessary first to cut them down: this is a heavy injury to the prosperity of Chiloe. — In the time of the Spaniards the Indians could not hold land; but a family after having cleared a piece of ground might be driven away & the property seized by Government. — The Chilian authorities are now performing an act of justice by making retribution to these poor Indians, by giving to each Cacique twelve quadras of land, to his widow six, to any man who has served in the militia the same number, & to the aged four. — The value of uncleared land is very little. Government

Breast ploughing at Chiloe, by T. Landseer after P. P. King (*Narrative* 1: 287).

gave Mr Douglas, the present surveyor, who was kind enough to give
me the above information, eight & a half square miles of forest near S.
Carlos in lieu of a debt, this he sold for 350 dollars or about seventy
pounds sterling.

[1] These last words are substituted for 'I should suppose at least as elevated as the Peak of
Teneriffe', and 'No' in the margin has been crossed out.
[2] Note added in margin by CD, 'V. Humboldt New Spain p 136 Vol I', refers to Alexander
von Humboldt, *Political essay on the kingdom of New Spain*, translated by John Black, 2 vols.
New York, 1811.
[3] Note in margin by CD: 'No'.

27th, 28th   We had the good luck to have these two days fine, & reached
by night the Isd of Quinchao. This neighbourhead is the centre of
cultivation; many of the islands are nearly cleared & a broard strip of
cleared ground follows the coast of the main Island. I was curious to
know the wealth of the Chilotans. — Mr Douglas tells me that no one
can be considered|497| to possess a regular income. Each person raises
enough for the consumption of his own family & a little more such as
hams & potatoes, which are sent in the rude country boats to S. Carlos,
where they are exchanged for such articles of clothing as they can not
themselves manufacture, & a few other luxuries. — One of the richest
landowners, in a long industrious life, might possibly accumulate as
much as a thousand pounds sterling; should this happen it would be
stowed away in some secret place, for each family generally possesses
a hidden jar or chest buried in the ground. —

29th & 30th   We reached on the Sunday morning Castro, the ancient
capital of Chiloe. — I never saw before so truly a deserted city. — The
usual quadrangular arrangement of Spanish towns was to be traced,
but the streets & Plaza were coated with fine green turf on which Sheep
were browzing. — A church, built by that all-powerful order of the
Jesuits shortly before their expulsion, is highly picturesque; (*Note in
margin:* It is not the Jesuits Church but the Parochial one.) it is entirely
built of plank, even to the roof: it seems wonderful that wood should
last for half a century in so wet a climate. — The arrival of our boats was
a rare event in this quiet retired corner of the world, nearly all the
inhabitants came to the beach to see us pitch our tents. They were very
civil & offered us a house; & one man even sent us a cask of cyder as a
present. — In the afternoon we paid our respects to the Governor; a
quiet old man, who in his appearance & manner of life was scarcely
superior to an English cottager. — |498|I afterwards went out riding, to

examine the geology of the neighbourhead. — The country rises to some height behind the town, it is partly cultivated & pleasant looking. — At night rain commenced, which was hardly sufficient to drive away from the tents the large circle of lookers on. — An Indian family, who had come to trade in a canoe from Caylen, bivouacked near us: & they had no shelter during the heavy rain: in the morning I asked a young Indian, who was wet to the skin, how he passed the night; — he seemed perfectly content & answered "Muy bien Signor". —

December 1st We steered for the Isd of Lemuy. — I was anxious to examine a reported coal mine, which turned out to be Lignite of little value in the tertiary Sandstones of which these Islands are composed. — During the day we passed many Chapels; the number of these all over Chiloe is remarkable; every collection even of a few houses has its Capella. When we reached Lemuy we had great difficulty in finding a place for the tents, owing to it being Spring tides & the land being universally wooded to the high water line. —

We were soon surrounded by a large group of the nearly pure Indian inhabitants. They were much surprised at our arrival & said one to the other, this is the reason we have seen so many Parrots lately; the Cheucau (an odd red-breasted little bird, which inhabits the thick forest & utters very peculiar noises) has not cried "beware" for nothing. —[1] They were|499| soon eager for barter. Money is scarcely worth anything, but their esteem & anxiety for tobacco was something quite extraordinary: after tobacco, indigo came next in value, then capsicum, old clothes & gunpowder; the latter article was required for a very innocent purpose; each parish has a public musket, & the gunpowder was wanted to make a noise on their Saint or Feast days. —

The people here live chiefly on shell-fish & potatoes; at certain seasons they catch also, in "Corrales" or hedges under water, many fish which are left as the tide falls dry on the mud-banks. — They occasionally possess fowls, sheep & goats, pigs, horses & cattle, the order in which they are mentioned expressing their frequency. — I never saw anything more obliging & humble than the manners of these people. — They generally begin with stating that they are poor natives of the place & not Spaniards & are in sad want of tobacco and other comforts. At Caylen, the most Southern island, we bought with a stick of tobacco, of the value of three half-pennies, two fowls (one of which the Indian stated had skin between its toes & turned out to be a fine duck): & with some cotton handkerchiefs worth three shillings, we procured three sheep & a large bunch of onions. — All these purchases were transacted under the denomination of money; the stick of tobacco was valued at one shilling & the proportion of a shilling to the half-pennies expresses the profit of the traders with these Islanders. —²|500|

The Yawl at this place was anchored some way from the shore & we had fears for her safety during the night. Our pilot, Mr Douglas, accordingly told the constable of the district that we always placed sentinels with loaded arms, & not understanding Spanish, if we saw any person in the dark, we should assuredly shoot him. The constable, with much humility, agreed to the propriety of this consequence & promised us that no one should stir out of his house during the night.

¹In his pocketbook (Down House Notebook 1.8, and see *CD and the Voyage* p. 229) CD noted this day: 'Chucao. 3 distinct noises I know. — Nest?'. In *Journal of Researches* pp. 351–2 he wrote: 'In all parts of Chiloe and Chonos, two very strange birds occur, which have many points of affinity with the Turco and Tapacola. One is called by the inhabitants "Cheucau" (*Pteroptochos rubecula*). It frequents the most gloomy and retired spots within the damp forests. Sometimes, although its cry may be heard close at hand, let a person watch ever so attentively, he will not see the cheucau; at other times, let him stand motionless, and the red-breasted little bird will approach within a few feet, in the most familiar manner. It then busily hops about the entangled mass of rotting canes and branches, with its little tail cocked upwards. I opened the gizzard of some specimens: it was very muscular, and contained hard seeds, buds of plants, and vegetable fibres, mixed with small stones. The cheucau is held in superstitious fear by the Chilotans, on account of its strange and varied cries. There are three very distinct kinds, — one is called "chiduco," and is an omen of good; another, "huitreu," which is extremely unfavourable; and a third, which I have forgotten. These words are given in imitation of its cries, and the natives are in some things absolutely governed by them. The Chilotans assuredly have chosen a most comical little creature for their prophet.' See *Journal of Researches* pp. 351–2.

²Followed by a deleted sentence: 'Immense as this profit is, our arrival was always hailed as a piece of high good fortune. —'

2nd   The day was calm & we only reached the South extreme of Lemuy.¹

¹In Down House Notebook 1.8 (and see *CD and the Voyage* pp. 229–30), CD wrote: 'I have mentioned having found at Lemuy on beach very much silicified wood—one piece pentrated by Teredo. Likewise I found many fragments on coast above Yal.[?]— *At last* I found *in the* yellow sandstone a great trunk (structure beautifully clear) throwing off branches: main stem much thicker than my body & standing out from weathering 2 feet.— Central parts generally black & vascular, & structure not visible.— This tree coetaneous (near in position) with the shells of above: it is curious chemical action, such a sandstone in sea, holding such silex in solution: vessels transparent quartz. This observation most important, as proof of general facts of petrified wood. For here the inhabitants firmly believe the process is now going on. —'

CD next found petrified trees at Villa Vicencio near Mendoza on 1 April 1835. See footnote 1 for that day (pp. 317–18) and *Journal of Researches* p. 406.

3rd   During our last visit, I fancied Chiloe never enjoyed such a day as this; I cannot imagine a more beautiful scene, than the snowy cones of the Cordilleras seen over an inland sea of glass, only here & there rippled by a Porpoise or logger-headed Duck. And I admired this view from a cliff adorned with sweet-smelling evergreens, where the bright colored, smooth trunks, the parasitical plants, the ferns, the arborescent grasses, all reminded me of the Tropics, neither did the temperature recall me to the reality. —

4th The weather was squally, but we reached P. Chagua: the general features of the country remain precisely the same: it is much less thickly inhabited: the whole of the large island of Tanqui has scarcely one cleared spot; the trees on every side extend their branches over the sea. — |501|

5th I noticed to day growing on the cliffs of soft sandstone some very fine plants of the Pangi, which somewhat resembles the Rhubarb on a gigantic scale. — The inhabitants eat the stalks, which are sub-acid, & with the root tan leather & prepare a black dye. (*Note in margin:* The stalks are called Nalca, so indeed is the plant sometimes.) The leaf is much indented in its margin & is nearly circular; the diameter of one was nearly 8 feet (giving a circumference of 24 feet!). The stalk rather more than a yard high: each plant throws out from four to six of these enormous leaves & a group of them hence has a very fine appearance. —

6th Reached Caylen, called "el fin del Christianitad". It is rather better inhabited.

7th In the morning we stopped for a few minutes at a house at the extreme North point of Is^{d} of Laylec. This was the last house; the extreme point of S. American Christendom; & a miserable hovel it was. — The latitude is about 43° 10', which is considerably to the South of the R. Negro on the Atlantic coast of America. The people were miserably poor & as usual begged for a little tobacco. — I forgot to mention an anecdote which forcibly shows the poverty of these Indians; some days since, we met a man who had travelled 3 & ½ days on foot, on bad roads, & had the same distance to return to recover the value of an axe & a few fish! How difficult it must be to buy the smallest article, where such trouble is taken to recover so small a debt. —

We had a foul wind & a good deal of swell|502| to struggle with, but we reached the Island of S. Pedro, the SE extremity of Chiloe, in the evening. When doubling the point of the harbor, Mrs Stuart & Usborne landed to take a round of angles. — A fox (of Chiloe, a rare animal) sat on the point & was so absorbed in watching their mænœvres, that he allowed me to walk behind him & actually kill him with my geological hammer.[1] We found the Beagle at anchor, she had arrived the day before & from bad weather had not been able to survey the outer coast of Chiloe. — The most singular result of the observations is that Chiloe is made 30 miles too long, hence it will be necessary to shorten the island $\frac{1}{4}$ of its received size. —

[1] In *Journal of Researches* p. 341, CD added: 'This fox, more curious or more scientific, but less wise, than the generality of his brethren, is now mounted in the museum of the Zoological Society.'

8ᵗʰ A party with Capt FitzRoy tried to reach the summit of San Pedro, the highest part of the islands. — The woods here have a different aspect from those in the North, there is a much larger proportion of trees with deciduous leaves. — the rock also being primitive Micaceous slate, there is no beach, but the steep sides of the hills dip directly down into the sea; the whole appearance is in consequence much more that of T. del Fuego than of Chiloe. — In vain we tried to gain the summit; the wood is so intricate that a person who has never seen it will not be able to imagine such a confused mass of dead & dying trunks. — I am sure oftentimes for quarter|503| of an hour our feet never touched the ground, being generally from 10 to 20 feet above it; at other times, like foxes, one after the other we crept on our hands & knees under the rotten trunks. In the lower parts of the hills, noble trees of Winters bark, & the Laurus sassafras (?) with fragrant leaves, & others the names of which I do not know, were matted together by Bamboos or Canes. — Here our party were more like fish struggling in a net than any other animal. — On the higher parts brushwood took the place of larger trees, with here & there a red Cypress or an Alerce. — I was also much interested by finding our old friend the T. del F. Beech, Fagus antarcticus; they were poor stunted little trees, & at an elevation of little less than a thousand feet. — This must be, I should apprehend from their appearance, nearly their Northern limit. — We ultimately gave up the ascent in despair. —

10ᵗʰ The Yawl & Whale-boat, with Mr Sulivan, started to continue their survey: & the next day (11ᵗʰ) we left S. Pedro in the Beagle. —

13ᵗʰ On the 13ᵗʰ we ran into an opening in the Southern part of the Guyatecas or Chonos Archipelago & soon found a good harbor. —

14ᵗʰ It is fortunate we reached this shelter. For now a real storm of T. del Fuego is raging with its wonted fury. White massive clouds were piled up against a dark blue sky & across them black ragged sheets of vapor were rapidly driven. The successive ranges of mountains appeared|504|

like dim shadows; it was a most ominous, sublime scene. — The setting sun cast on the woodland a yellow gleam, much like the flame of spirits of wine on a man's countenance. The water was white with the flying spray; & the wind lulled & roared again through the rigging: the gale in all its features was complete. It was curious to notice the effect which the spray had on a bright rain-bow; instead of approaching to a semicircle, the ring was nearly complete, for the prismatic colors were carried on the surface of the water on both sides to the Ships stern; & hence formed a circle. — [1]

[1] The entry for this day has mostly been written in the present tense, and afterwards altered into the past, as it appears in *Journal of Researches*, p. 342.

15th–17th The weather continued bad; to me it did not much signify, because the land in all these islands is next thing to impassable; the coast is rugged & so very uneven that it is one never ceasing climb to attempt to pass that way; as for the woods, I have said enough about them; I shall never forget or forgive them; my face, hands, shin-bones all bear witness what maltreatment I have received in simply trying to penetrate into their forbidden recesses. —

18th Stood out to sea. — Mr Stokes, the day before, was despatched in a Whale-boat with three weeks provisions to survey the Northern part of the Archipelago & there meet us. — We have now three boats away; which is something for a ten gun-brig to say. — The Jonas is out of the Ship (whoever he may be);

20th the Beagle had a fair wind to the extreme Southern point where it was necessary to proceed; & when at Noon|505| on the 20th, we bid farewell to the South & put the Ships head to the North, the wind continued fair. — From C. Tres Montes we ran pleasantly along this lofty weather-beaten coast. It is remarkable by the bold outline of the hills & the thick covering, even on the almost precipitous sides of [the] forest. —

Sunday 21st Found an harbor, which on this unknown & dangerous coast might be of great utility to a distressed vessel. It can be easily recognized by the most perfectly conical hill I ever saw; it quite beats the famous Sugar-loaf at the entrance of Rio de Janeiro harbor.

22nd On the Monday I succeeded in reaching the summit (1600 ft. high); it was a laborious undertaking; the ascent being so steep as to make it necessary to use the trees like a ladder. (*Note in margin:* Great thickets of Fushza.) In these wild countries it gives much delight to reach the summit of any high hill; there is an indefinite expectation of meeting something very strange, which however often it is baulked, never with me failed to recur. — Every one must know the feeling of triumph or pride which a great & extensive view from a height communicates to the mind. — In this case there is joined to it a little vanity of distinction, that you perhaps are the first man who ever stood on this pinnacle, or admired this view. — There is always a strong desire to ascertain whether any body has previously visited the place where one may happen to be. — A bit of wood with a nail in it is picked up & studied as if it was covered with hieroglyphics. Owing to this feeling, I was much interested by finding on a wild piece of the coast, a bed made of grass, beneath|506| a ledge of rock; close by it there had been a fire, & the man had used an axe. — The fire, bed & situation were chosen with the dexterity of an Indian, but it could scarcely be an Indian. — We subsequently found traces of a sealing vessel having been in here; yet I cannot help having some misgivings that the solitary man who had made his bed on this wild spot, was some poor shipwrecked sailor, trying to travel up the coast. If so, probably before this, he has laid himself down & died. —

23rd Stood out to sea, but bad weather coming on from the Northward, we ran back again & anchored in another cove.

24th I was here much interested by finding quantities of Lava & other Volcanic products. —

28th At last the weather barely permitted us to run out; our time has hung heavy on our hands, as it always does when we are detained from day to day by successive gales of wind. — Our Christmas day was not such a merry one as we had last year at Port Desire. — Between 30 & 40 miles of coast was surveyed & in the afternoon we found an excellent harbor. — Directly after anchoring we saw a man waving a shirt. A boat was sent & brought two men off. — They turned out to be N. American seamen, who from bad treatment had run away from their vessel when 70 miles from the land. The party consisted of five men & the officer of the watch; who together in the middle watch had lowered a boat & taken a weeks provisions with them, thinking to go along the coast to Valdivia; The boat on their first landing had been|507| dashed into pieces. — This happened 15 months ago; since which time the poor wretches have been wandering up & down the coast, without knowing which way to go or where they were (they knew nothing of Chiloe). What a singular piece of good fortune our happening to discover this harbor at the very time they were in it. Excepting by this chance they might have wandered till they had been old men & probably would not have been picked up. — This explains the bed in the last harbor; the party had separated when this was used. — They were now all together & the boat subsequently brought off three more. —one man had fallen from a cliff & perished. — I never saw such anxiety as was pictured in the mens faces to get into the boat. —before she landed, they were nearly jumping into the water. They were in good condition, having plenty of seals-flesh which together with shell-fish had entirely supported them. — In the evening we paid a visit to their little hut made of reeds; a few days since, they had killed nine seals; they cut the flesh into pieces & secured it on sticks which they place cross-wise over the fire & thus preserve it. — They had some few clothese, a book (well thumbed), 2 hatchets & knives; with these they had hollowed out two trees to make canoes, but neither answered. — The difficulties they encountered in trying to travel up the coast were dreadful; it was in passing a head-land the man was lost; some of the Bays gave them 5 days walking to reach the head. Latterly they appear to have given up in despair their attempt at reaching Valdivia!|508| And well they might. — They had one comfort in having always plenty of firewood; they managed to make a fire by placing a bit of tinder with a spark from a steel & flint between two pieces of charcoal, & by blowing this was sufficient to ignite it. — There are no Indians. — Their treatment on board the Whaler does not appear to have been so very bad; but their remedy, probably from ignorance of the dangers, has been a most desperately perilous one. I am very glad the Beagle has been the means of saving their lives. — Considering what they have undergone, I think they have kept a very good tally of the time; they making this day to be the 24th instead of the 28th. —

29th   Ran along the Coast till we came to an anchor at Yuche Island, a little to the North of the Peninsula of Tres Montes.

30th   In the morning went on shore; to our great surprise we found the Island well stocked with fine wild Goats. The sportsmen soon killed eight, which have given us two days fresh meat. I should think these Goats must originally have been turned out by some of the old Spanish Missionary expeditions. Others besides us have visited this place; I found marks of trees long ago cut down, an old fire, & remains of a sort of Shed. — I presume it has been one of the prowling tribe of Sealers. — In the evening changed our anchorage to a snug cove at the foot of some high hills.

31ˢᵗ After breakfast the next morning, a party ascended one of the
highest viz. 2400 ft. elevation. — The scenery was very remarkable; the
chief part of the|509| range is composed of grand solid abrupt masses of
granite, which look as if they had been coeval with the very beginning
of the world. — The granite is capped with slaty gneiss, & this in the
lapse of ages of time has been worn into strange finger-shaped points.
These two formations, thus differing in their outlines, agree in being
almost destitute of vegetation; and this barrenness had to our eyes a
more strange appearance, from being accustomed to the sight of an
almost universal forest of dark green trees. I took much delight in
examining the structure of these mountains. — The complicated & lofty
ranges bore a noble aspect of durability — equally profitless however to
man & to all other animals. Granite to the Geologist is a classic ground:
from its wide-spread limits, its beautiful & compact texture, few rocks
have been more anciently recognised. Granite has given rise perhaps to
more discussion concerning its origin than any other formation. — We
see it generally the fundamental rock, & however formed, we know it
to be the deepest layer in the crust of this globe to which man is able to
penetrate. — The limit of mans knowledge in every subject possesses a
high interest, which is perhaps increased by its close neighbourhood to
the realms of imagination.

January 1ˢᵗ, 1835. The new year is ushered in with the Ceremonies
proper to it in these regions: — she lays out no false hopes; a heavy NW
gale with steady rain bespeaks the rising year. Thank God we shall not
here see the end of it; but rather in the Pacific, where |510|a blue sky
does tell one there is a heaven, a something beyond the Clouds, above
our heads. —

4ᵗʰ The NW winds continued to prevail & we only managed to cross a
sort of great bay & anchored in an excellent harbor. — This is the place
where the Anna Pink, one of Lord Ansons squadron, found refuge
during the disasters which beset him. —[1] A boat with the Captain went
up to the head of the bay. The number of the Seals was quite astonishing; every bit of flat rock or beach was covered with them. They appear
to be of a loving disposition & lie huddled together fast asleep like pigs:
but even pigs would be ashamed of the dirt & foul smell which
surrounded them. Often times in the midst of the herd, a flock of gulls
were peaceably standing: & they were watched by the patient but
inauspicious eyes of the Turkey Buzzard. — This disgusting bird, with
its bald scarlet head formed to wallow in putridity, is very common on

this West Coast. Their attendance on the Seals shows on the mortality of what animal they depend. —

We found the water (probably only that of the surface) nearly fresh; this is caused by the number of the mountain torrents which in the form of cascades come tumbling over the bold Granite rocks into the very sea. — The fresh-water attracts the fish & this brings many terns, gulls & two kinds of cormorant. — We saw also a pair of the beautiful black-necked swans; & several small sea-otters, the fur of which is held in|511| such high estimation. In returning we were again amused by the impetuous manner in which the heap of seals, old & young, tumbled into the water as the boat passed by. They would not remain long under, but rising, followed us with outstreched necks, expressing great wonder & curiosity. —

The entire absence of all Indians amongst these islands is a complete puzzle. That they formerly lived here is certain, & some even within a hundred years; I do not think they could migrate anywhere; & indeed, what could their temptation be? For we here see the great abundance of the Indians highest luxury, seals flesh; I should suppose the tribe has become extinct; one step to the final extermination of the Indian race in S. America. —

[1] The small merchantman or pink *Anna* accompanied Anson's fleet on the voyage in which after capturing the Spanish galleons laden with treasure from Manila, he completed the circumnavigation of the globe. Earlier in the voyage, after rounding Cape Horn in April 1741, the ships were scattered in a severe storm, and the badly damaged *Anna* was repaired here at Port Refuge before proceeding to the rendezvous with Anson at Juan Fernandez Island. A copy of Richard Walter's account of Anson's voyage, published in London in 1748, was in the *Beagle* library, and CD had recorded that he had finished reading it on 22 May 1832.

5th   The Barometer says we shall have fine weather; & although we have at present a foul wind & plenty of rain, we stand out to sea. —

6th   The Captains faith is rewarded by a beautiful day & Southerly wind. — After noon, the ship was hove to, & the Captain ran in his boat to reconnoitre some harbors. We passed a dead whale; it was not very putrid; the barnacles & great parasitical crabs being alive; the skin of this great mass of flesh & blubber was quite pink; I suppose owing to partial decomposition. (*Note in margin:* Outer thin skin having been removed.) In one of the harbors in P. Tres Montes, we found another cast up on the beach & of the same color. — A sight of a Whale always puts me in mind of the great fossil animals; he appears altogether too big for the present|512| pigmy race of inhabitants. He ought to have coexisted with his equals, the great reptiles of the Lias epoch. —

During our absence, a French Whaler bore down on the Beagle & here we found her Captain on board. — He had lately been at anchor when two other great ships; one of which was commanded by our old friend Le Dilly, who was wrecked in the Falklands. — So that the French government are not tired of their expensive school to make Sailors. —

7th We ran on during the night. The French ship most pertinaciously followed us; she supposed we were making for some Harbor; & a harbor on this lee-shore is a prize which a Whaler dare not herself look for. We found Mr Stokes had arrived a week before at this (Lowes Harbor) our rendevous. — The islands here are chiefly of the same Tertiary formation as at Chiloe, & are beautifully luxuriant: The woods come down to the beach in precisely [the] same manner as an evergreen shrubbery over a gravel walk. We found here a Periagua from Caylen; the Chilotans had most adventurously crossed in their miserable boat the open space of the sea which separates Chonos from Chiloe. — I think this place will soon be inhabited; there is a great abundance of fine muscles & oysters; wild potatoes grow in plenty, one which I measured was oval, & its longest diameter two inches. — Mr Stokes & his party cooked & ate them & found them watery but good. — The Chilotans expected to catch fish, & the very great numbers of|513| sea-otters shows to be the case.

We enjoyed from the anchorage a splendid view of four of the great snowy cones of the Cordilleras; the most Northern is the flat-topped Volcano, & next to this comes "el famoso Corcovado". — The range itself is almost hidden beneath the horizon.

8th–14th Our week in this port passed rather heavily; the climate is so very bad & the country so very uniform in its character. —

15th & 16th On the 15th we sailed & steered for the SW point of Chiloe; the next day it was attempted to survey the coast, but the weather again becoming bad, we bore up & run to an anchorage under Huafo. We had the misfortune to lose our best Bower anchor, which parted in bringing up the ship. — I went on shore in the evening, & extracted from the rock a good many fossil shells. — There are here some large caverns; one which I could by no means see the length of, had been inhabited some long time ago. — During the night it rained as if rain was a novelty; the rain in this country never seems to grow tired of pouring down. —

17th We ran along during the next day the Southern part of [the] outer coast of Chiloe; The country is similar to that on the inside coast viz a thickly wooded plain & white cliffs facing the sea: further to the North the coast becomes bolder. — We made during the night a good run &

18th by noon on the Sunday reached S. Carlos. — We found Mr Sulivan
with the Yawl & Whaleboat, who had made a prosperous cruize.|514|

19th Early in the morning the ship ran out to sound on the English bank.
A boat put me on shore on P. Tenuy, where I found some very
interesting geology. In the evening we returned to our old anchorage at
P. Arena. — During this night the Volcano of Osorno was in great
activity; at 12 oclock the Sentry observed something like a large star,
from which state it gradually increased in size till three oclock when
most of the officers were on deck watching it. — It was a very magnificent sight; by the aid of a glass, in the midst of the great red glare of
light, dark objects in a constant succession might be seen to be thrown
up & fall down. — The light was sufficient to cast on the water a long
bright shadow. — By the morning the Volcano seemed to have regained
its composure. —

22nd Capt FitzRoy being anxious that some bearings should be taken on
the outer coast of Chiloe, an excursion was planned that Mr King &
myself should ride to Castro & from thence across the Island to the
Capella de Cucao, situated close to the West coast. — Having hired
horses & a guide, we set out on the morning of the 22nd. We had not
proceeded far, before we were joined by a woman & two boys, who
were bent on the same journey. Every one on this road acts on a "Hail
fellow well met" fashion: and one may here enjoy the priviledge, so rare
in S America, of travelling without fire arms. — In the first part the road
lies across a succession of hills & valleys; nearer|515| to Castro it crosses
a plain. — The road is a very singular affair as I have formerly said, is
almost entirely composed of logs of wood. — These are either broard
slabs laid longitudinally or smaller ones transversely to the direction of
the road. — Being summer time & fine weather the road is not so very
bad; but in winter, when the wood is slippery with rain, by all accounts
the travelling becomes quite dangerous. It is remarkable how active
custom has made the Chilotan horses; in crossing bad parts of the road
where the logs are displaced, the horse skips from one to the other with
quickness & certainty of a dog. — In winter the road on each side of the
line of logs is a perfect swamp & is in many places overflowed; so that
the logs are fastened down by transverse poles, which are pegged into
the earth on each side. — These same pegs render a fall from a horse
more dangerous as the chance of alighting on one is not small. — On
either hand of the road we have the forest of lofty trees, their bases
matted together by the Canes. — When occasionally a long reach of this
avenue could be seen, it presented a curious scene of uniformity; the

white line of logs, narrowing in perspective, became hidden by the one
colored forest, or it terminated in a zig zag line which ascended some
steep hill. — The first opening of this road must have cost considerable
labor. — I was told that many people had lost their lives in attempting
to cross the forest, & that the first who succeeded was an Indian who
cut his way through the canes in 8 days & reached S. Carlos. — He was
rewarded by the Spanish government by a large grant of land. — The
distance in a straight line is only 12 sea-leagues, yet from the nature of
the forest the labor must have been excessive. — During the summer
time many of the Indians wander about the woods, chiefly in the higher
parts where it is not quite so thick, in search of half wild cattle, which
live in the forest on the leaves of the Cane & various trees. It was one of
these Indians who by chance found a few years since an English ship
which had been wrecked on the West coast, the crew of which was
beginning to fail in provisions: it is not probable [that] without the aid
of this man, they would have been able to extricate themselves. —as it
was, one of the men died of fatigue on their march. — The Indians in
these excursions steer by the Sun & are very expert in finding their way;
if however they have a continuance of cloudy weather, they cannot
travel; This reminds one of the state which navigation must have been
in before the invention of the compass. —

The road to Castro will before very long become inhabited; we now
meet 3 or 4 cleared spots, each with its house, in the interval between
the two inhabited ends. — It is at this time of year much frequented;
chiefly however by foot men, who carry on their backs heavy loads of
corn &c &c & buy at S. Carlos clothese, Capsicum &c to sell in the
country. These men perform the journey in less than two days.

The day was beautiful; the number of trees which were|517| in full
flower perfumed the air; yet even this could scarcely dissipate the
gloomy dampness of the forest. The number of dead trunks, which
stand like great white skeletons, never fails to give these primeval
woods a character of solemnity which is wanting in those of countries
long civilized, such as England. —

I noticed in some particular tracts that nearly all the large trees were
dead. — I cannot give any reason for this. — My guide cut the matter
short by saying that a " bad wind" had killed them! Shortly after sunset,
we bivouaced for the night. Our female companion was rather good
looking; she belonged to one of the most respectable families in
Castro. — She rode, however, without shoes or stockings & cross-leg-
ged. — I was surprised at the want of pride shown by both her & her

brother; they brought food with them, but at all our meals sat watching Mr King & myself eating, till out of shame they compelled us to feed the whole party. — The night was cloudless; we enjoyed, & it is an high enjoyment, whilst lying in our beds the sight of the multitude of stars which brightened the darkness of the forest. —

23rd We started early in the morning & reached the pretty quiet town of Castro at 2 oclock. The governor who was here on the former occasion was dead, & in his place was a Chileno. — We had a letter of introduction to him; he had formerly been in much better circumstances, but was now very poor, & his Governorship only confers honor but no pay. — We found Don Pedro most exceedingly hospitable & kind; & a degree of|518| disinterestedness which I believe to be as common in the Spanish character, as assuredly it is most rare in the present Creole race (i.e. in Chili). —

24th Don Pedro procured us fresh horses & offered himself to accompany us. We proceeded to the South, generally following the coast. We passed through several hamlets, each with its large barn-like chapel built of wood. — Near Castro we saw a remarkably pretty waterfall: it was very small: but the water fell in a single sheet into a large circular basin; around which stately trees from 100 to 120 feet high cast a dark shadow. — (Note in margin: Trees were measured here of this height & some appeared higher.) At Vilèpilli Don Pedro asked the Commandante to give us a guide to Cucao: The old gentleman offered to come himself; but for a long time he could not believe that anything could induce two Englishmen to go to such an out of the way place as Cucao. He repeatedly asked "but where are you really going?" & when Don Pedro answered to Cucao—He replied "a los infiernos, hombre;—what is the good of deceiving me?"— We thus were accompanied by the two greatest aristocrats in the country; as was plainly to be seen in the manner of all the poorer Indians. — But yet, it must not for a moment be imagined that either of these men had at all the air of a gentleman. —

At Chonchi we struck off across the island and followed intricate winding paths, sometimes passing through magnificent forest & then|519| opening into pretty cleared spots, abounding with corn & potatoe crops. In this undulating woody country, partially cultivated, there was something which brought to mind the wilder parts of England, & hence to my eye wore a most fascinating aspect. — On the road we met a small herd of cattle which had just been collected at a "Rodeo" in the Pampas or Chili, where many hundreds are collected by a few men: here there were more men than cattle!— The cattle are hunted by dogs, which like our bull-dogs seize & hold them by the ears & nose, till men with lazos can come up & secure them. — At Vilinco, which is situated on the borders of the great lake of Cucao, only a few fields are cleared out of the forest, & the inhabitants appear all Indian. — This lake is twelve miles long & runs in an East & West direction: from local circumstances, the sea breeze blows very regularly during the day & during the night it falls calm. — This has given rise to strange exaggerations; for the phenomenon as described to us at S. Carlos was quite a prodigy.

The road to Cucao was so very bad, that we determined to embark in a periagua. The Commandante in the most authoritative manner ordered six Indians to get ready to paddle us over & without deigning to tell them whether they would be paid. The periagua is a strange rough boat, but the crew were still stranger: I do not think six uglier little men ever were in a boat together. — They pulled however very well & cheerfully; the stroke oar gabbled Indian & uttered|520| strange crys, much after the fashion a pig-driver drives pigs. — We started with a light breeze against us, but yet reached after night fall the Capella de Cucao; having pulled at the rate of three miles an hour. — The country on each side of the lake is one unbroken forest. In the same Periagua with us, a cow was embarked; it would seem a puzzle how to get a cow into a small boat, but the Indians managed it in a minute. They brought the cow along side the boat, & heeling the gunwale towards her, placed two oars under her belly & resting on the gunwale; with these levers they fairly tumbled the poor animal heels over head into the bottom of the boat. — At Cucao we found an uninhabited hovel (which is the residence of the Padre when he pays this Capella a visit) where lighting a fire, we cooked our supper & were very comfortable. —

The district of Cucao is the only inhabited part on the whole West coast of Chiloe. It contains about thirty or forty Indians, who are scattered along four or five miles of the shore, and without a single Spanish resident. — They are very much secluded from the rest of Chiloe & have scarcely any sort of commerce, excepting sometimes a little oil which they get from seal blubber. They are pretty well dressed in clothes of their own manufacture, & they have plenty to eat. — They seemed however discontented, yet humble to a degree which it was quite painful to behold. The former feeling is I think chiefly to be attributed to the harsh & authoritative manner in which they are treated by their rulers. Our companions, although so very civil to|521| us, behaved to the poor Indians as if they were slaves rather than free men. They ordered provisions, & the use of their horses, without ever condescending to say how much, or indeed if the owners should at all be paid.

25th In the morning being left alone with the Indians, we soon ingratiated ourselves by presents of cigars & matte: a lump of white sugar was divided between all present & tasted with the greatest curiosity. — The Indians ended all their complaints by saying "& it is only because we are poor Indians & know nothing, but it was not so when we had a King". — I really think a boats crew with the Spanish flag might take the island of Chiloe.

The next day after breakfast we rode to P. Huantamò, a little way to the Northward; the road lay along a very broard beach, on which even after so many fine days a terrible surf was breaking. I am assured that after a gale the roar can be heard at night even at Castro, a distance of no less than twenty one sea miles across a hilly and wooded country. We had some difficulty in reaching the point owing to the intolerably bad paths; for every where in the shade, the ground in Chiloe soon becomes a perfect quagmire. The point itself is a bold rocky hill; it is covered by a plant allied I believe to the bromelias, with little recurved hooks on the leaves, and which the inhabitants call Chepones. In scrambling through the beds, our hands were very much scratched; I was amused by seeing the precaution our Indian guide took, in *turning up* his trowsers thinking them more delicate than his hard skin. — This plant bears a fruit, in shape like an Artichoke; in it a number of seed-vessels are packed together|522| which contain a pleasant sweet pulp & are here much esteemed. I saw at Lowes Harbor the Chilotans making Chichi or cyder with this fruit; so true is it, that everywhere man finds some means to make intoxicating drink. —

The coast to the Northward of P. Huantamò is exceedingly rugged & broken & is fronted by many breakers on which the sea is eternally roaring. — Mr King & myself were anxious to return, if it had been possible, on foot along this coast; but even the Indians say it is quite impracticable. — We were told that men have crossed by striking into the Woods from Cucao to S. Carlos, but never by the Coast. — On these expeditions the Indians only carry with them toasted corn; & of this they eat sparingly but twice a day.

I made some enquiries concerning the history of the Indians of Chiloe. — They all speak the same language which is the Birliche or Williche: is different from that of the Araucanians; yet their method of address is nearly the same; the word being "Mari-Mari", which signifys "good morning". They recognize amongst themselves certainly some divisions: do not believe that the Ragunias or Chahues come (as M[r] Douglass states) from the North, & only recognized the former name. They say the Bybenies formerly spoke quite a distinct language; the Commandante believes they came from the South. The Indian word, to the S. of C. Tres Montes, for the Potatoe is Aquina, here they have quite a distinct name. — These Indians of Cucao are said to have originally belonged to Isd [of] Huafo, & to have been brought over by the Missionaries. — In a |523|similar manner the Missionaries finding the passage to the Chonos Islands difficult & dangerous tempted by presents the Inhabitants to come & live in Caylen. This agrees with what the Chilotans said in Lowes Harbor & it perfectly accounts for the deserted state of that Archipelago. — I before heard that the few remaining Bybenies chiefly lived in Caylen. Is it not probable that these are the original inhabitants of Chonos? —

I understand since the time of the Patria, the Caciques have been entirely done away with. —

26th   We again embarked in the Peragua & crossed the lake: & then took to our horses. — The whole of Chiloe took advantage of this week of unusually fine weather to clear the ground by burning: in every direction volumes of smoke were curling upwards: although the inhabitants were so assiduous in setting fire to every part of the wood, I did not see a single one which they had succeeded in making extensive. — We dined with our friend the Commandante & did not reach Castro till after dark. — I cannot give a better idea of the poverty of Castro, than the fact that we had great difficulty to buy a pound of sugar; & a knife which we wanted was quite out of the question. — (*Note in margin:* No Watch or Clock, strike the Bell by guess!) Don Pedro gave as a reason for this; that there being no money, goods could only be taken in exchange, so that a trader must at the same time be a merchant. A man wanting to buy a bottle of wine, carrys on his back an Alerce board! —

27th   We left Castro early in the morning; after having entered for some time the forest, we had from a steep brow of a hill, (& it is a rare thing in this road)|524| an extensive view of the great forest; over the horizon of trees the Volcanoes of Corcovado & Lagartigas stood out in proud preeminence; Scarcely another peak of the Cordilleras showed their snowy tops. I hope it will be long before I forget this farewell view of the magnificent Cordilleras of Chiloe. — At night we again bivouaced with a cloudless sky,

28th & rising before day-break reached S. Carlos in the morning. We arrived on the right day, for in the evening heavy rain commenced. — I have now well seen Chiloe, having both gone round it & crossed it in two directions. —

February 4th We sailed from P. Arena; but from dirty weather were obliged to return & anchor in English harbor. — In this last week I made some short excursions; one was to see a bed of oysters, out of which large forest trees were growing at an elevation of 350 feet. — Another was to P. Huechucucuy. I had with me a good Vaqueano, who pertinaciously told me the Indian name for every little point, rivulet & creek. In the same manner as in T. del Fuego, the Indian language appears singularly well adapted for attaching names to the most minute divisions of land. —

I believe every one is glad to say farewell to Chiloe. Yet if we forget the gloom & ceaseless rain of winter, Chiloe might pass for a charming island. — There is also something very attractive in the simplicity & humble politeness of all the cottagers; when we look however to their morality, there is, as in the weather, a dark as well as bright point of view.|525|

5th We steered along the coast, but owing to thick weather did not reach Valdivia till the night of the 8th.

8th The forest is no where cleared away; the geological structure being evidently the same with that of the central parts of Chiloe, the external features are the same. We have everywhere on the coast bold rocky points, which more inland are covered up by Tertiary plains of different altitudes.

9th The morning after our arrival in the Port, two boats were sent to the town of Baldivia.[1] This is seated on the banks of a river 9 or 10 miles distant from the anchorage. — At the latter place there are only a few cottages & some strong fortresses. — I ought rather to say which were formerly strong; for now most of the guns have been carried to Valparaiso. — This port is well known from Lord Cochranes gallant attack when in the service of La Patria. —[2] We followed the course of the river; occassionally passing a few hovels & cleared patches of ground, & sometimes meeting a canoe with an Indian family. The scenery otherwise is one unbroken forest. The town of Valdivia is seated on the low banks of the river: it is completely hidden in a wood of Apple trees; the streets are merely paths in an orchard. —[3] I never saw this fruit in such abundance. — There are but few houses; even I think less than in S. Carlos; they are entirely built of Alerce planks. The manners & habits of the upper classes are evidently superior to what we meet|526| with at poor Chiloe. There is perhaps also more pure Spanish blood. — Beyond this, there is little to show that Baldivia is one of the most ancient colonies on this West coast of America. —

Our first impression on seeing this quiet little town certainly has been a pleasing one. — There are several Englishmen residing here (as indeed in every corner of S. America); their number has lately been increased by an addition of seven run-away convicts from Van Diemen's land. They stole (or made) a vessel & ran straight for this coast; when some distance from the land they sunk her & took to their boats. — They all took wives in about a weeks time; & the fact of their being such notorious rogues appears to have weighed nothing in the Governors opinion, in comparison with the advantage of having some good workmen. — In all these Spanish colonies, it appears to me that the committal of enormous crimes lessens but very little the public estimation of any individual; that is, as long as they remain unpunished. The Chilians in St Jago think it very hard that the Englishmen cease to hold communication with any of their countrymen who may have acted dishonorably. — This must partly be the consequence of their absolving, forgiving religion. — I am afraid however, this Christian charity, both of the public & the Church, is chiefly extended to the rich. —

[1] Pencil note in margin: 'Read over &c &c'.
[2] While in command of the Chilean navy in December 1819, Lord Cochrane landed at Valdivia with 300 men and overcame the strongly fortified Spanish base with very few casualties of his own.
[3] Pencil note in margin: 'Apple story'.

11th I set out on a short ride, in which however I managed to see singularly little either of the geology of the country or of the inhabitants. — There is not much cleared land near Valdivia; after crossing a river at the distance of a few miles, we entered the forest, & then only passed one miserable hovel before reaching our sleeping place for the night. — The short difference in latitude of 150 miles has given to the forest, as compared to that of Chiloe, another aspect. This is owing to a slightly different proportion in the kinds of trees; the evergreens do not appear to be quite so numerous; & the forest in consequence is coloured by a brighter & more lively green. — As in Chiloe, the lower parts are matted together by Canes; here also another kind, about twenty feet high and which strictly resembles in form the bamboos of Brazil, grows in clusters: the banks of some of the streams are thus ornamented in a very pretty manner. — It is with this plant that the Indians make their Chusas, or long tapering spears. — Our resting house was so dirty I preferred sleeping outside; the first night is generally an uncomfortable one, because ones body is not accustomed to the tickling & biting of the fleas: I am sure in the morning there was not the space of a shilling on my legs which had not its little red mark where the flea had feasted. —

12th We continued to ride through the uncleared forest; & only occassionally met an Indian on horseback, or a troop of fine mules bringing Alerce planks or corn from the Southern plains. In the afternoon one of the horses tired; we were then on the|528| brow of a hill which commanded a fine view of the Llanos. The view of these open plains was very refreshing, after being hemmed in & buried amongst the wilderness of trees. The uniformity of a forest soon becomes very wearisome; this West coast makes me remember with pleasure the free, unbounded plains of Patagonia; yet with the true spirit of contradiction, I cannot forget how sublime is the silence of the forest. The Llanos are the most fertile & thickly peopled parts of the country: they possess the immense advantage of being nearly free from trees; before leaving the forest we crossed some flat little lawns, around which single trees were encroaching in the manner of an English park. — It is curious how generally a plain seems hostile to the growth of trees: Humboldt found much difficulty in endeavouring to account for their presence or absence in certain parts of S. America; it appears to me that the levelness of the surface very frequently determines this point; but the cause why it should do so I cannot guess. — In the case of Tierra del Fuego the deficiency is probably owing to the accumulation of too much moisture; but in Banda Oriental, to the North of Maldonado, where we have a fine undulating country, with streams of water (which are themselves fringed with wood) is to me, as I have before stated, the most inexplicable case. —

On account of the tired horse I determined to stop close by at the Mission of Cudico; to the Friar of which I had a letter of introduction. — Cudico is an intermediate district between the forest & the Llanos: there are a good many cottages with patches of corn & potatoes nearly all|529| belonging to Indians. The Plank-built Chapel is small & in sad decay;

the Government is building a school for the Indian children. — The Padre tells me they are very easily taught any subject, & that the school will be the means of doing a great deal of good. — All the Indians belonging to Valdivia are "reducidos & Christianos"; they are divided into tribes, & have their Caciques: their quarrels & crimes are superintended by Spanish authorities, & I do not quite understand what power the Cacique has, excepting that of oppressing his subjects. The Indians to the North, about Imperial & Arauco, are yet very wild & not converted; they all have however much intercourse with the Spaniards. — There are 26 tribes more or less dependant on Valdivia; each of these have Spanish residents, called "Capitanes delos amigos", whose office is to interpret & plead for their respective tribes with the Governor of Valdivia. The Caciques of three or four of the tribes, who have remained very faithful & have been of service during the wars, receive a pension of 30 dollars a year (6 pounds sterling); a sort of bribe with which they are well satisfied to remain quiet. — Some of the tribes are large, one is supposed to have 3–4000 Indians. — The Padre says that the Indians do not much like coming to mass, but otherwise show much respect to religion; the greatest difficulty is in making them observe the ceremonies of marriage. — The wild Indians take as many wives as they can support; & a Cacique will sometimes have more than ten: — on entering his house, the number|530| can be told by that of the separate fires. This last plan must be a good one to prevent quarrelling. The wives live each a week in turns with the Cacique; but all are employed in weaving Ponchos &c for his advantage: to be the wife of a Cacique is an honor much sought after by the Indian women. — The besetting sin with all is that of drunkedness; it seems wonderful that they are able to drink enough of our sour weak cyder to make themselves drunk. — But it is certain that they remain in this state for whole days together & are then very dangerous & fierce. — The Indian temperament, all over the Americas, seeks with singular eagerness the excitation produced by Spirituous liquors. —

The common Indian dress to the South of Valdivia is a dark woollen Poncho, beneath which they wear nothing, & short tight trousers & leggings. To the North, they wear a garment folded round their bodies in the manner of the Chilipa of the Gauchos. This alone will immediately point out from which side any Indian comes. — They all wear their long hair bound by a red band, & without covering to their heads. Both of which tastes are constantly seen in the Indians on the other side of the Cordilleras. Some of the women wear curiously shaped & very large plates of silver in their ears; & I saw one man with a similar necklace; which at a distance looked like a white ruff. — It appears to me that these Indians have a slightly different physiognomy from any which I have seen; they are more swarthy, their hair is not so straight & in greater profusion, |531| their cheek bones are very prominent: they are good sized men. — The expression of their faces is generally grave & even austere; & possesses much character; this may either pass for goodnatured bluntness or for fierce determination. — On the road a traveller meets with none of that humble politeness so universal in Chiloe; some however gave their "Mari-Mari" (good morning) with promptness. — The resemblance very likely is imaginary, but the long hair, the grave & much lined features, & dark complexion, called to my mind old portraits of Charles the First. — The independence of manners of these Indians is probably a consequence of the long & victorious wars which they have fought with the Spaniards. — At present all the Southern Indians seem in a fair way of continuing subjects of Chili. — They are said to be very good horsemen; they do not much use the lazo, or the Bolas, & this latter only to the North. — The Chusa is the proper Weapon of the country. — It is odd what difficulty is found in ascertaining even the most simple question from the Spaniards. I was assured by what would appear excellent authority, that the Indian language of Chiloe is quite distinct from that of these Araucanians: yet I now am convinced they are the same. — The greater part of the latter talk some Spanish.

I spent the evening very pleasantly, talking with the Padre. — He was exceedingly kind & hospitable; & coming from St Jago had contrived to surround himself with some few comforts. |532| Being a man of some little education, he bitterly complained of the total want of society; — with no particular zeal for religion, no business or pursuit, how completely must this mans life be wasted. —

13th I found nothing worth staying for or for proceeding, so again returned through the forest. — We met seven very wild Indians, amongst whom were some Caciques who had just received their yearly stipend. They were fine upright men, but rode one after the other, with gloomy looks. An old Cacique who headed them, I suppose had been more excessively drunk than any of the rest, for he seemed both extremely grave & crabbed. — Shortly before this two Indians joined us, who were travelling from a distant Mission to Valdivia concerning some law suit. — One was a good humored old man, but from his wrinkled beardless face looked more like an old Woman. I frequently presented both with cigars; though ready to receive them & I daresay grateful, they would hardly condescend to thank me:— A Chilotan Indian would have taken off his hat & given his "Dios le pagé" (may God repay you).— My guide talked the Indian language fluently; so that I heard plenty of their conversation. It is entirely free from guttural sounds; none of the words proceeding from the throat.— We reached before night-fall a sort of warehouse for the reception of muleteers; the other of the two houses in the whole line of road.— The travelling was very tedious, from heavy rain of the preceding night; another great|533| difficulty is the number of large trees which have fallen across the road.— If they are so big that the horse cannot leap them, it is often necessary to go fifty yards on one or the other side.—

14th We reached Valdivia by noon & had the good fortune to find boats from the Beagle, so that I got on board the same evening.— I forgot to mention as a proof how congenial this climate is to the Apple tree, that in several places in the forest I found trees which must have been sown by chance. An old man illustrated his motto that "Necessidad es la Madre del invencion" by giving an account of how many things he manufactured from apples: After extracting the cyder from the refuse, he by some process procured a white & most excellently flavoured spirit (which many of the officers tasted); he also could make wine.—by a distinct process he produced a very sweet & well tasted treacle or as he called it honey.— None of these processes require much attention.

18th I crossed over to the Fort called Niebla, which is on the opposite side of the bay to the Corral where we are at anchor.— The Fort is in a most ruinous state; the carriages of guns are so rotten that Mr Wickham remarked to the commanding officer, that with one discharge they would all fall. The poor man trying to put a good face on it, gravely replied, "No I am sure Sir they would stand two!" The Spaniards must have intended to have made this place impregnable. There is now lying in the middle of the court-yard a little mountain of mortar, which rivals in hardness the rock on which it lies. — It was brought from Chili|534| & cost seven thousand dollars. The revolution breaking out prevented its being applied to any purpose; but now it remains a monument to the fallen greatness of Spain. — I wanted to go to a house about a mile & half distant; my guide said it was quite impossible to penetrate the wood in a straight line; but he offered to lead me by the shortest way, following obscure cattle tracks: after all, the walk took no less than three hours! This man is employed in hunting strayed cattle; yet well as he must know the woods, he was not long since lost for two whole days & had nothing to eat. These facts convey a good idea of the impracticability of the forest of these countries. — A question often occurred to me, how long does any vestige of a fallen tree remain? This man showed me one which a party of fugitive Royalists had cut down fourteen years ago. — Judging from the state in which it was I should think a bole a foot and a half in diameter in thirty years would present a mere ridge of mould. —

20th This day has been remarkable in the annals of Valdivia for the most severe earthquake which the oldest inhabitants remember. — Some who were at Valparaiso during the dreadful one of 1822, say this was as powerful. — I can hardly credit this, & must think that in Earthquakes as in gales of wind, the last is always the worst. I was on shore & lying down in the wood to rest myself. It came on suddenly & lasted two minutes (but appeared much longer). The rocking|535| was most sensible; the undulation appeared both to me & my servant to travel from due East. There was no difficulty in standing upright; but the motion made me giddy. — I can compare it to skating on very thin ice or to the motion of a ship in a little cross ripple.

An earthquake like this at once destroys the oldest associations; the world, the very emblem of all that is solid, moves beneath our feet like a crust over a fluid; one second of time conveys to the mind a strange idea of insecurity, which hours of reflection would never create. In the forest, a breeze moved the trees, I felt the earth tremble, but saw no consequence from it. — At the town where nearly all the officers were, the scene was more awful; all the houses being built of wood, none actually fell & but few were injured. Every one expected to see the Church a heap of ruins. The houses were shaken violently & creaked much, the nails being partially drawn. — I feel sure it is these accompaniments & the horror pictured in the faces of all the inhabitants, which communicates the dread that every one feels who has *thus seen* as well as felt an earthquake. In the forest it was a highly interesting but by no means awe-exciting phenomenon. — The effect on the tides was very curious; the great shock took place at the time of low-water; an old woman who was on the beach told me that the water flowed quickly but not in big waves to the high-water mark, & as quickly returned to its proper level; this was also evident by the wet sand. She|536| said it flowed like an ordinary tide, only a good deal quicker. This very kind of irregularity in the tide happened two or three years since during an Earthquake at Chiloe & caused a great deal of groundless alarm. — In the course of the evening there were other weaker shocks; all of which seemed to produce the most complicated currents, & some of great strength in the Bay. The generally active Volcano of Villa-Rica, which is the only part of the Cordilleras in sight, appeared quite tranquil. — I am afraid we shall hear of damage done at Concepcion. I forgot to mention that on board the motion was very perceptible; some below cried out that the ship must have tailed on the shore & was touching the bottom. —

21st We moved our anchorage to one nearer the mouth of the harbor. — during the last week there has been an unusual degree of gaiety on board. — The Intendente paid us a visit one day & brought a whole boat full of ladies: bad weather compelled them to stay all night, a sore plague both to us & them. — They in return gave a ball, which was attended by nearly all on board. Those who went returned exceedingly well pleased with the people of Valdivia. — The Signoritas are pronounced very charming; & what is still more surprising, they have not forgotten how to blush, an art which is at present quite unknown in Chiloe. — |537|

22nd We finally sailed from Valdivia & continued the survey up the coast. About thirty miles to the Northward the country becomes lower & more level, neither is it quite so concealed in forest. — We saw much cattle; & several groups of Indians on horseback appeared to watch with interest our movements. Seeing us so close to the land, they perhaps hoped we should be wrecked; a fate which happened not long since to a French Whaler, the crew of which were robbed of every single thing in a very short time. — It is said that the country of these Araucanians is the most fertile in Chili; my friend the Padre at Cudico bitterly regretted that it should be so wasted & wished with *Christian* humanity, that all the provinces would unite & make a complete end of the Indian race. —

This is a dangerous coast; shoal water extends to some distance in the offing, & a heavy swell is constantly setting right on shore. A ship in a calm in such a situation is most awkwardly placed. — The swell lost us an anchor & 16 fathoms of cable; We only anchored for an hour, & in heaving up, the jerks were so violent that the cable snapped in two. — This is the sixth anchor since leaving England!

23rd We have not been very lucky with the survey; during part of each day there has been a fog: I suppose this fog is heavy rain in Chiloe; we now are in a land of blue skys. —

24th In the evening came to an anchor under the lee of the island of Mocha: we had an unusual|538| spectacle in seeing five ships under sail at once. They are Whalers cruizing for fish. —

Remains of the cathedral at Concepcion, by S. Bull after J. C. Wickham (*Narrative* **2**: 405).

25ᵗʰ   This island of Mocha has been an island of trouble to us. This day we sailed round it making a plan; in the evening the swell prevented us anchoring. — A gale from the North followed, & the wind, instead of changing to the South, continued in this unlucky point. —

27ᵗʰ   The Captain at last effected a landing through a heavy surf; during his observations the tide fell, & it was found impossible to launch the boat again.

28ᵗʰ   On the next day, by noon, there was less surf & she was brought off. — Early in the last night a strong jerk was felt at our anchor, more cable being veered she remained all fast; but this morning we found the anchor snapped right in two. — This is most unfortunate; we have now only one anchor; & instead of being able to survey the coast, it will be necessary after touching at Concepcion to run up to Valparaiso to purchase a fresh stock.

March 1ˢᵗ & 2ⁿᵈ   To crown our ill fate, we now have a light foul wind. Nobody, but those on board a ship can know how vexatious these petty misfortunes are. —

3ʳᵈ   We felt, on board, a very smart shock of an earthquake: some compared the motion to that of a cable running out, & others to the ship touching on a Mud bank. — Capt. FitzRoy heard when on Mocha that the Sealers had experienced a succession of shocks during the last fortnight.

4ᵗʰ   As soon as the ship entered the harbor of Concepcion, I landed|539| on the island of Quiriquina, & there spent the day, whilst the ship was beating up to the anchorage. The Major domo of the estate rode down to tell us the terrible news of the great Earthquake of the 20ᵗʰ: — "That not a house in Concepcion or Talcuhano (the port) was standing, that seventy villages were destroyed, & that a great wave had almost washed away the ruins of Talcuhano". — Of this latter fact I soon saw abundant proof; the whole coast was strewed over with timber & furniture as if a thousand great ships had been wrecked. Besides chairs, tables, bookshelves &c &c in great numbers, there were several roofs of cottages almost entire, Store houses had been burst open, & in all parts great bags of cotton, Yerba, & other valuable merchandise were scattered about. During my walk round the island I observed that numerous fragments of rock, which form the marine productions adhering to them must recently have been lying in deep water, had been cast high up on the beach: one of these was a slab six feet by three square & about two thick. —

The Island itself showed the effects of the Earthquake, as plainly as the beach did that of the consequent great wave. Many great cracks which had a North & South direction, |540| traversed the ground; some of these near the cliffs on the coast were a yard wide; & many enormous masses in every part had fallen down; in the winter when the rain comes, the water will cause greater slips. The effect on the underlying hard slate was still more curious; the surface being shattered into small fragments. — If this effect is not confined, as I suppose it is, to the upper parts, it appears wonderful that any solid rock can remain in Chili. — For the future when I see a geological section traversed by any number of fissures, I shall well understand the reason. I believe this earthquake has done more in degrading or lessening the size of the island, than 100 years of ordinary wear & tear.

5th I went on shore to Talcuana, & afterwards rode with the Captain to Concepcion. — The two towns presented the most awful yet interesting spectacle I ever beheld. — To any person who had formerly known them it must be still more so; for the ruins are so confused & mingled & the scene has so little the air of an habitable place that it is difficult to understand how great the damage has been. — Many compared the ruins to those of Ephesus or the drawings of Palmyra & other Eastern towns; certainly there is the same impossibility of imagining their former appearance & condition. In Concepcion each house or row of houses stood by itself a heap or line of ruins: in Talcuhano, owing to the great wave little more was left than *one* layer of bricks, tiles & timber, with here & there part of a|541| wall yet standing up. From this circumstance Concepcion, although not so completely desolated, was the more terrible, & if I may so call it, picturesque sight. The Earthquake took place, as we have seen at Valdivia, at half past eleven. It is generally thought if it had happened in the night, at least $\frac{3}{4}$^{s} of the inhabitants would have perished. It is probable that not more than 100 have met their deaths; yet many must still lie buried in the ruins. The earthquake came on with tremendous violence & gave no notice; the *constant* habit of these people of running out of their houses *instantly* on perceiving the *first* trembling only saved them. The inhabitants scarcely passed their thresholds before the houses fell in. This is thought to be the worse Earthquake ever known in Chili; it is however hard to tell, for the worst sorts happen only after long intervals from 60 to 100 years. Indeed several degrees worse would not signify, for the desolation is now complete. After viewing the ruins of Concepcion, I cannot understand how the greater part escaped unhurt; the houses in many places have fallen outwards on each side into the street, so that it is frequently necessary to pass over little hillocks several feet high. In other places the houses fell in; in a large boarding school, the beds were buried 8 feet beneath bricks, yet all the young ladies escaped. — How dreadful would the slaughter have been, if as I said it had happened at night. Mr Rous, the English Consul, told us he was|542| at breakfast; at the first motion he ran out, but only reached the middle of his little court-yard when one side of his house came thundering down; he retained presence of mind to remember that if he once got on the top of that part which had already fallen, he should be safe; not being able, from the motion of the ground, to stand on his legs he crawled up on his hands & knees; no sooner had he ascended this little eminence, than the other side of the house fell in; the great beams sweeping close in front of his head. — The sky became dark from the dense cloud of dust; with his eyes blinded & mouth choked he at last reached the street. Shock succeeded Shock at the interval of a few minutes; no one dared approach the shattered ruins; no one knew whether his dearest friends or relations were perishing from the want of help. The thatched roofs fell over the fires, & flames burst forth in all parts; hundreds knew themselves ruined & few had the means of procuring food for the day. — Can a more miserable & fearful scene be imagined? —

I shall never again laugh when I see people running out of their houses at a trifling shock; nor will any on board who now has seen what an Earthquake is. The earthquake alone is sufficient to destroy the prosperity of a country; if beneath England a volcanic focus should reassume its power; how completely the whole country would be altered. What would become of the lofty houses, thickly packed cities, the great manufactories, the beautiful|543| private & public buildings? If such a Volcanic focus should announce its presence by a great earthquake, what a horrible destruction there would be of human life. — England would become bankrupt; all papers, accounts, records, as here would be lost: & Government could not collect the taxes. — Who can say how soon such will happen? — Talcuana is built on a low flat bit of ground at the foot of some hills; a great wave, so common an occurrence with Earthquakes, entirely flowed over the whole town; after the houses had been shaken down, the destruction caused by the water can be well imagined. Few of the inhabitants were drowned; for the unbroken swell was seen travelling onwards at the distance of 5 or 6 miles. The people ran for the high land; — as soon as the swell came close on shore it broke & is believed to have risen 23 ft higher than the

Spring tides; it was followed by two other lesser ones; in the retreat of the water many things which could float were carried out to sea; hence the wreck on Quiriquina. The force of the wave must have been very great, for in the fort a gun & carriage, which some of the officers thought weighed about 4 tuns, was removed 15 ft upwards. — 200 yards from the beach & well within the town there is now lying a fine Schooner, a most strange witness of the height of the wave. Before the swell reached the town it was seen tearing up all the cottages which were scattered around the Bay; some boats pulled out to meet it, the men knowing well that they would be|544| safe if they reached the wave before it broke. In the confusion a little English boy 4 or 5 years old & an old woman got into a boat, but with nobody to pull them to seaward; the surf in consequence carried the boat with immense force into the town, where striking against an anchor it was cut into two; the old woman was drowned but the little boy clinging to the broken boat was carried out to sea, & was picked up some hours afterwards quietly seated on the thwart. The Ships at anchor were whirled about; two which were near each other had their two cables with three turns; although anchored in 36 ft water, they were for some minutes aground. — In another part of the harbor a vessel was pitched high & dry on shore, was carried off, was again driven on shore & again carried off!— The wave is said to have come from the South and in its road sadly devastated the Is^d of St Mary; it is certain that it entered this harbor by the entrance nearest to the South. The permanent level of the land & water is, I believe, altered, but this Capt. FitzRoy will investigate when we return. —[1]

At this present time there are pools of sea water in the streets of the town; & the children making boats with old tables & chairs, appear as happy as their parents are miserable.— I must however say it is admirable to see how cheerful & active every-body is. Mr Rous remarked that it makes a wonderful difference the misfortune being universal: a man is not humbled, he has no reason to suspect|545| his friends will look down on him & this perhaps is the worst part in loss of wealth. Mr Rous has a few Apple trees in his garden. He & a large party lived there for the first week & were as merry as if it had been a pic-nic. Some heavy rain after that period added much to their misery; many, Mr Rous for one, being absolutely without any shelter. Almost every one has now made a hut with planks. The hovels built of sticks & straw which belonged to the poorest class of people, were not shaken down, & they are now hired at a high price by the richest people. We saw many pretty ladies standing at the doors of such Ranchos. Those who have estates have gone there: The town is in such complete ruins that it is not yet decided whether it will not be better to change the situation, although at the loss of the close neighbourhead of the materials. —

Heavy misfortunes are well known to make the bad worse; & here there were many robbers; there was a mixture of religion in their depredations which we should not see in England; at each little trembling of the ground, with one hand they beat their breasts & cried out "Miserecordia", & with the other continued to filch from the ruins. The necessity of every man watching what he contrived to save, added much to the trouble of the more respectable inhabitants. —

With respect to the extent of the earthquake, we know it was severely felt at Valdivia; at Valparaiso they had|546| a sharp shock but it did no damage. — All the towns, Talca, Chillan, &c &c between Concepcion & St Jago have been destroyed, till we reach S. Fernando, which has only been partially destroyed. We may imagine the shock at this place & at Valdivia to have had the same degree of force, & looking at the map, they will be found to be nearly equally distant; hence Concepcion may be supposed to be about the centre of the disturbance; The length of coast which has been *much* affected is rather less than 400 miles. Mr Rous thinks the vibration came from the East, & this would appear probable from the greater number and & longest cracks having a N & S direction, which line would correspond to the tops of the undulations. — The Volcano of Antuco, which is a little to the North of Concepcion is said to be in great activity. The people in Talcuana say that the Earthquake is owing to some old Indian Woman two years ago being offended, that they by witchcraft *stopped the Volcano*, & now comes the Earthquake. This silly belief is curious because it shows that experience has taught them the constant relation between the suppressed activity of volcanoes & tremblings of the ground. It is necessary to apply the Witchcraft to the point where their knowledge stops, & this is the closing of the Volcanic Vent.

The town of Concepcion is built, as is usual, with all its streets at rt angles; one set runs (SW by W & NE by E) & the other (NW by N & SE by S). The walls which have the former|547| direction certainly have stood better than those at right angles to them; If, as would seem probable Antuco may be considered as the centre it lying rather to the Northward of Concepcion,[2] the concentric lines of undulation would not be far from coincident with NW by N & SE by S walls: this being the case the whole line would be thrown out of its centre of gravity at the same time & would be more likely to fall, than those which presented their ends to the shock. The different resistance offered by the two sets of walls is well seen in the great Church. This fine building stood on one side of the Plaza: it was of considerable size & the walls very thick, 4 to 6 ft & built entirely of brick: the front which faced the NE forms the grandest pile of ruins I ever saw; great masses of brick-work being rolled into the square as fragments of rock are seen at the base of mountains. — Neither of the side walls are entirely down, but exceedingly fractured; they are supported by immense buttresses, the inutility of which is exemplified by their having been cut off smooth from the wall, as if done by a chisel, whilst the walls themselves remain standing. There must have been a rotatory motion in the earth for square ornaments placed on the coping of this wall are now seated edgeways. — Generally in all parts of the town arched doorways & windows stood pretty well; an old man however, who was lame, had always been in the custom of running to a certain doorway; this time however it fell & he was crushed to pieces. — |548|

With my idea of a vibration having come from Antuco, the Northward of E, I cannot understand the wave travelling from the South. The cause however of an earthquake causing one, two, or three great waves does not to me appear very clear. —

The effect of so violent a shock on the springs was of course considerable; some poured out much more water than usual, some were closed: in one place black hot water flowed from a crack & it is said bubbles of gas & discoloured water were seen rising in the Bay. Many geological reasons have been advanced for supposing that the earth is a mere crust over a fluid melted mass of rock & that Volcanoes are merely apertures through this crust. When a Volcano has been closed for some time, the increased force (whatever its nature may be) which bursts open the orifice might well cause an undulation in the fluid mass beneath the earth; at each successive ejection of Lava a similar vibration would be felt over the surrounding country; these are known gradually to become less & less frequent, & with them probably the earthquakes, till at last the expansive force is counterbalanced by the pressure in the funnel of the Volcano. — Where Earthquakes take place without any volcanic action, we may either imagine that melted rock is injected in the inferior strata, or that an abortive attempt at an eruption has taken place beneath the Volcano. — On the supposition of an inferior fluid mass there is no difficulty|549| in understanding that gases, the results of the Chemical action of the great heat, should penetrate upwards through the cracks; or water that had percolated deep near to the

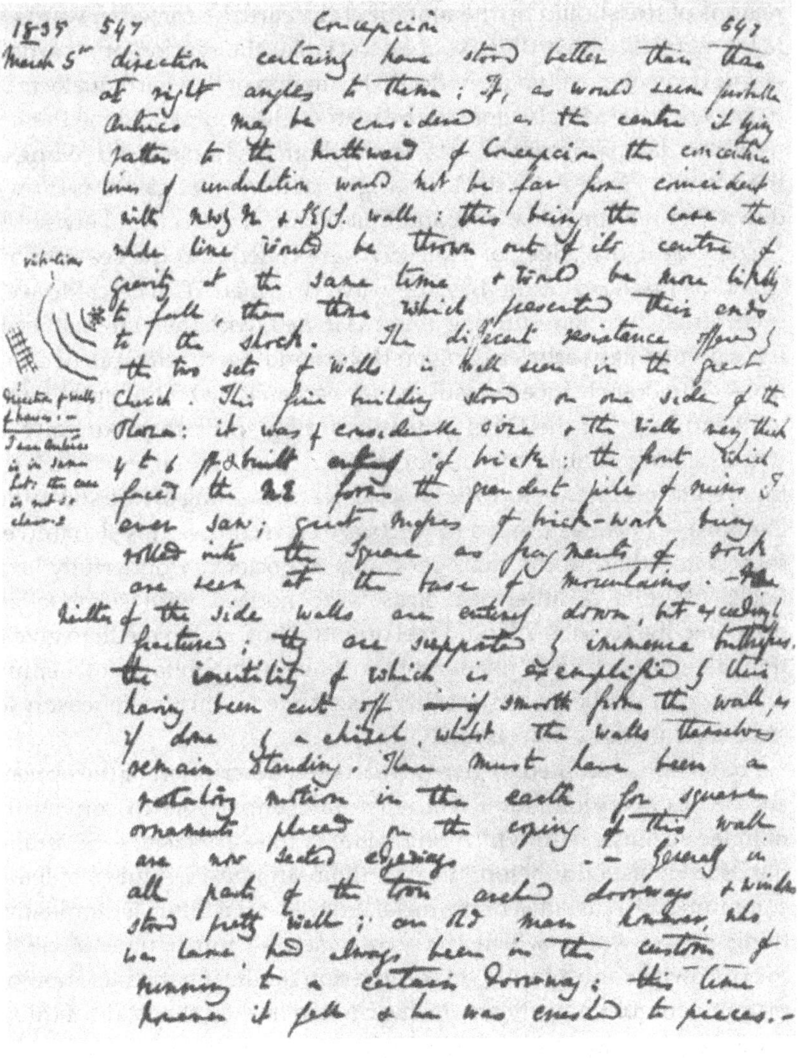

A page of the entry for 5 March 1835 in the MS diary.

regions of fire should by the motion of the earth be forced upwards. —
Most certainly an earthquake feels very like the motion of a partially
elastic body over a fluid in motion. The motion of this Earthquake must
have been exceedingly violent; the man at Quiriquina told me the first
notice he had of the shock, was finding both his horse & self rolling on
the ground. He rose, hardly knowing what it was, & again was thrown
down, but not the horse a second time; some of the cattle likewise fell,
& some near the edges of the cliffs were rolled into the sea. On one
island at the head of the Bay the wave drowned 70. The cattle were
exceedingly terrified, running about as if mad, with their tails in the air.
It is said that light articles lying on the ground, were fairly pitched to &
fro. — The French Vice Consul mentioned a fact which if authentic is
very curious, that the Dogs generally during an Earthquake howl, as
when hearing military music, but that this time they all quietly left the
town some minutes before the shock & were standing on the surrounding hills. — I believe other such facts are on record. — It is also universally stated that on the same morning at 9 oclock, wonderfully large
flocks of gulls & other sea birds were noticed with surprise|550|
directing their course inland. I feel doubtful how much credit to give to
this statement: I have not forgotten that the inhabitants of Lemuy,
when we in the boats arrived there, exclaimed, "this is the reason we
have seen so many parrots lately". —

I have not attempted to give any detailed description of the appearance of Concepcion, for I feel it is quite impossible to convey the
mingled feelings with which one beholds this spectacle. — Several of
the officers visited it before me; but their strongest language failed to
communicate a just idea of the desolation. — It is a bitter & humiliating
thing to see works which have cost men so much time & labour
overthrown in one minute; yet compassion for the inhabitants is almost
instantly forgotten by the interest excited in finding that state of things
produced at a moment of time which one is accustomed to attribute to
a succession of ages. — To my mind since leaving England we have
scarcely beheld any one other sight so deeply interesting. The Earthquake & Volcano are parts of one of the greatest phenomena to which
this world is subject.

[1] FitzRoy wrote: 'Besides suffering from the effects of the earthquake and three invading
waves, which, coming from the west round both points of the island, united to overflow
the low ground near the village, Santa Maria was upheaved nine feet. It appeared that the
southern extreme of the island was raised eight feet, the middle nine, and the northern
end upwards of ten feet. The Beagle visited this island twice—at the end of March and in
the beginning of April: at her first visit it was concluded, from the visible evidence of dead

shell-fish, water-marks, and soundings, and from the verbal testimony of the inhabitants, that the land had been raised about eight feet. However, on returning to Concepcion, doubts were raised; and to settle the matter beyond dispute, one of the owners of the island, Don S. Palma, accompanied us the second time. An intelligent Hanoverian, whose occupation upon this island was sealing, and who had lived two years there and knew its shores thoroughly, was also passenger in the Beagle.'

'When we landed, the Hanoverian, whose name was Anthony Vogelborg, showed me a spot from which he used formerly to gather "choros", a large kind of muscle, by diving for them at low tide. At dead low water, standing upon the bed of "choros", and holding his hands up above his head, he could not reach the surface of the water: his height is six feet. On that spot, when I was there, the "choros" were barely covered at high spring-tide.'

'Riding round the island afterwards, with Don Salvador and Vogelborg, I took many measures in places where no mistake could be made. On large steep-sided rocks, where vertical measures could be correctly taken, beds of dead muscles were found ten feet above the recent high-water mark. A few inches only above what was then the spring-tide high-water mark, were putrid shell-fish and seaweed, which evidently had not been wetted since the upheaval of the land. One foot lower than the highest bed of muscles, a few limpets and chitons were adhering to the rock where they had grown. Two feet lower than the same muscles, chitons and limpets were abundant.' See *Narrative* 2: 413–14.

[2] Note and diagram in margin: 'I see Antuco is in same Lat: the case is not so clear'. (See p. 301.)

6th I crossed the Bay to Linguen to see the best coal-mine of Concepción: as all the rest which I have seen, it is rather Lignite than Coal & occurs in a very modern formation. — The mine is not worked, for the coal when placed in a heap has the singular property of spontaneously igniting, it is certain that several vessels have been set on fire. — I found Capt. Walford, a Shropshire man, |551| residing in a nice quiet valley. — Linguen is a short distance from Penco; the former port of Concepcion, which was destroyed by an Earthquake & consequent wave in the year 1751. — From what I could see at the distance, the overthrow could not have been so complete as now at Talcuana. How strange it is with this example before their eyes, people should build houses & massive churches with bricks. —

I am much disappointed with the scenery of Concepcion; the outline of the land is very tame & no part of the Cordilleras or intermediate high mountains are in view: In the general aspect of the vegetation there is a greater similarity to Valparaiso than the damp forests of the South; yet here in the valleys there is plenty of wood. — I could see none of the Park-like scenery mentioned by Capt. Basil Hall[1] on the road to Concepcion: it might perhaps wear that aspect to a person who had just returned from the sterile sands of Peru. —

In the course of the day I felt two smart Earthquakes & there was a third which I did not notice; the first was sufficient to make a heap of tiles rattle. — Yet I believe this day has been freer from shocks than almost any one since the great & first Earthquake. They expect to feel small tremblings for some weeks to come. — These shocks render the searching for property amongst the ruins very dangerous; for there must always be a great probability of the shattered walls falling in. |552|

[1] See Basil Hall, *Extracts from a journal written on the coasts of Chili, Peru and Mexico for the years 1820, 1821, 1822.* 2 vols. Edinburgh, 1824.

7th   After the last three active days we made sail for Valparaiso. — Mr Stokes & Usborne are left on shore with tents to work at the Charts. It is a great convenience to many of the inhabitants our proceeding & returning directly from Valparaiso. — there is a great derth of money, & we shall be able to bring a supply. — The Captain took on board a Padre, whom we found houseless. — We had known him at Chiloe. — The wind being Northerly, we only reached the mouth of the harbor after it was dark; a heavy fog coming on & being very near the land we dropped the anchor. Presently a large American Whaler appeared close a long side of us: we heard the Yankee swearing at his men to keep quiet whilst he listened where the breakers were: The Captain hailed him in a loud clear voice to anchor where he then was. The poor man must have thought the voice came from the shore, such a Babel of cries at once issued from the ship; every one hollowing out, "Let go the anchor, veer cable, shorten sail"; it was the most laughable thing I ever heard; if the ship had been full of Captains & no men to work, there could not have been such an uproar of orders. We afterward found the Mate stuttered: I suppose all hands were assisting him in giving his orders. —

11th   After a succession of calms we reached Valparaiso in the evening. — On the next day I moved into Mr Corfields house in the Almendral. |553|

14th   Set out for St Jago in one of the covered gigs or Birloches which travel between the two places; sleeping at the Post house at the foot of the Rado, reached the city early in the day.

15th   Mr Caldcleugh most kindly assisted me in making all the little preparations for crossing the Cordilleras & on the 18th started by the Portillo Pass for Mendoza.

18th   I took with me my former companion, Mariano Gonzales, & an Arriero with ten mules & the Madrina. The Madrina is a mare with a little bell round her neck; she is a sort of step-mother to the whole troop. — It is quite curious to see how steadily the mules follow the sound of the Bell, — if four large troops are grazing together during the night, the Muleteers in the morning have only to draw a little apart each Madrina & tinkle the Bell, & immediately the mules although 2–300 together, will all go to their proper troop. — The affection of the mules for the Madrina saves an infinity of trouble; if one is detained for several hours & then let loose, she will like a dog track out the Troop or rather the Madrina, for *she* seems the chief attraction. — Six of the mules were for riding & four for Cargoes; each taking turn about. — We carried a good deal of food, in case of being snowed up, as the season was rather late for passing by the Portillo. — Leaving St Jago in the morning we rode over the great burnt up plain till we arrived at the mouth of the valley of the Maypo. This is one of the principal rivers in Chili; the valley is bounded by the high mountains of the first Cordilleras; it is not broard, but very fertile. The numerous cottages are surrounded with Grapes, Apples, Nectarines &|534|[1] Peaches; the boughs of the latter were bending & breaking with the weight of the beautiful ripe fruit. In the evening we passed the Custom house, where our boxes were examined. The frontier of Chili is better guarded by the Cordilleras than by so much sea; the mountains on each side of the few narrow valleys where there are Custom-houses, are far too steep & high for any beast of burden to pass over. — The officers were very civil, partly owing to my carrying a strong passport from the President of Chili. But I must express my admiration of the politeness of every Chileno. In this instance the contrast is strong with the same class of officers in England. — I may mention an anecdote which at the time struck me: we met in Mendoza a very little, fat, poor Negress, with so enormous a goitre, that ones eyes almost involuntarily were fixed with surprise; but I noticed my two companions after looking for a short time took off their hats as an apology. Where would one of the lower classes in Europe show such feeling politeness to a poor & miserable object of a degraded race?— We slept at a cottage; our manner of travelling is delightfully independent; in the inhabited parts, we hire pasture for the animals, buy a little firewood, & bivouac in the corner of the field; carrying our cooking apparatus, we eat our supper under the cloudless sky & know no troubles. —

[1] There is an error in the pagination of the manuscript at this point, since CD has written 534 instead of 554, and continues with 535 on the next page, so that there are two sets of pages numbered 534–553.

19th We rode during this day to the last or highest house in the valley. — The number of inhabitants became scanty, but wherever water could be brought on the land it was very fertile. The valley is very narrow &|535|

consists of a plain of shingle, generally elevated some hundred feet above the river. The Maypo is rather a great mountain torrent than a river: the fall is very great, & the water the color of mud; the roar is very like that of Sea, as it rushes amongst the great rounded fragments. — Amidst the din, the noise of the stones rattling one over the other is most distinctly audible. — The hills on each side are I suppose 3–5000 ft high; their faces are very steep & bare, the color generally purple & stratification of the rocks is very striking; but the forms are not wild. — If the scenery is not very beautiful, it is remarkable & grand. — We met during the day several troops of Cattle which had been driven down from the higher valleys; this sign of the approaching winter hurried our steps more than was convenient for geology. Our sleeping place was about a league below where the Maypo divides into R. del Valle del Yeso & R. del Volcan; the valley leaving here its Southerly course enters more directly the main Cordilleras. — The house also is at the foot of the mountain, in the top of which are the mines of S. Pedro de Nolasko: the ascent to which is so spiritedly described by Capt. Head. —[1] We saw the mules creeping up the zigzag track. — Even at this advanced season of the year, there were some small patches of Snow on the summit; — the height must at least be 10,000 ft. —Capt. Head wonders how mines in such extraordinary situations are discovered. In the first place, metallic veins are here generally harder than|536| the surrounding strata, hence during the gradual degradation of the hills, they project above the surface of the ground. — Secondly almost every labourer, especially in the Northern parts of Chili, understands something about the appearances of ores. — In the great mining provinces of Coquimbo & Copiapo, firewood is very scarce & men are employed in searching for it over every hill & dale; by this means nearly all the richest mines have been discovered. — Chanuncillo, from which silver to the value of many hundred thousand pounds has been raised in the course of a few years, was thus discovered by a man picking up a stone to throw at his loaded donkey, which afterwards it struck him was very heavy, & again picking it up, he found it was nearly pure Silver. — The vein occurred at no great distance standing up like a wedge of Silver. — The miners also, on Sundays, taking a crowbar often sally out on such discoveries. — In the South part of Chili, the men who drive cattle into the Cordillera & who hunt out every ravine where there is a little pasture, are the usual agents. —

[1] See Francis Bond Head, *Rough notes taken during some rapid journeys across the Pampas and among the Andes.* London, 1826.

20th   As we ascended the valley the vegetation became exceedingly scanty; there were however a few very pretty Alpine plants. — Scarcely a bird or insect was to be seen. — The lofty mountains, their summits marked with a few patches of snow, stood well separated one from the other; The valleys are filled up with an enormous thickness of Alluvium. In the scenery of the Andes, the parts which strike me as contrasted with the few other mountain chains which I have seen, are; — the flatness of the valleys, the narrow plain being composed of shingle, through which the river |537| cuts a channel. Geological reasons induce me to believe that this gravel &c was deposited by the ocean when it occupied these ravines, & that the agency of the rivers is solely to remove such rubbish. If such be the case the elevation of the Andes, being posterior to most other mountains, accounts for these fringes[1] still remaining attached to the sides of the valleys. Again, the bright colors, chiefly red & purple, of the utterly bare & steep hills; — the great & continuous wall-like dykes; — the manifest stratification, which where nearly vertical causes the wildest & most picturesque groups of peaks, where little inclined we have massive unbroken mountains; these latter occupy the outskirts of the Cordilleras, as the others do the more central & lofty parts. And lastly the vast piles of fine & generally bright colored detritus. These decline from the sides of the mountains at a high angle into the bottom of the valley. These smooth & unbroken conical piles must often have an elevation of 2000 ft.

I have often noticed that where snow lies long on the ground, the stones seem very apt to crumble, and in the Cordilleras, *Rain* never falls. Hence the quantity of degraded rock. —[2] It occassionally happens that in the Spring, a quantity of such rubbish falls over the drift snow at the base of the hills; & so forms for many years a natural Ice-house: We rode over one of these: the elevation is far beneath the line of perpetual snow. — During the day, in a very desert & exposed part of the valley, we passed the remains of some Indian houses; I shall have occasion to mention this subject again. —

As the evening was drawing on, we reached the |538| Valle del Yeso. This is a very singular basin which must once have been a large lake. — The barrier is formed by what deserves the name of a mountain of Alluvium, on one side of which the river has cut a gorge. The plain is covered by some dry pasture, & we had the agreeable prospect of herds of cattle. The valley is called Yeso from a great bed, I should think nearly 2000 ft thick, of white & in many parts quite pure Gypsum. — We slept

with a party of men who were employed in loading mules with this
substance & who had come up for the Cattle. —

[1] Note in margin: 'what *fringes*?'
[2] Note in margin: 'Scoresby Spitzbergen'.

21st We began our march early in the day; we followed the course of the
river, which by this time was small, till we arrived at the foot of the ridge
which separates the waters which flow into the Pacifick & Atlantic
Oceans. Untill now our road had been good & the ascent steady but
very gradual; now commenced the steep zigzag track. — The Cordilleras in this pass consist of two principal ridges, each of which must be
about 12000 ft high; the first called Puquenes forms the division of the
waters & hence of the Republics of Chili & Mendoza; to the East of this
we meet an undulating track with a gentle fall & then the second line of
the Portillo; through this, some way to the South the intermediate
waters have a passage. —

We began the tedious ascent, & first experienced some little difficulty
in the respiration. The mules would halt every fifty yards & then the
poor willing animals would after a few seconds of their own accord start
again. — The short breathing from the rarified air is called by the
Chilenos, Puna. They have most|539| ridiculous ideas respecting its
nature; some say "all the waters here have Puna" others that "where
there is snow there is Puna" & which no doubt is true. — It is considered
a sort of disease, & I was shown the crosses of several graves where
people had died "*Punado*". I cannot believe this, without perhaps a
person suffering from organic disease of the Chest or Heart: or very
likely any one dying from whatever cause would have unusual difficulty in breathing. The only sensation I experienced was a slight
tightness over the head & chest; a feeling which may be known by
leaving a warm room & running violently on a frosty day. — There was
a good deal of fancy even in this, for upon finding fossil shells on the
highest ridge, in my delight I entirely forgot the Puna. Certainly the
labor of walking is excessive, & in breathing *deep* & difficult; & it is
nearly incomprehensible to me how Humboldt (& others subsequently)
have reached 19000 ft. No doubt a residence of some months in Quito,
10000 ft high would prepare the constitution for such an exertion. Yet
in Potosi, strangers, I am told, suffer for about a year. —

When about halfway up, we met a large party of seventy loaded
mules & passengers; it was a pretty sight to see the long string
descending, & hear the wild cries of the Muleteers; they looked so

diminutive; no bushes, nothing but the bleak mountains with which to compare them. — Near the summit the wind, as is almost always the case, was violent & very cold; on each side of the ridge we had to pass over broard bands of Snow, which is perpetually there & now would soon be covered by a fresh layer. — I there first observed the substance described by the|540| Arctic navigators as Red Snow.[1] Subsequently I found under the microscope it consists of groups of minute red balls, the diameter of which is $\frac{1}{1000}$th of an inch, & having several envelopes. — The snow was only tinged where crushed by the mules hoofs & where the thaw had been rapid. —

When we reached the crest & looked backwards, a glorious view was presented. The atmosphere so resplendently clear, the sky an intense blue, the profound valleys, the wild broken forms, the heaps of ruins piled up during the lapse of ages, the bright colored rocks, contrasted with the quiet mountains of Snow, together produced a scene I never could have imagined. Neither plant or bird, excepting a few condors wheeling around the higher pinnacles, distracted the attention from the inanimate mass. — I felt glad I was by myself, it was like watching a thunderstorm, or hearing in the full Orchestra a Chorus of the Messiah.[2]

We descended into the intermediate district & took up our quarters for the night: the elevation cannot be much short of 10,000 ft, in consequence the vegetation is very scanty & there are no bushes; the roots of a certain plant which are thick, serve for bad fuel. — It was piercingly cold, & I having a headache went to bed. — During the night the sky suddenly became clouded; I awakened the Arriero to know if there was any danger, but he told me, without thunder & lightning there is no risk of a bad Snow storm. The peril is imminent, & the difficulty of subsequent escape great, to a person caught in a|541| heavy storm between the two Cordilleras. There is only one place of shelter, a cave, where Mr Caldcleugh, who crossed on the very same day of the month, took refuge for some time. From this cause the Portillo pass in the Autumn is so much more dangerous than the other one, where there are Casuchas built. —

Under the diminished pressure, of course water boils at a lower temperature; in consequence of this the potatoes after boiling for some hours were as hard as ever; the pot was left on the fire all night, but yet the potatoes were not softened. I found out this, by overhearing in the morning my companions discussing the cause; they came to the simple conclusion that "the cursed pot (which was a new one) did not choose to boil potatoes". —

[1] In Down House Notebook 1.13, CD's observations on the red snow are recorded in greater detail. He wrote: 'Red snow. Seen on both of the highest ridges—above limit of perpetual snow—little spores rather more than twice their diameter apart—appeared like bits of brown dirt scattered over snow—partly optical deception seen through the globule of ice: appeared of all sizes, to about 1/8 of inch. — When picked up appear to disappear: Examined in lens are groups of 20–40 little circular balls & through both lens appear like eggs of small molluscous animals:— Crushed, stain fingers & paper. Noticed by hoofs of mules & where thawed: thought it dust of Brecc & Porph, although remembering Miers: Colour where mules have trod, beautiful rose with slight touch of brick red: Examine paper.' See CD and the Voyage p. 233. He wrote a month later to Henslow: 'I send with this letter my observations & a piece of Paper on which I tried to dry some specimens. If the fact is new, & you think it worth while, either yourself examine them or send them to whoever has described the specimens from the North, & publish a notice in any of the periodicals. —' See Correspondence 1: 433. In Journal of Researches pp. 394–5, the red snow was later identified as arising from the plant Protococcus nivalis.

[2] Followed by a deleted sentence: 'This one view stands distinct in my memory from all others.'

22nd After eating our potatoe-less breakfast, we travelled across the intermediate tract to the foot of the Portillo range. In the very middle of summer cattle are brought up here to graze, but they had now all been removed, even the greater number of the guanaco had decamped, they knowing well that if overtaken by a snow storm they would be caught in a trap.

We had a fine view of a mass of Mountains called Tupungato, the whole clothed with unbroken snow; from one peak my Arriero said he had once seen smoke proceeding; I thought I could distinguish the form of a large crater.— In the maps Tupungato flourishes as a single mountain; this Chileno method of giving one name to a tract of mountains is a fruitful source of error.— In this region of snow there was a blue patch; no doubt a glacier.— A phenomenon which is not thought to occur in these mountains.—|542| Again we had a heavy & long climb similar to that up the Puquenes range. On each hand were bold conical hills of red Granite. We had to pass over still broader pieces of perpetual snow; this by the action of the thaw had assumed the form of numberless pinnacles, which as they were close together & high rendered it difficult for the Cargo Mules to pass. (Note in margin: Mem. Icebergs Arctic Regions.) A frozen horse was exposed, sticking to one of these points as to a pedestal, with its hind legs straight up in the air; the animal must have fallen into a hole head downmost & thus have died. — When nearly on the ridge we were enveloped in a cloud which was continually falling in the shape of minute frozen spiculæ. This was very unfortunate as it continued the whole day & quite intercepted the view. This pass of the Andes takes the name of the Portillo from a narrow cleft in the crest of this range, through which the road passes. — From this point on a clear day the great plains are to be seen. —

We descended to the first vegetation, & found good quarters under the shelter of some large fragments of rock. — We have found some passengers, who made anxious enquiries about the state of the Roads. — Shortly after it was dark, the clouds suddenly cleared away; the effect was quite magical, the great mountains, bright with the full-moon, seemed impending over us from all sides, — as if we had been at the bottom of some deep crevice. — I saw the same striking effect one morning very early. — Now that the clouds were dispersed it froze severely: but as there was no wind we were very comfortable. The increased brilliancy|543| of the moon & stars at this elevation is very striking & is clearly owing to the great transparency of the Air. — All travellers have remarked on the difficulty of judging of heights & distances in mountainous districts & generally attribute it to the want of objects of comparison. — It appears to me that it is full as much owing to the extreme transparency, confounding different distances; & partly likewise to the novel degree of fatigue from a little exertion opposing habit to the evidence of the senses. I am sure this transparency gives a peculiar aspect to the landscape; to a certain extent all the objects are brought in one plane as in a drawing. The cause of this state of the atmosphere is, I presume, owing to the equal dryness. The skin & some of the flesh of the Carcases of dead animals are preserved — articles of food, such as bread & sugar, become very hard — woodwork shrinks, as I found with my Geological hammer. All of which shows the extreme dryness. Another curious effect is the facility with which Electricity is excited. My flannel waistcoat appeared in the dark when rubbed as if washed with Phosphorus — every hair on a dogs back crackled, the sheets & leather gear of the saddle in handling all sent out sparks. —

23rd The descent is much shorter & therefore steeper than that on the other side. — that is, the Cordilleras rise more abruptly from the plains than from the Alpine country of Chili. At some depth beneath our feet there was extended a level & brilliantly white sea of clouds which shut out from our view the equally level Pampas. We soon entered the band of clouds|544| & did not again emerge from them. At one oclock finding pasture & bushes for firewood at a spot called Los Arenales, we stopped there for the night. This is nearly the uppermost limit of bushes. I should apprehend the elevation to be about 7000 ft. —[1] I was surprised at the general difference of the vegetation in the valleys on this side & those of the other; & still more so with the close identity in the greater part of all the living productions with Patagonia. I recognised here many of the thorny bushes & plants which are common on those sterile plains, & with them we have the same birds & peculiar insects. It has always been a subject of regret to me that we were unavoidably compelled to give up the ascent of the S. Cruz river before reaching the mountains. I always had a latent hope of meeting with some great change in the features of the country; I now feel sure it would only have been pursuing the plains of Patagonia up an ascent. —

[1] The text for 23 March in *Journal of Researches* pp. 399–400 follows the *Diary* closely up to this point, but CD then inserts a significant comment on the question of geographical distribution of species that has been a recurrent theme in his notebooks. He writes: 'I was very much struck with the marked difference between the vegetation of these eastern valleys and that of the opposite side: yet the climate, as well as the kind of soil, is nearly identical, and the difference of longitude very trifling. The same remark holds good with the quadrupeds, and in a lesser degree with the birds and insects. We must except certain species which habitually or occasionally frequent elevated mountains; and in the case of the birds, certain kinds, which have a range as far south as the Strait of Magellan. This fact is in perfect accordance with the geological history of the Andes; for these mountains have existed as a great barrier, since a period so remote that the whole races of animals must subsequently have perished from the face of the earth. Therefore, unless we suppose the same species to have been created in two different countries, we ought not to expect any closer similarity between the organic beings on opposite sides of the Andes, than on shores separated by a broad strait of the sea. In both cases we must leave out of the question those kinds which have been able to cross the barrier, whether of salt water or solid rock.
Footnote: This is merely an illustration of the admirable laws first laid down by Mr Lyell of the geographical distribution of animals as influenced by geological changes. The whole reasoning, of course, is founded on the assumption of the immutability of species. Otherwise the changes might be considered as superinduced by different circumstances in the two regions during a length of time.'
'A great number of the plants and animals were absolutely the same, or most closely allied with those of Patagonia. We here have the agouti, bizcacha, three species of armadillo, the ostrich, certain kinds of partridges, and other birds, none of which are ever seen in Chile, but are the characteristic animals of the desert plains of Patagonia. We have likewise many of the same (to the eyes of a person who is not a botanist) thorny stunted bushes, withered grass, and dwarf plants. Even the black slowly-crawling beetles are closely similar, and some, I believe, on rigorous examination, absolutely identical. It had always been a subject of regret to me, that we were unavoidably compelled to give up the ascent of the St Cruz River before reaching the mountains. I always had a latent hope of meeting with some great change in the features of the country; but I now feel sure, that it would only have been following the plains of Patagonia up an ascent.'
The interpolated passage 'founded on the assumption of the immutability of species' was written after CD had already opened his first notebook on the Transmutation of Species in July 1837, but little hint of his impending change of view is given to the reader even in the 1845 edition of the *Journal of Researches*.

24*th* Early in the morning climbed up one side of the valley: & had a most extensive view of the Pampas. This was a spectacle to which I had always looked forward to with interest, but I was disappointed; it was no ways superior to that from crest of the Sierra Ventana. — At the first glance there was a strong resemblance to the ocean; but to the North many irregularities were distinguishable. The rivers were the most striking part of the scene, these facing the rising sun glittered like silver threads till lost in the immense distance. — We descended until reaching a hovel, where an Officer|545| & three soldiers were posted to examine passports. One of these men was a thorough bred Pampas Indian. — He was kept much for the same purpose as a blood-hound, to track out any person who might pass by secretly either on foot or horseback. Some years ago, a passenger had endeavoured to escape detection by making a long circuit over a neighbouring mountain; the Indian happening to cross his track followed it for the whole day over dry & very stony parts, till at last he discovered his prey hidden in a gully. — We heard that the silvery clouds which we had admired from the bright region above had poured down torrents of rain. — The valley from this point gradually opened, & the hills became mere water-worn hillocks as compared to the giants behind; it soon expanded into a gently sloping plain of shingle, covered with low trees and bushes. — This talus, although it looks of no breadth, must be nearly ten miles wide before it blends into the apparently dead level Pampas. — We had already passed the only house in this neighbourhead, the Estancia of Chaquaio, & at sunset we pulled up in the first snug corner & there bivouacked.

25*th* I was reminded of the Pampas of Buenos Ayres by seeing the disc of the rising sun intersected by an horizon level as that of the ocean. — During the night a heavy dew had fallen, a thing we did not experience within the Cordilleras. The road proceeded for some distance due East across a low swamp, then meeting with the dry plain it turned up North to Mendoza. — The distance is two very long days' journey. Our first day was called fourteen leagues to Estacado, & second seventeen to Luxan, near Mendoza. The whole is a level, sterile plain, with only|546| two or three houses: — we scarcely met a single person. — The sun was exceedingly powerful & the ride devoid of all interest. There is very little water in this Traversia; in the whole of the second day there was only one little pool. — The small streams which flow from the mountains are dried up, or rather absorbed before they reach this distance, although we generally were only from 10–15 miles from the first range. — The ground is in many parts encrusted by a saline efflorescence, hence we see the same salt-loving plants which are so common near B. Blanca. As I have already remarked respecting the Eastern valleys, there is in this Traversia also a great resemblance to the plains of Patagonia. — There is one character of landscape from the Sts of Magellan to some distance North of B. Blanca; it would appear that this kind of country extends in a sweeping line to about S. Luis de la Punta, & that to the East of this is the basin of the damp & green plains of B. Ayres. The dry & sterile Traversia of Mendoza & Patagonia is a formation of pebbles, worn smooth & deposited by a former sea; whilst the Cienegas, or plains of grass, is a deposition of fine mud from a former æstuary of the Plata, which was then bounded by a coast the line of which is pointed out by the two sorts of country. The Zoology of these plains is also similar to those near the Atlantic; we have here the Ostrich, Guanaco, Agouti or Hare, Bizcatcha, same Foxes, Lions, the four species of Armadillo; the same sorts of Partridges, Carrion hawks, Butcher-birds &c &c. —

26th After our tedious days ride, it was refreshing to|547| see in the distance the rows of Poplars & willow trees & green gardens around the village of Luxan. — Shortly before arriving at this place, we saw to the South a large ragged cloud of a dark reddish brown color. — For some time we were convinced that it was heavy smoke from a large fire in the Pampas; it afterward turned out to be a Pest of Locusts. They were travelling due North with a light breeze & overtook us, I should think, at the rate of 10—15 miles an hour. — The main body reached from 20 to perhaps 2000 to 3000 ft above the ground. The noise of their approach was that of a strong breeze passing through the rigging of a Ship. The sky seen through the advanced guard appeared like a Mezzotinto engraving, but the main body was impervious to sight; they were not however so thick but what they could escape from the waving backwards & forwards of a stick. — When they alighted they were more numerous than the leaves in a field and changed the green into a reddish colour: — the swarm having once alighted the individuals flew from side to side in any direction. This is not an uncommon pest in this country; already during the season several smaller swarms had come up from the sterile plains of the South & many trees had been entirely stripped of their leaves. — Of course this swarm cannot even be compared [to] those of the Eastern world, yet it was sufficient to make the well known descriptions of their ravages more intelligible. I have omitted perhaps the most striking part of [the] scene, namely the vain attempts of the poor cottagers to turn the stream aside; many lighted fires, & with the smoke, shouts, & waving|548| of branches, endeavoured to avert the attack.

We crossed the river of Luxan; this is a considerable body of water, its course however toward the sea coasts is but very imperfectly known: They either are dried up in the plains or form the R. Sauce & R. Colorado. — We slept in the village, it is a small place, 5 leagues South of Mendoza, & is the S. limit of the fertile territory of that Province. At night I experienced an attack, & it deserves no less a name, of the Benchuca, the great black bug of the Pampas.[1] It is most disgusting to feel soft wingless insects, about an inch long, cawling over ones body; before sucking they are quite thin, but afterwards round & bloated with blood, & in this state they are easily squashed. They are also found in the Northern part of Chili & in Peru: one which I caught at Iquiqui[2] was very empty; being placed on the table & though surrounded by people, if a finger was presented, its sucker was withdrawn, & the bold insect began to draw blood. It was curious to watch the change in the size of the insects body in less than ten minutes. There was no pain felt. — This one meal kept the insect fat for four months; In a fortnight, however, it was ready, if allowed, to suck more blood.

[1] The entry in Down House Notebook 1.13 runs as follows: 'At night, good to experience everything once. — Chinches, the giant bugs of the Pampas; horribly disgusting, to feel numerous creatures nearly an inch long & black & soft crawling in all parts of your person — gorged with your blood. —' In CD and the Voyage p. 236, the word 'Chinches', which is the Spanish term in general use for bed bugs, has been incorrectly transcribed as 'Chindass'. CD was slightly in error when he wrote 'Benchuca' here, for the insect Triatoma infestans is known today as the 'Vinchuca' bug. It is the vector for Trypanosoma cruzi, the causative agent of Chagas's disease, which then as now was endemic in this and neighbouring parts of South America. However, if a component of the illness from which CD suffered from 1839 onwards was indeed trypanosomiasis, it seems unlikely that it was on this occasion that he picked it up, since there is no mention in the following days of the period of fever that characteristically accompanies the initial infection. It remains possible, nevertheless, that he had already been infected six months earlier, as discussed under footnote 1 for 27 September 1834 (p. 263).
[2] Since the Beagle did not reach Iquique until 12 July 1835, this passage must actually have been written about three and a half months after the events described.

27th   We rode on to Mendoza; the country was beautifully cultivated & resembled Chili. — From the number of houses it was almost one straggling village; the whole is celebrated for its fruit, & certainly nothing could appear more flourishing than the orchards of Figs,

peaches, vines & olives. We bought water melons nearly twice as large
as a mans head, most deliciously cool & well flavoured for a halfpenny
a piece; & for a Medio (3^d), half a wheel-barrow full of Peaches. — |549|

The cultivated & enclosed part of the province of Mendoza is very
small: being chiefly what lies between Luxan & the capital. Beyond this
we have a plain such as we have seen more or less sterile; where there
is Water there is pasture for cattle. — The cultivated land, as in Chili,
owes its fertility to artificial irrigation; & it really is wonderful, when one
reflects how abundantly productive an utterly barren Traversia can be
made by this simple process. — The inhabitants have the reckless
lounging manners of the Pampas, as also the same dress, riding-gear
&c &c. They appear however a dirty drunken race, of mixed Indian &
Negro blood. —

28th We reached Mendoza early & staid the ensuing day there. The
prosperity of Mendoza has much declined of late years; the inhabitants
say "it is a good place to live in, but a poor one to grow rich in". To me
it had a forlorn & stupid air; to those however coming from B. Ayres,
after having crossed the monotomous Savannahs of grass, the gardens
& Orchards around the town are very pleasing. — Capt. Head talking
about the inhabitants, says, "They eat their dinner, & it is so very hot
they go to sleep — & what could they do better?" I quite agree with Capt
Head, the happy doom of the Mendozinos is to eat, sleep & be idle.
Neither the boasted Alameda or the scenery is at all comparable with
that of St Jago. —

29th We had to cross this day a long & most sterile Traversia of 15
leagues. — There is no water, & of course beyond the outskirts of
Mendoza, not a single house. |550| On this plain although elevated from
2 to 3000 ft above the sea, the sun is excessively powerful; this together
with the clouds of fine dust renders the travelling very irksome. — We
continued riding all day nearly parallel to, or rather gradually approaching to the chain of mountains; at last we entered one of the wide valleys
or bays which open on the plains; this soon narrowed into a ravine, a
little way up which is the house of Villa Vicencio. — As we had ridden
all day without any water, we were very thirsty, & looked out anxiously
for the stream which flows down the valley. — It was curious how
gradually the water made its appearance; on the edge of the grand plain
of shingle the course was quite dry, by degrees it became damper, till
there were puddles of water; these soon were connected, & at Villa
Vicencio there was a small running brook. —

30th   The solitary hovel which bears the imposing name of Villa Vicencio has been mentioned by every traveller. I staid half a day to examine the geology. In the evening rode a few leagues on to Hornillos, where I stopped the ensuing day.

April 1st[31st]   There are here a few miserable houses & there is a Tracpiche for Gold ores.[1]

[1] For the next few days, as noted by CD on 5 April, his dates are wrong by one day. The correct dates are shown in square brackets.

2nd[1st]   On the 2nd we crossed the Uspallata range of mountains; these correspond in their position & probably in their age to the Portillo range, but are of very inferior height. they are separated from the main range by a level plain of the same appearance & nature as those basins described in Chili. On this barren plain which has an altitude of nearly 6000 ft |551|are the houses of the Estancia of Uspallata. We slept here at night; it is the custom house & the last inhabited place on this side of the Cordillereas. The Uspallata mountains are deficient in water & quite barren; on the road shortly before reaching the plain there is a very extraordinary view; there are quite white, red, purple & green sedimentary rocks & black Lavas; these strata are broken up by hills of Porphyry of every shade of Brown & bright Lilacs. All together they were the first mountains which I had seen which literally resembled a coloured Geological section. −[1]

[1] In the course of his ride from Hornillos over the Uspallata range, CD made an important geological discovery. The long entry in Down House Notebook 1.13 contains the following passage: 'Looking for Silicified wood found in broken escarpment of green sandstone (58) [this figure refers to a specimen now in the Sedgwick Museum in Cambridge] 11 silicified trees & 50 or 60 columns, (Lots wife) of Sulph. of Barytes: drusy cavities: form completely kept either entire silex or Barytes: nearly all same diameter, little more or less 18 inches: in silicified centre of tree evident & all the rings: impression of bark in Sandstone: in Barytes only analogy makes me know what they are: the 11 are within 60 yards of each other; & the most remote not above 120. *No where else* did I see a trace: The strata incline 20°–30° WSW: *All* the trees incline about 70° to ENE: I except 2 silicified pieces as thick as my arm & smooth, which are embedded: horizontal: some trees only a yard apart, many two or three: appear vertical: Barytes one traced seven feet: silicified 4½ ft: Sandstone consists of many layers in colour & texture which embrace trees:'

This finding of the grove of fossilized trees was also described in similar terms in CD's letter to Henslow of 18 April 1835 (see *Correspondence* 1: 440–5). In *Journal of Researches* pp. 405–7, CD concluded: 'It required little geological practice to interpret the marvellous story, which this scene at once unfolded; though I confess I was at first so much astonished that I could scarcely believe the plainest evidence of it. I saw the spot where a cluster of fine trees had once waved their branches on the shores of the Atlantic, when that ocean (now driven back 700 miles) approached the base of the Andes. I saw that they had sprung from a volcanic soil which had been raised above the level of the sea, and that this dry land, with its upright trees, had subsequently been let down to the depths of the ocean. There it was covered by sedimentary matter, and this again by enormous streams of submarine lava—one such mass alone attaining the thickness of a thousand feet; and these deluges of melted stone and aqueous deposits had been five times spread out alternately. The ocean which received such masses must have been deep; but again the subterranean forces exerted their power, and I now beheld the bed of that sea forming a chain of mountains more than seven thousand feet in altitude.' Yet the era of plate tectonics was still in the distant future!

The site of the grove beside the main road from Mendoza to Santiago is now marked by a marble plaque erected in 1959. The silicified trunks of the trees have been removed by later visitors, but some of the cavities in the sandstone are still to be seen.

April 3ʳᵈ [2ⁿᵈ] We left the houses at noon & crossed the plain, which is so extensive that to the North nothing can be seen over its level horizon. — Our road lay along the side of the mountain torrent which we had crossed by the village of Luxan, it was here a furious & quite impassable stream, & similarly to the case of V. Vicencio appeared larger than in the plain. — We followed the valley which trends very Southerly, & slept at a place called the Pulvadera. —

4ᵗʰ [3ʳᵈ] In the morning when we arose, we had much difficulty in saddling the mules owing to a gale of wind, this brought such clouds of dust that we soon were convinced the name was properly applied. — In the evening we reached the R. de las Vacas, which is about the worst stream in the Cordilleras; it was deemed prudent to take up our nights lodging on the side. As all the water in the rivers proceeds from the melted snow, & the course being short & rapid, the hour of day makes a|552| considerable difference in the difficulty of crossing, in the evening the stream is muddy & full, about an hour after day-break it is both both clearer & much less impetuous. And this we found to be the case on the ensuing morning. — The scenery during the whole of the ascent is very uninteresting as compared to the pass of the Portillo, little can be seen beyond the grand valley with its broard base which the road follows up to the very crest of the chain. — It is moreover very sterile; during this & the previous night the poor mules had eaten nothing. Besides a few low resinous bushes, there are very few plants. During the last day we have crossed some of the worst passes in the Cordilleras. I have been quite surprised at the degree of exaggeration concerning the danger & difficulty. These are not only Travellers tales, for I was told in Chili that if I attempted to pass on foot my head would turn giddy, that there was no room to dismount &c &c. Now I did not see a place which I would not walk backwards over & get off on either side of my mule. One of the bad passes, called Las Animas (the Souls), I had crossed, & did not find out till a day afterwards that it was one of the *awful* dangers. — No doubt in very many places if the Mule should fall you would be hurled down an enormous precipice; in a like manner if a Sailor falls from aloft, it is probable he will break his neck; (& by the latter way many more in proportion have lost their lives). — I daresay in the Spring time, the Laderas or roads which each year are formed anew across the|553| piles of fallen detritus, are much worse; but from what I have seen I believe the real danger is nothing, & the apparent very little. With Cargo mules the case is rather different, the loads project so far beyond the animals sides that they occassionally run against another mule or overhanging point, & losing their balance are lost. — With respect to the rivers, I can well believe the difficulty amounts to every degree till it is impossible to cross them. In the Autumn, at this season, there is no trouble, but in the summer they must be very bad. I can quite imagine what Capt Head describes, the different expression of countenance of these who *have* passed & those who *are* passing these torrents. I never heard of any man being drowned, but of plenty of Cargo mules. The Arriero tells you to show your mule the best line & then allow her to take her own manner in crossing. The Cargo mule takes a bad line & then with its great load is lost. —

5th [4th] From the Rio de las Vacas to the Puente del Inca, half a days journey; here was a little pasture for the mules; & some interesting geology for me, so we bivouaced for the night. When one hears of a Natural bridge, one pictures to oneself some deep & narrow ravine across which a bold mass of rock has fallen, or a great archway excavated. Instead of all this the Incas bridge is a miserable object. The bottom of the valley is nearly even & composed of a mass of Alluvium; on one side are several hot mineral springs, & these have deposited over the pebbles|554| a considerable thickness of hard stratified Tufa; The river running in a narrow channel, scooped out an archway beneath the hard Tufa; soil & stones falling down from the opposite side at last met the over hanging part & formed the bridge. The oblique (D) junction of the stratified (A) rock & a confused mass is very distinct & this latter is different from the general character of the plain (B). — This Inca's bridge is truly a sight not worth seeing. —

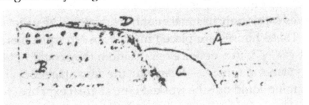

Near to this place are some ruins of Indian buildings; they consist now merely of the vestiges of walls; I saw such in several other stations; the most perfect were the Ruinas del Tambillos. —[1] The rooms were small & square & many huddled together in distinct groups; some of the doorways yet stood, these were formed of a cross slab of stone & very low, not more than 3 ft high. — The whole were capable of containing a good many people. Tradition says they were the halting places for the Incas when they crossed the Cordilleras, & these Monarchs would probably travel with a large Retinue. The situation of Tambillos is utterly desert & that of the Puente only a shade better. Traces of Indian buildings are common all over the Cordilleras; those mentioned in the Portillo pass probably were not only used as lodging houses in the passage; because if so, there would have been others, & the situation is by no means central. — Yet the Valley is now quite useless & destitute of vegetation. — In the ravine of Jajuel near Aconcagua I frequently heard of numerous remains situated at a great elevation, & of course both cold & sterile; — there is no pass in that part. — |555| I at this time imagined these might have been places of refuge on the first arrival of the Spaniards. Subsequently what I have seen has led me almost to suspect there has been a change of Climate in these Latitudes.[2] In very many places, indeed in all the ravines, in the Cordilleras of Copiapo remains of Indian houses are found; in these they find bits of woollen articles, instruments of precious metals, Indian corn, & I had in my possession the head of an arrow made of Agate, of precisely the same figure as those in T. del Fuego. — It is the opinion of the people of the country that the Indians resided in these houses; Now I am assured by men who have passed their lives in travelling the Andes, that these ruins are found at the greatest elevations, almost on the limit of perpetual snow, in places where there are no passes, where the ground produces nothing, & what is more extraordinary where there is no water. In the "Despoblado" (ininhabited) valley near Copiapò at a spot called Punta Gorda, I saw the remains of seven or eight square little rooms; they were of a similar form with those at the Tambillos but chiefly built of mud instead of stone, & which mud the people of the country cannot imitate in hardness: there was no water nearer than 3 or four leagues & this only in small quantity & bad. — The valley is utterly desert. — These houses are placed in the most conspicuous spot in a broad flat valley & in a defenceless position; they could not therefore have been places of refuge. — Even with the advantage|556| of beasts of burden, a mine could only be worked here at great expense; yet former

Indians chose it out as a place of residence. A person who has never seen such countries will not readily understand how entirely unfit they are for human habitations. If however a few showers were to fall annually, in the place of one in several years, so as to make a small rill of water, by irrigation such spots would be highly fertile. — All these facts strongly incline me to suspect that some change for the worse has taken place since the period when the ruins were inhabited. — (*Note in margin:* The Indians in the Quebrada of [*illegible*] had built an extensive Azequi or Conduit with the *hard* mud.)

I have certain proof that the S. part of continent of S. America has been elevated from 4 to 500 feet within the epoch of the existence of such shells as are now found on the coasts. It may possibly have been much more on the sea-coast & probably more in the Cordilleras. If the Andes were lowered till they formed (perhaps 3–4000 ft) a mere peninsula with outlying Islands, would not the climate probably be more like that of the S. Sea Islands, than its present parched nature — At a remote Geological æra, I can show that this grand chain consisted of Volcanic Islands, covered with luxuriant forests, some of the trees one of which, 15 feet in circumference, I have seen silicified & imbedded in marine strata. — If the mountains rose slowly, the change of climate would also deteriorate slowly; I know of no reason for denying that a large part of this may have taken place since S. America was peopled. — We need not be surprised|557| at the remains of stone & hardened mud walls lasting for so many ages as I imagine; it will be well to call to mind how many centuries the Druidical mounds have withstood even the climate of England. — I may also remark that the above conjecture explains the *present* elevation of the ruins; I am aware that the Peruvian Indians chose stations so lofty that a stranger is affected with Puna, but I am assured there are "muchissimas" houses where during the whole long winter snow lies. Surely no people would found a village under such circumstances. — When at Lima[3] I was conversing with a civil engineer, Mr Gill, about the number of Indian ruins & quantity of ground thrown out of cultivation in that province, & he told me that the conjecture about a change of climate had sometimes crossed his mind; but generally he thought that the present sterility where there was formerly cultivation was chiefly owing to neglect or subterranean movements injuring the Conduits or subterranean passages, which the Indians had formed on so wonderful a scale to bring water for the purposes of irrigation. — As an illustration he told me one very curious fact, that travelling from Casma to Huaraz he found a plain covered with ruins &c &c & now quite bare; near to it was the dry course of a considerable river; in its bed there were pebbles & sand, & in one spot solid rock to the depth of 8 feet & about 40 yards wide had been cut through. (*Note in margin:* The fall in perpendicular ft. about 40–50.) From its appearance he could not tell that the river had not followed this|558| line within a few years; but upon following *up* the course for a short distance, to his astonishment he afterwards found it going down hill; that is the bed of the river was arched; this could, of course, only happen after some subterranean movement which would throw the water back on itself untill some new lateral line of drainage was opened.[4] The inhabited plain from that year would necessarily be deserted. —

[1] The huge area ruled by the Incas reached its southern limit at Tambillos, near which remnants of the paved Inca path can still be seen, and where the buildings of three Inca staging posts have recently been excavated.

[2] CD was probably wrong in concluding that the climate had changed to any appreciable extent since the time of the Incas. Their posts were always located close to a source of water, though they might have to be moved if a stream dried up, and were not necessarily occupied during the winter.

[3] CD did not arrive in Lima until 19 July, which confirms that this section of the Diary was written at least three and a half months after the events described.

[4] A pencil note in the margin, probably in Hensleigh Wedgwood's hand, says 'This account is not very clear.' Although small shifts in the levels of the elaborate systems of irrigation canals constructed by the Incas that might be caused by earthquakes could often be repaired, large ones or movements of river beds during floods and earthquakes could not, and have been claimed to account in some cases for the abandonment of settlements that at one time were agriculturally productive.

6*th* [5*th*]   A long days march across the central ridge down to the Ojos del Agua. This was near to the lowest Casucha on the Western slope. — These Casuchas are round little towers, with the floor elevated above the ground & steps outside to reach it. — There are eight in number; formerly in the time of the King of Spain, stores were kept in them, & the Couriers took with them in the Winter master keys. Now they only answer the purpose of caves & are miserable dungeons; seated on some little eminence in the wild valleys, they are not ill suited to the surrounding desolation. The zigzag ascent of the Cumbre or partition of the waters is very steep & tedious: the road does not pass over any perpetual snow, but there are patches on either hand. — The wind on the summit was very piercing; but it was impossible not to admire again & again the intense color of the Heavens & the brilliant transparency of the Air. — The scenery moreover was grand; to the West there was a fine Chaos of huge mountains divided by profound ravines. — By this time of the year there have generally fallen a few snow storms, &|559|

not infrequently the Cordilleras are shut up; but we were favoured with the brightest fortune, & everything happened well: the sky was cloudless, excepting sometimes a few round little masses of Vapour floated around the highest peaks; such little islands in the sky are seen from a distance, when the Cordilleras are beneath the horizon to mark their position. — We met during the day several parties & Cargo troops; the road is well frequented, I suppose during the whole passage we met at least ten different parties. — (NB The six foregoing days have wrong dates owing to there being 31 days & I only counted 30 for March) —

6th In the morning we found some thief had stolen one of our mules & the bell from the Madrina; we only rode a short distance to the remains of the old Guard House. —

7th I staid here the ensuing day in hopes of finding the mule, which the Arriero thought had been hidden in the mountains. — The valley here had assumed the air of a Chilian landscape; certainly the lower parts of the hills dotted over with the pale evergreen Quillay tree & the great candlestick-like Cactus are much prettier than the bare Eastern valleys. Yet I can hardly understand the admiration expressed by some travellers at this view; the extreme delight is, I suspect, chiefly owing to escaping from the cold regions & the prospect of a good fire. I am sure I participated in such feelings. —

8th We left the valley of the river of Aconcagua by which we had descended, & reached in the evening a cottage near the Villa de St Rosa.|560| The fertility of this plain was extremely delightful; the Autumn being well advanced the leaves of many of the fruit trees were falling, & the labourers were all busy in drying on the roofs of their cottages, figs & peaches; while others were gathering the grapes from the vineyards. It was a pretty Scene; but there was absent that pensive stillness[1] which makes the autumn in England indeed the evening of the year.

[1] Followed by the deleted words '& the song of the Robin at dusk'.

9th We were now on the high road to St Jago, & crossing the Cuesta of Chacabuco reached at night the village of Colina. From this day till I reached Valparaiso, I was not very well & saw nothing & admired nothing.

10th We reached St Jago by the middle of the day, having been absent 24 days, & being well repaid for my trouble.

15th Started for Valparaiso, was two days & a half on the road endeavouring to geologize. —

17th   At Valparaiso I lived with my good friend Mr Corfield. — On the 23rd the Beagle called off the port. — I went on board. —[1] The survey of the coast to the South was concluded, & in the evening the Beagle continued her progress to Coquimbo.

[1] FitzRoy wrote: 'At noon, on the 23rd, we hove-to off Valparaiso, and sent boats ashore. Mr Darwin came on board, and among other pieces of good news, told me of my promotion. I asked about Mr Stokes and Lieut. Wickham, especially the former; but nothing had been heard of their exertions having obtained any satisfactory notice at head-quarters, which much diminished the gratification I might otherwise have felt on my own account.' See *Narrative* 2: 425. FitzRoy had held the rank of Commander since 1828, and now became a full Captain in rank as well as title.

27th   I set out on a journey to Coquimbo, from thence through Guasco to Copiapo, where Capt. FitzRoy offered to call for me. — The distance in a straight line is only 420 maritime miles, but as I travelled I found the journey a very long one. — I took with me the same man, Mariano Gonzales, four horses & two mules. — We travelled in the usual independent manner, cooking our own meals & sleeping in the open air. — |561| As we rode towards the Viño del Mar, I took a farewell view of Valparaiso & admired its picturesque appearance.[1] For geological purposes I made a detour from the high-road to the foot of the Bell Mountain: we passed through a highly auriferous district to the neighbourhead of Limache, where we slept. The country is covered with much Alluvium & this by the side of each little rivulet has been washed for gold. This employment supports the inhabitants of numerous scattered hovels, who like all those who gain by chance, are unthrifty in their habits. —

[1] For a fine watercolour view of the Bay of Valparaiso looking towards Viña del Mar, attributed to Conrad Martens (CM No. 261), but possibly painted by J. M. Rugendas, see *Beagle Record* p. 228.

28th   Passed Limache & Umiri, villages in one of the broard, level & fertile valleys & lodged at a cottage at the South foot of the Bell Mountain. It has been discovered on board the Beagle by angular measurements that the hill is 6200 ft. high; as there are others of the same & many of little less height, it will be evident how truly an Alpine country Chili is. The Volcano of Aconcagua, the magnificent appearance of which I have so often admired in this journal, actually attains the enormous height of 23000 ft—!!! The inhabitants of the cottage were freeholders, which is not very common in Chili; they support themselves on the produce of a garden & little field, but are very poor. — So deficient is Capital, that they are obliged to sell their green corn when standing in the field, in order to buy necessaries; wheat is in consequence dearer here in the very district where it is produced, than in the town of Valparaiso where the Contractors live.— Having failed in my geological pursuit, I took|562| the road for Quillota, which we reached by the middle of the day.—

29th During the night a very light shower of rain fell; this is the first since the heavy rain of Septemb. 11th & 12th which detained me a prisoner at Cauquenes. The interval is 7 months & a half; but the rain this year in Chili is rather late.—

30th We passed the Cerro of Chilicauquen at the same place as in my trip last year. At a pretty little village called Plazilla, we joined the Coquimbo road.— The surrounding country is barren & uninteresting.— I think the view from Chilicauquen of the valley of Quillota with the distant Cordilleras now thickly covered with new snow, is one of the most beautiful in Chili. With this may be ranked the view from the hills behind Valparaiso, the basin of S. Filipe or Aconcagua, & the plains of St Jago & Rancagua. Thus none of the finest scenery in Chili is very distant from the Capital city & its port.—

May 1st Catapilco to valley of Longotomo—a few small inhabited valleys—trees are becoming scarcer & are replaced by a *large* plant which has leaves like a Pineapple & long flowering stem like a Yucca.

2nd Longotomo to Quilimar.— As yesterday, the road generally runs at no great distance from the sea coast. The country on a small scale singularly broken & irregular: abrupt little peaks rise out of small plains or basins: the bottom of the neighbouring sea, studded with breakers, & the indented coast would if converted into dry land, present similar forms.—|563|

3rd Quilimar to Conchalee.— The Country becomes more & more barren; the valleys have so little water that there is scarcely any irrigation; of course the intermediate country is quite useless & will not even support goats.— In the Spring after the winter rains there is a rapid growth of thin pasture & cattle are then brought down from the Cordilleras to graze. It is rather curious the manner in which the Vegetation *knows* how much rain to expect;[1] one shower at Copiapo produces an equal effect with a couple at Guasco & 3 or 4 at Coquimbo, whilst at Valparaiso torrents of rain fall. Travelling North from the latter place, the quantity does not decrease in a regular proportion to the distances. At Conchalee which is not half-way between Valparaiso & Coquimbo (being only 67 miles to the North of the former) they do not

Robert FitzRoy as a young man, drawn by Philip Gidley King in 1838
(original in the Mitchell Library, Sydney).

expect rain till end of May, whereas at Valparaiso generally early in April; the quantity likewise which falls is proportionally small to the later time it comes.

I heard of the Beagle surveying all these ports; all the inhabitants were convinced she was a Smuggler, they complained of the entire want of confidence the Captain showed in not coming to any terms; each man thought his neighbour was in the secret—I had even difficulty in undeceiving them. — By the way, this anecdote about the smuggling shows how little even the upper classes in these countries understand the wide distinction of manners. A person who could possibly mistake Capt. FitzRoy for|564| a smuggler, would never perceive any difference between a Lord Chesterfield & his valet. —

[1] In *Journal of Researches* p. 417, this sentence becomes 'It is curious to observe how the seeds of the grass seem to know, as if by an acquired instinct, what quantity of rain to expect.' In the 1845 edition it is altered again to 'It is curious to observe how the seeds of the grass and other plants seem to accommodate themselves, as if by an acquired habit, to the quantity of rain which falls on different parts of this coast.'

4th Conchalee—Illapel. The country near the coast possessed little Geological interest, & otherwise the rocky barren hills were very monotomous; so I determined to strike in the country to the mining town of Illapel. — It was a long days journey & we had to cross a Cuesta I should think at least 2000 ft high. — The valley of Illapel is like all the others, dead level, broad, bordered by gravel cliffs or mountain sides, & very fertile. — Above the *straight* line of the upper irrigating ditch, all is as brown as a turnpike road, all beneath is Alfarfa (a kind of Clover) green as Verdigris—the contrast is singular. — Illapel is a very regular & pretty little town, its flourishing condition depends on the numerous mines, chiefly Copper, in the vicinity. —

5th On account of my animals I staid the day here.

6th & then travelled on to Los Hornos, which is a "Mineral" or particular district abounding with mines; the principal hill was so drilled with excavations that it was a magnified edition of a large Anthill. The Miners in Chili are a peculiar race of men; in their habits they somewhat resemble men-of-war sailors; living for weeks together in the most desolate spots, when they descend on the feast days to the villages, there is no excess or extravagance into which they do not run. They sometimes gain a considerable|565| sum & then like Sailors with prize money, they try how soon they can possibly squander it. They drink excessively, buy quantities of clothes & in a few days return penniless to their miserable abodes there to work harder than beasts of burden. —

Their dress is peculiar & rather picturesque; they wear a very long shirt of some dark coloured baize & leathern apron; around the waist there is also a broard & gayly coloured Senador (like the red silk woven band of officers); their trowsers are very broard & their heads are covered by little scarlet caps. — We met a party of these Miners on horseback in full costume, carrying for burial the body of one of their companions. — They marched at a very quick trot; four men on foot carried the corpse; each set running as hard as they could for about 200 yards, were relieved by four others who had previously dashed on ahead on horseback & so on. — They encouraged each other by wild crys; altogether it formed a most strange funeral.

7th & 8th Staid here a day on account of Geology & then rode on to Combarbala, a very pretty little town at the foot of the main Cordilleras. Country very mountainous & desolate. —

9th & 10th There are so very few inhabited spots & the roads so obscure we had some difficulty in finding our way—during these last days there was nothing of interest, & the travelling sufficiently wearisome. — We passed the Mineral of Punitague, from which much Copper & Gold has been|566| extracted; there are also Quicksilver mines which are not worked. — We reached Ovalle, a small town on the R. Limari, late in the evening. — Before arriving there we had to cross some extensive sterile plains or Traversias, which extend from the coast many leagues in the interior. —

11th Next day crossing the river, passed over plains to some hills where the copper mines of Panuncillo are seated. They belong to Mr Caldcleugh of St Jago. —

12th I staid here a day in order to see the mines. The mine is not a very rich one, the ore being the common yellow Copper pyrites; its value may be from 30000 to 40000 dollars, (Note in margin: £6000–£8000.) yet when the English first came into the country, Mr Caldcleugh bought it for the association for one ounce ($3^£..8^s$). The mine had been abandoned when full of this kind of ore, the inhabitants not believing it possible to reduce it. Likewise from ignorance piles of scoriæ abounding with particles of copper & fused pyrites were sold at about the same scale of profit. — Yet with all these possibilities the mining associations contrived to lose great sums. The folly of the commissioners & the shareholders amounted to madness: such enormous salaries, libraries of well bound geological books, $1000^£$ per annum to entertain the Authorities;

bringing out miners for particular metals before such were known to exist; their contracts with the workmen to find them with so much milk &c &c every day; their machinery, where such could not be used; |567|& a hundred similar things bear witness to their absurdity & afford amusement to the natives. — Yet there can be no doubt the same Capital employed in working mines in the country method would have given an immense return. A confidential man of business & a practical miner & assayer would have been quite sufficient. — The English & Chilian miners were tried against each other at this place, & I believe the latter fairly laughed at our countrymen, being so entirely victorious. —

Capt. Head has described the wonderful load which the "Apires", truly beasts of burden, carry up from deep mines. — I confess I thought the account exaggerated; so that I was glad to take the opportunity of weighing one of the loads, which I picked out by chance. When standing straight over it, I could just lift it from the ground, the weight was 197 pounds (equal [to] a 14 stone man). — The Apire had carried this up 80 perpendicular yards, by a very steep road, & by climbing up a zigzag nearly vertical notched pole. — He is not allowed to halt to breathe, excepting the mine is more than 600 ft deep. — The average weight is rather more than 200 £. (Nearly equal 22 & ½ stone.) — I have been assured that 300 £ have been carried for a trial from the deepest mines.

In this mine they bring up the above load on their backs 12 times in the day, that is 2400 £ from 80 yards deep to the surface. These men work nearly naked; their bodies are not very muscular; but excepting from accidents, |568| they are healthy and they appear cheerful. They rarely eat meat once a week & never oftener & then only the hard dry Charqúi. — Knowing that the labor is voluntary, it is yet quite revolting to see the state in which they reach the mouth of the mine. — their bodies bent forward, leaning with their arms on the steps; their legs bowed, the muscles quivering, the perspiration streaming from their faces over their breasts, the nostrils distended, the corners of the mouth forcibly drawn back, & the expulsion of their breath most laborious: each time from habit they utter an articulate cry of ay-ay, which ends in a sound rising from deep in the chest, but shrill like the note of a fife. — After staggering to the pile of ore, they empty the "Carpacho" — in two or three seconds recovering their breath, they wipe the sweat from their brows & apparently quite fresh descend the mine again at a quick pace. — This appears to me a wonderful instance of the amount of labor which habit, for it can be nothing else, will teach a man to endure. —

The Mayor-domo of these mines, Don Joaquin Edwards, is a young man & the son of an Englishman, but till some years old did not learn English. — Talking with him about the number of foreigners in all parts of the country, he told me he recollected being at school in Coquimbo, when a holiday was given to all the boys to see the Captain of an English Ship, who came on some business from the Port to the city. He believes that nothing would have induced any body in the school, including himself, |569|to have gone close to the Englishman; so fully had they been impressed with all the heresy, contamination & evil to be derived from contact with such a person. To this day they hand down the atrocious actions of the Buccaniers; one of them took the Virgin Mary out [of] the Church & returned the ensuing year for St. Joseph, saying it was a pity the Lady should not have a husband. I heard M\(^r\) Caldcleugh say that sitting by an old lady at a dinner in Coquimbo, she remarked how wonderfully strange it was that she should live to dine in the same room with an Englishman. — Twice as a girl, at the cry of "Los Ingleses" every soul carrying what valuables they could had taken to the mountains. —

13th From Panuncillo to Tambillos, where there are some old Copper Mines. They are now in possession of some Englishmen, who came out as Mechanicks, but have accumulated by their industry some thousand pounds & have bought this mine, which they hope [to] empty of its water. Slept at a spot called the Punta; it is the point of [a] range of hills which abuts on an extensive plain, precisely in the same manner as a headland in the sea.

14th Over the plain & Traversia we had to cross to the port of Coquimbo. — We found the Beagle in the little harbor of Herradura a league to the South. — All hands were living on shore under tents; the ship undergoing a thorough refit before the long passage of the Pacifick.

15th I staid one day on board & on the 16th hired|570| with Capt. FitzRoy lodgings in the city of Coquimbo, which is distant 11 miles from the Beagles anchorage. Coquimbo is said to contain 6000 to 8000 inhabitants, it is remarkable for nothing but the extreme quietness which reigns in all parts; like the other towns in the North of Chili, it depends, but in a less degree, for its support on the mines.

17th In the morning it rained lightly for about five hours; the first time this season; with this the farmers would break the ground, with a second plant their corn; & if a third shower fell, would in the spring reap a good harvest. It was curious to witness the effect of this trifling amount of moisture; the ground apparently was scarcely damp 12 hours afterwards, yet on the 27th an interval of 10 days, all the hills were tinged green in patches, the grass being sparingly scattered in hair-like fibres a full inch long. — Before this *every* part was as destitute of Vegetation as a turnpike road. In the evening I dined with Mr Edwards, during dinner there was a smart shock of an Earthquake. I heard the forecoming rumble, but from the screams of the ladies, the running of servants & the rush of several of the gentlemen to the doorway I could not distinguish the motion. Some of the women afterwards were crying with *fear* & one person said he should not be able to sleep all night or if he did, it would only be to dream of falling houses. — The father of this gentleman had lost all his property at Talcuhuana, & he himself only just escaped from a falling roof at Valparaiso in 1822; a curious coincidence|571| happened; he with a party were playing at cards when a German remarked he never would sit in a room in these countries with the door shut, as he had with difficulty escaped in the Copiapò earthquake, accordingly it was opened. No sooner was this done than the famous shock commenced, & the whole party effected their escape by this coincidence. The danger in an Earthquake is not the time lost in opening a door, but the chance of its being jammed by the movement of the walls. — It is impossible to be much surprised at the fear which Natives & old Residents experience. — I think the excess of panic may be partly owing to a want of habit in governing fear; the usual restraint, shame, being here absent. — Indeed the natives do not like to see a person indifferent. I heard of two Englishmen who, sleeping in the open air near to some houses during a smart shock, knowing there was no danger did not rise — the natives cried out indignantly "Look at those hereticks, they will not even get out of their beds". —

19th   I walked a little way up the valley & saw those step-like plains of shingle described by Capt. B. Hall, the origin of which has been discussed by Mr Lyell.[1] The same phenomenon is found in the valley of Guasco in a more evident manner; in places there [are] as many as seven perfectly level & unequally broad plains, ascending by steps on one or both sides the valley. — There can be no doubt that during the rise of the

land each line of cliff was for a period the beach of a large bay. — At Coquimbo marine shells were embedded in strata near the surface; independent of this |572|proof, the explanation of the successive breaking down of the barrier of a lake adduced by Capt. Hall is quite inapplicable. — The appearance of these steps, especially in Guasco, is sufficiently remarkable to call the attention of any one who is not at all interested concerning the causes of the present forms of the land. The number of parallel & horizontal lines, of which many have exactly corresponding ones on the opposite side of the valley, is rendered more conspicuous by the irregular outline of the neighbouring mountains. —

[1] See Charles Lyell, *Principles of Geology, being an attempt to explain the former changes of the earth's surface, by reference to causes now in operation*. Vol. III, p. 131. John Murray, London, 1833.

21st   I set out on a short excursion with Don José Maria Edwards, a pleasant young Anglo-Chilian, to the famous silver Mineral of Arqueros & from thence up the valley of Elque or Coquimbo; passing through a fine alpine country we reached his Fathers mine after it was dark. I enjoyed my nights rest here from a cause the force of which will not be understood in England—there were no fleas! The rooms in Coquimbo swarm with them; at an elevation of about 3000 ft they will not live, & if brought there, as for instance to these mines, they will dye. It can scarcely be the trifling diminution in temperature, 22°, but some other cause which is destructive to these troublesome insects.

22nd   I spent half of the ensuing day in examining the mines. — The Mineral extends over a few miles of hilly country, & abounds with Silver mines, the ore of which always occurs with white Sulp of Barytes. — The Mineral was only discovered a few years since by a wood-cutter, although the veins project beyond the surface & are very abundant. |573| The mines are now in a bad state; they formerly yielded about 2000 pounds weight of Silver a year. It has been said "a person with a Copper mine will gain, with Silver he may gain, but with Gold is sure to lose". This is not true, all the large Chilian fortunes have been made by mines of the richer metals. — The other day Dr Seward returned to England from Copiapò taking with him the profits of a share of a Silver mine & this amounted to 120000 dollars.[1] — No doubt a Copper mine with care is a sure game, whereas the other is gambling, or rather taking a ticket in a lottery. The owners lose great quantities of rich ores, no care can prevent robbery. I heard of a man laying a bet with another that one of his men should rob him before his face. The ore when brought out of the mine is broken into pieces by men who sit down & separate the pieces of useless stone, which are rolled over the side of the mountain. Two of the men as if by accident pitched two pieces of stone away at the same moment & then cried out for a joke, let us see which rolls furthest. — The owner who was standing by bet a cigar with his friend on the race. The Miner by this means watched the very point amongst the rubbish, where the stone lay; in the evening he picked it up & carried it to his master, showing a rich mass of Silver ore & saying "This was the stone you won a cigar by its rolling so far". — Some of the|574| Mine owners say to their men, "We know you manage to steal, but why not sell us the ore, that we may receive the usual large profits?", and this contract has sometimes been made. —[2] In the afternoon, we left Arqueros & descended into the valley (the region of fleas) and slept at the first Rancho we came to. —

[1]Note in margin: '£24000'.
[2]Followed by a deleted sentence which runs: 'Near Copiapò I met three mules travelling by night loaded with rich ore; the robbers bribe every one whom they think would betray them; they very quietly gave a fine specimen to a guide whom I had with me.'

23rd We followed up the fertile valley, geologising by the way till we reached an Hacienda, the owner of which was a relation of Don Josè.

24th We staid here the ensuing day. The Signora was a very pretty girl not 17 years old, yet the mother of two children & would soon add another to the family of Salzera. It is strange what little attention the Hacienderos, who correspond to our country gentlemen, pay to comfort. — This house furniture &c was in no ways superior to a second rate English farm house; although the lady was dressed most elegantly & the gentleman with the usual respectability. — The scenery here was exceedingly beautiful, — truly Chilian in its character. — it reminded me of some of the views in the Annuals of Alpine Scenery. —

25th Leaving Don Josè behind I travelled a days ride further up where the R. Claro joins the Elque. — I had heard of petrified shells & beans, the former turned out true, the latter small white quartz pebbles.|575| We passed through several small villages; the valley was beautifully cultivated & the whole scenery very grand. We were here near the main Cordillera, the surrounding hills being very lofty. In all parts of Northern Chili, the fruit trees produce much more abundantly at a considerable elevation near the Andes. The figs & grapes of Elque are famous for their superiority & are cultivated to a great extent. This valley is perhaps the most productive one to the North of Quillota: I believe it contains, including Coquimbo, 25 thousand inhabitants. —

26th Having seen what I wanted – returned to the Hacienda &

27th the following day, Don Josè & I reached Coquimbo late in the afternoon. –

June 2nd Set out for the valley of Guasco, taking with me a guide for the road. – The Beagle was to sail for Valparaiso a few days afterwards, from thence to Copiapò to pick me up & then to Peru. Capt FitzRoy hired a small vessel & left a party under the command of Mr Sulivan to survey the Northern coast of Chili & to rendezvous at Lima. – We rode this day to a solitary house, called Yerba buena, where pasture for the animals can be bought. – On the whole road we passed only one other house or inhabited spot. – There are two roads from Coquimbo to Guasco, one near the sea coast, the other in the interior; in this latter there is nothing for the animals to eat during the whole time. I therefore followed the former line. – The shower alluded to a fortnight ago had reached about half way to Guasco, we had therefore in this first part a slight tinge|576| of green, just sufficient to remind me of the freshness of the turf & budding flowers in the Spring of England. – Travelling in these countries, like to a prisoner shut up in gloomy courts, produces a constant longing for such scenes. –

3rd Yerba-buena to Carizal. – During first part of day crossed a mountainous rocky desert, like near Conchalee, then a long deep sandy plain covered with broken marine shells. – There is very little water & that little saline; the few streamlets are bordered on each side by white encrustations, amongst which the succulent, salt-loving plants grow. The whole country from the coast to the Cordillera is a desert & uninhabited. – I saw only traces of one living animal in abundance; this was a Bulimus, the shells of which were collected together in extraordinary numbers in the driest parts. In the Spring, & at the dawn of day when the ground is damp with the dew these animals are crawling about in all parts. As they are never seen excepting in the early morning, the Guassos think they are born from the dew. – I have noticed in other places that the driest & very sterile districts are most favourable to an extraordinary increase of land-shells. – At Carizal there are a few cottages, some brackish water & a trace of cultivation; with difficulty we purchased a little corn & straw for the Horses.

4th Carizal to Sauce. – Continued to ride over a desert plain tenanted by some large herds of Guanaco. – This plain is crossed by the valley of Chañeral, the most fertile one between Guasco|577| & Coquimbo. – it is however very narrow & although green produces very little. – Pasture for animals at this time of year could not be procured. — At Sauce we found a civil old gentleman superintending a Copper-smelting furnace. — as an especial favor he allowed me to purchase at a high price an armfull of dirty straw, which was all the poor horses had for supper after their long days work. — There are very few smelting furnaces still employed in Chili; it is found more profitable to ship the ore to England, owing to the extreme scarcity of fire-wood & the loss of metal from the clumsy Chilian method of reduction. — The poor Chilenos think that England is quite dependant for her Copper to Chili; they will scarcely believe that all the quantity which is imported there must again be exported to other countries.

5th   This Sauce is a little way out of the direct road — from it to Freyrina we had to cross some mountains, — every days march to the Northward the vegetation becomes more scanty; here a few tiny bushes were coated by a filamentous green Lichen & the large Candlestick-like Cactus was succeeded by a much smaller species. During the winter months both in Chili & Peru a thin but uniform stratum of clouds hangs at no great height over the Pacifick. — From the above hill we had a striking view of this great white & brilliant field; from it arms entered all the valleys, leaving Islands & promontories precisely in the same manner as the sea intersects the land in the Chonos Archipelago. — We reached Freyrina|578| early in the day to the great joy of ourselves & poor horses. — In the valley of Guasco, beginning at the mouth we have the little village at the port — a spot entirely desert & without water immediately at hand. — 5 leagues up is the village of Freyrina, consisting of one long straggling street; the houses white washed & generally decent. — 10 leagues further up is the principal town — Ballenar. — And again near the Cordilleras there is Guasco alto, an agricultural or rather Horticultural village, famous for its dried fruit. — Ballenar & Freyrina depend chiefly on the mines. —

6th & 7th   I staid here on account of my animals two days, & lived with Mr Hardy, an owner of Copper mines. — One of the days I rode down to the Port. — On a clear day the view up the valley is very fine; the opening being nearly straight is terminated at a great distance by the distinct outline of the snowy Cordillera; on each side an infinity of crossing lines blend together in a beautiful haze. The foreground is singular from the number of parallel & extensive terraces; & the included strip of green valley abounding with its willow bushes is contrasted on each hand by the naked hills. — From Capt. B. Halls description I had expected a valley luxuriant as those at C. de Verds, but it appears to me that all Capt. Halls beautiful descriptions require a little washing with a Neutral tint[1] — it may partly destroy their charm but I am afraid will add to their reality. — But it may be well|579| imagined how bare the hills must have been, since a shower had not fallen for 13 months. The inhabitants heard with the greatest envy of the rain in Coquimbo: From the looks of the weather they had strong expectations of equally good fortune and a fortnight afterwards this was verified. I was at Copiapo at the time, & there the people with equal envy talked of the abundant rain at Guasco. After two or three very dry years, that is perhaps with not more than one shower during the whole time, a rainy year generally follows, & this does more harm than even the drought. — The river swells & covers with gravel & sand the narrow strip of ground which alone is fit for cultivation; the flood also injures the irrigating ditches: great devastation had thus been caused three years ago. I called in the evening at the house of the "Governador"; the Signora was a Limerian & affected blue-stockingism & superiority over her neighbours. Yet this learned lady never could have seen a Map. Mr Hardy told me that one day a coloured Atlas was lying on a Pianoforte & this lady seeing it exclaimed, "Esta es contradanca". This is a country dance! "que bonita" how pretty! — On the other hand, the good people at Valdivia hearing so much about our making Charts thought everything a map. As they mistook a Sextant & artificial horizon, doubtless a piece of Music would have gone under the same name. (*Note in margin:* NB I have exaggerated this story.)

[1] Pencil note in margin, probably made by Hensleigh Wedgwood, runs: 'a very happy expression'.

8th   Rode up to Ballenar; as the rocky mountains on each side were concealed by clouds, the terrace-like plains caused the valley to have a very similar appearance to that of S. Cruz in Patagonia. The quantity of cultivated ground is small. — |580|

9th   I staid the day here. — Ballenar is a considerable town, nearly as large as Coquimbo & well built; it is only sprung up in late years & owes its prosperity entirely to the Silver mines. The produce of the valley is not sufficient to support the inhabitants. Ballenar takes its name from Ballenagh in Ireland, the birthplace of the family of O Higgins, who were presidents & generals in Chili. Freyrina likewise takes its name from a General Freyre. — [1] Ballenar is rather a nice & pretty town: the valley & indeed all the valleys in Chili are well worth visiting. —

[1] Ramon Freire Serrano was Director-General of Chile, and commanded the expedition that in 1826 freed Chiloe from the Spaniards.

10th Instead of going from this place direct to the town of Copiapò, I determined to take a guide & fall into the valley higher up. — We rode all day over an uninteresting country. — I am tired of repeating the epithets barren & sterile. — These words, however, as commonly used, are comparative. I have always applied them to the plains of Patagonia, yet the vegetation possesses spiny bushes & some dry prickly grasses, which is luxuriant to anything to be seen here. — There are not *many* spots where in 200 y^{ds} square, some little bush, plant, Cactus or Lichen can not be discovered, & in the ground seeds lie buried ready to spring up during the first rainy winter. — In Peru absolute deserts are to be met with over a large extent of country.

In the evening we came to a little valley in which the bed of a little streamlet was damp — following this up for a mile we came to water & that not very bad. During the night the stream flows a league lower down than in day, before it is evaporated & absorbed.|581| There was plenty sticks for firewood, so that for us it was a good place of bivouac; but the poor animals had not a mouthful to eat. Even here there were two cottages of Indians with a troop of *donkeys*, employed in carrying firewood &c &c to the mines; these donkeys are without any inaccuracy supported on the *stumps* of the dry twigs of the Bushes. There is not a plant of any sort for them to eat. I believe every now & then they are taken to feed for a short time in the valleys of the Cordilleras, but generally, what I have stated is their sole support. — The fact of the gnawed stumps proved the truth & quite astonished me. —

11th Rode for 12 hours without stopping, till we reached a spot where there was water & firewood. Formerly there had been a smelting furnace here. — Our horses again had not anything to eat, being shut up in an old Corrall. — The whole line of road was hilly; any view of the distant landscape was interesting from the various colors of the bare mountains & splendid weather. — It is a pity to see the sun so constantly bright over so useless a country; such shining days ought only to brighten a prospect of fields, cottages & gardens. —

12th By noon we arrived at the Hacienda of Potrero Seco in the Valley of Copiapò. I was heartily glad of it; it is most disagreeable to hear whilst you are eating a good supper, your horse gnawing the post to which he is tied & to know that you cannot relieve his hunger. — the whole journey is a source of anxiety to see how fast you can cross the Traversia. To all *appearance*|582| however the horses were quite fresh & no one could have told they had not eaten for the last 55 hours. —

This Hacienda belonged to one of the British associations; their affairs when bankrupt were purchased by some English merchants; one of these, Mr Bingley, came out as managing agent. — I had a letter of introduction to him & by good luck he came from the town this day to the Estate. — To his credit, there is no land in the whole valley which looks in such good order. — At one time the whole was rented at 500 dollars[1] now one mere part is let 1800, & another 400. — He reserving the best part for himself. This is a specimen of the management of those mad associations. — This estate is called 14 leagues (perhaps 25 English miles) long: it is of course very narrow, seldom a mile & often of no breadth, that is the valley in some parts cannot be irrigated. — Generally it has a width of two fields. The whole is cultivated with the Clover of the country for the Pasturage of Mules. Mr Bingleys business is the shipment of Copper ores, but everything depends on the carriage of the ores to the Port. — There is so little land in the whole valley, that mules sufficient for the mines cannot be pastured. It would sound odd in England, the whole value of a mining business depending on the quantity of pasturage to be obtained by any individual. — The scarcity of cultivated land does not depend so much on the inequality or unfitness for irrigation, as on|583| the little water. The river this year is remarkably full; at this Hacienda it reaches up to a horses belly, is about 15 yards wide & rapid; of course it grows gradually decreasing till it reaches the sea. This however happens rarely; for a period of 30 years not a drop ever entered the Pacifick. The inhabitants watch a storm in the Cordilleras with great interest; one good fall of snow secures water for the ensuing year. — This is of infinitely more consequence than rain in the lower country. With the latter, which often does not fall for two & even three years together, they are enabled to pasture the mules & cattle for some time in the mountains; but without Snow in the Andes desolation extends over the whole valley. — It is on record that three times nearly all the inhabitants have been obliged to emigrate to the South. The valley is said to contain 12000 souls, but its produce is sufficient only for three months in the year; the rest being drawn from Valparaiso & the South. —[2] Before the discovery of the famous Silver Mineral of Chanuncillo, Copiapò was in a rapid state of decay; now it is in a very thriving condition. The town which was completely overthrown by an Earthquake has been rebuilt. —

The valley of Copiapò runs in a very Southerly direction, so that it is of considerable length to its source in the Cordillera; it forms a mere green ribbon in a desert. Both the|584| valleys of Guasco & Copiapò may be considered as islands to the Northward of Chili, separated by deserts in the place of Salt water. Beyond these, there is one other very miserable valley called Paposo, which contains about 200 people. Then we come to the real desert of Atacama, a far worse barrier than the most turbulent sea. At the present time there is plenty of water & every man irrigates his land as much as he likes; when it is scarce guards are sent to the sluices of all the Azequis to see they do not take more than their allotted number of hours in the week. — In consequence of this abundance & the rich nature of soil, which is much less gravelly than in the other valleys, the stripe of Vegetation is very luxuriant. But when the latitude 27° is considered, & that it is nearly in the same parallel with St Catherines on the coast of Brazil, it is surprising that there is no trace of a Tropical character in the Vegetation. —

[1] Note in margin: '100$^{\ell}$'.
[2] Followed by deleted sentence: 'It is therefore manifest how entirely this place depends on the mines'.

13th & 14th Staid here two days, employed in geologizing the huge surrounding mountains. —

15th Proceeded up the valley towards the Cordilleras; its course was however very oblique, running about SSE instead of East. — I dined at a hospitable old Spaniard, Don Eugenio Matta, at whose house General Aldonati was staying. he was governor of Chiloe during the time of Capt. Kings Voyage & well known to the officers. — I found him the pleasantest gentleman I have met in Chili.[1] The valley continued much as I have described it; & always pleasant to behold. — |585| At night fall we reached the Hacienda of las Amolanas to the owner of which Don Benito Cruz I had a letter of introduction.

[1] Colonel Aldunate was the first Governor of Chiloe after its liberation from the Spaniards in 1826. See *Narrative* 1: 298–9.

16th I staid there the ensuing day & found him most hospitable & kind; indeed I defy a traveller to do justice to the goodnature with which strangers are received in this country. —

17th I hired mules & a guide to penetrate a little way in the Cordilleras. A few leagues beyond the Hacienda, the valley of Copiapò is divided into three branches; the Southern one, Mamflas, has a long course skirting the Cordilleras & is inhabited during much of its length; — the other two arms each only have one or two houses. — I entered the one called Jolquera, it was very barren & uninhabited, slept where there was a little pasture.

18ʰ Pursued our course, the valley becoming more fertile; we passed only one house. — At midday seeing the valley ran in a very Northerly direction for a long distance, I did not think it worth while to proceed; so we turned back & again chose a snug spot to bivouac. — This ravine is one of the passes of the Cordilleras. — At night it appeared like an approaching rain storm. We experienced a trifling shock of an Earthquake. — We were at a considerable elevation although the ascent from the sea is only just perceptible. With a clear sky it froze sharply every night. —

19ᵗʰ Returned down the ravine to las Amolanas.

20ᵗʰ Staid there the following day. I found an abundance of petrified shells & wood. It is amusing to find the same subject discussed here as formerly amongst the learned of Europe concerning the origin of these shells, |586| whether they really were shells or were thus "born by Nature".[1] At night a stranger came in & asked permission to sleep there: it turned out he had been lost & wandering about for the last 17 days. he started from Guasco alto, with baggage mules & servants, expecting to find (without a guide) his way in two days to the valley of Copiapo. Missing his track he became involved in a labyrinth of mountains & could not escape. Some of his mules fell over the precipes & if it had not been for the good fortune of meeting a herd of cattle he would have been obliged to have killed his mules to eat. — They could not fairly leave the mountains, on account of not knowing in the more level country the few spots where water is found. — I mention this as a proof of the impracticable nature of the country; It is a constant subject of surprise to me, whenever I reflect about it, how the Spanish soldiers, who at the time of the Conquest marched, & many on foot, from Peru to Chili, did ever survive the dangers of these deserts.[2] That many perished is well known, but enough escaped to continue a war with numerous tribes of the native Indians. —

[1] Note in margin: 'My general method of explanation God made them'.
[2] Note in margin: 'Always took Indian guides perforce'.

21ˢᵗ Returned to the Hacienda of Potrero Seco, & from there a long days ride to the town of Copiapò.

22ⁿᵈ The lower part of the valley is broarder & near to the town it is a fine plain resembling that of Aconcagua or Quillota. — I staid three days here with Mʳ Bingley. — Copiapò covers a considerable space of ground, each house possessing more or less Garden. |587|

23rd–25th It is however a miserable looking place; I never saw so few houses furnished with any comforts. — Every soul appears to be endeavouring to make money & see how soon (& in this they are quite right) they can leave it. — Every person is more or less directly concerned with mines — & mines & ores are the sole subjects of conversation. Necessaries of all sorts are very dear. The town being 18 leagues from the sea port & the land carriage so expensive alone would nearly cause this. — A Fowl costs from 5 to 6 shillings; the fire wood or rather sticks are brought on donkeys from two & three days journey in the Cordilleras. Meat is nearly as dear as in England & pasturage for Animals a shilling per day; this for South America is wonderfully exorbitant. —

26th I hired a Vaqueano & 8 mules to take me into the Cordilleras by a more direct line than last time. As the country in this direction was utterly desert I took with me a cargo & half of Barley & Straw. About two leagues above the town, a broad valley called the "Despoblado" or uninhabited, branches off from the one by which I descended. This at first runs very Northerly, but then proceeds well Easterly & ends in a good pass to the other side. This valley is a very large one, both of great breadth & depth; it is however quite dry, perhaps with the exception of a few days during some very rainy winter. The sides of the crumbling mountains are but little furrowed with ravines & the bottom of the main valley level. — No considerable river ever could have poured its waters over the bed of shingle, without leaving a channel similar to what|588| is found in other valleys. — I feel no doubt as we now see it, so it was left by the gradually retiring sea; The dry valleys, mentioned by Travellers in Peru probably in the greater number of instances owe their present form to the same origin. —

We rode till an hour after sunset till we reached a side ravine with a small well called "Agua-amarga". — The water deserves its name, for besides being saline, it is most offensively putrid & bitter; I suppose the distance is about 25–30 (English) miles from the river of Copiapo; in this distance there is not a drop of water & the country almost deserves the name of an absolute desert. Yet it is about half where the Indian houses at Punta Gorda are situated. I also noticed in front of some of the small side valleys which enter into the main one, two piles of stones a little way apart & in a direction to point up the valley. My companions knew nothing about them & only answered my queries by their "Quien sabe". —

27th Set out early in the morning, by midday reached the ravine of Paypote, where there is a tiny rill of water, a little vegetation on its borders & some Algarroba (a Mimosa) trees. On this latter account formerly there was a smelting furnace here; we found a solitary man in charge of it, his sole occupation was hunting Guanacoes with a pack of large dogs. — At night it froze sharply, but we had plenty of firewood to make a good fire.

28th We continued gradually ascending as we followed the valley; this became more contracted & is called near to the Cordilleras Maricongo. We saw|589| during the day some Guanacoes & the track of the Vicuna; also a great many Foxes; I presume these latter animals prey on the small gnawing animals which manage to find sustenance, & abound in the most sterile & dry spots. — The scenery on all sides showed desolation brightened & made palpable by a clear, unclouded sky. Custom excludes the feeling of sublimity & this being absent, such scenery is rather the reverse of interesting. — We bivouacked at the foot of the "primera linea" or the first line of the partition of the waters; the streams however on the other side do not flow to the Atlantic, but into an elevated undulating district, in the middle of which there is a large Salina or salt-lake. Besides this ridge, there are two others to pass before arriving at the descent on the Eastern slope. — The outline of the Cordilleras in this part is very tame. — I climbed up on foot to very near the crest; from the *Puna* I experienced, I cannot suppose the elevation is less than 8000 to 10000 ft; There was a good deal of snow, which however only remains here in the winter months. The winds in these districts obey very regular laws; every day a fresh breeze blows up the valley & at night, an hour or two after sunset, the air from the cold regions above descends as through a funnel. — This night it blew a gale of wind, & the temperature must have been considerably below the freezing point, for water in a short time became a block of ice. No clothes seemed to oppose any obstacle to|590| the air; I suffered much from the cold, so that I could not sleep, & in the morning rose with my body quite dull & benumbed.

In the Cordillera further Southward people lose their lives from snow-storms, here it sometimes happens from another cause. My guide, when a boy of fourteen years old, was passing with some others the Cordillera in the month of May, & while in the central parts a furious gale of wind arose, so that the men could hardly stick on their mules, & stones were flying along the ground; the day was quite cloudless & not a speck of Snow fell, but the temperature was low. — It is probable that the thermometer would not have stood very many degrees below the freezing point, but the effect on their bodies, ill-protected by clothing, would be in proportion to the rapidity of the current of cold air. — The gale lasted for more than a day; the men began to lose all their strength & the mules would not move onwards. — My guide's brother tried to return but he perished & his body was found two years afterwards, lying by the side of his mule near the road, with the bridle still in his hand. Two other men in the party lost their fingers & toes, & out of two hundred mules & thirty cows only fourteen of the former escaped alive. Many years ago a large party all perished from a similar cause; but their bodies to this day have never been discovered: the union of a cloudless sky, low temperature, & a furious gale of wind, I should think must be in all parts of the world an unusual occurrence.

June 29th & 30th We gladly travelled down the valley to our former nights lodging; from the thence to near the "Agua amarga", where there is a bitter little well. — |591|

July 1st Reached the valley of Copiapò; the smell of the fresh clover was quite delightful after the scentless air of the dry sterile Despoblado. —

July 2nd & 3rd Staid in the town at Mr Bingleys house. —

4th Set out for the Port, which is called 18 leagues distant. — I slept at a cottage beyond the halfway. There is very little cultivation below the town; the valley expands & is covered with a wretched coarse kind of grass, which scarcely any animal will touch. The soil *appears* both rich & damp; its poorness in productive powers must be owing to the abundance of saline matter; in some spots there are layers several inches thick of white & pure Salts, which consist chiefly of the Carbonate & Sulphate of Soda. The whole line of road is only inhabited in a few places. —

5th We reached the port at Noon. — It is a miserable little assemblage of a few houses, situated at the foot of some sterile plains & hills. — At present, from the river reaching the sea they enjoy the advantage of fresh water within a mile & a half. — On the beach there were large piles of merchandize & the little place had an air of bustle & activity. — I found the Beagle had arrived on the 3rd. — Capt. FitzRoy was not on board: at Valparaiso he joined the Blonde to assist as Pilot in taking off the coast of Chili, South of Concepcion, the crew of H.M.S. Challenger, which had there been wrecked. —[1] I felt very glad to be again on board

the Beagle. — In the evening I gave my "adios" with a hearty goodwill
to my companion, Mariano Gonzales, with whom I had ridden so many
leagues in Chili. — |592|

[1] In a letter to Caroline Darwin, CD wrote: 'When I reached the port of Copiapo, I found
the Beagle there, but with Wickham as temporary Captain. Shortly after the Beagle got
into Valparaiso, news arrived that H.M.S. Challenger was lost at Arauco, & that Capt
Seymour a great friend of FitzRoy & crew were badly off amongst the Indians. — The old
Commodore in the Blonde was very slack in his motions, in short afraid of getting on that
lee-shore in winter; so that Capt FitzRoy had to bully him & at last offered to go as Pilot. —
We hear that they have succeeded in saving nearly all hands, but that the Captain &
Commodore have had a tremendous quarrel; the former having hinted some thing about
a Court-Martial to the old Commodore for his slowness. — We suspect, that such a taught
hand, as the Captain is, has opened the eyes of every one fore & aft in the Blonde to a most
surprising degree. We expect the Blonde will arrive here in a very few days & all are very
curious to hear the news; no change in state politicks ever caused in its circle more
conversation, that this wonderful quarrel between the Captain & the Commodore has
with us. —' See Correspondence 1: 458. A detailed account of the loss of the Challenger and
of his role in rescuing her captain and crew is given by FitzRoy in Narrative 2: 428–73.

6th   In the middle of the day the Beagle made sail: on the 10th we crossed
the Tropic of Capricorn: on the 12th in the evening came to an anchor at
the port of Iquique.

12th   The coast was here formed by a great steep wall of rock about 2000
feet high; the town containing about a thousand inhabitants, stands on
a little plain of loose sand at the foot of this barrier. The whole is utterly
desert; the fine white sand is piled up against the mountains to more
than a thousand feet high, & neither it nor the rocks produce one single
plant. In this climate a light shower only falls once in many years; hence
the ravines are filled up with loose detritus & the whole mountains
appear crumbling. At this season of the year, a heavy bank of clouds
parallel to the ocean seldom rises above the wall of coast rocks. — The
aspect of the place was most gloomy; the little port with its few vessels
& the small group of wretched houses, seemed overwhelmed & out of
all proportion with the rest of the scene. — The inhabitants live like
those on board a ship, everything comes from a distance. The water is
brought from Pisagua, about 40 miles off, in boats, & is sold at 9 Riales[1]
an eighteen gallon cask. — a wine-bottle full cost 3[d]. — In a like manner
firewood & of course every article of food is imported. The latter chiefly
from Arica where there is a stream & fertile valley. — Of course very few
animals can|593| be maintained in such a place; I with difficulty hired for
the morning two mules & a guide to go to the Saltpetre works.[2] These
are the present support of Iquique; during one year the value of 100
thousand pounds sterling was exported to France & England. It is
however of much less value than true Saltpetre, this being the Nitrate

of Soda, mixed with some common Salt. Formerly there were two exceedingly rich silver mining districts; at the present day they produce little. —

Our arrival in the offing caused some little apprehension; Peru is at present in a complete state of Anarchy; & each party having demanded a contribution, the poor town of Iquique was in tribulation, thinking that the evil hour was come. — They also have their domestic troubles; three French carpenters during one night broke open & robbed two Churches; subsequently from intimidation one confessed & the plate was recovered. The two convicts were sent to Arequipa (200 leagues distant) for punishment, but the chief man there thought it a pity to shoot such useful workmen who could make all sorts of furniture, & they were pardoned. — Things being in this state, the Churches were again broken open & the plate stolen; but this second time no traces can be discovered (some suspect the Cura!); the inhabitants were dreadfully enraged & declaring none but hereticks would "eat God Almighty", proceeded to torture some Englishmen, with the intention of afterwards shooting them. At last the|594| authorities interfered & peace was established.

[1] Note in margin: $4^s.6^d$.
[2] Note in margin: 'Paid 4£ Sterling. Mention this amount'.

13th In the morning I started for the Saltpetre works, a distance of 14 leagues. — Our ascent by a zig-zag sandy track up the steep coast line of mountain (1900 ft. Barom:) was very tedious. — We soon came in view of the Minerales of Guantajaya & S. Rosa: These two small villages are placed at the very mouths of the mines; if Iquique had a desolate appearance, these perched up on a hill had a still more unnatural air. — We did not reach the Saltpetre works till after sunset; the road crossed an undulating country; a complete & utter desert. The road was strewed over with the bones and skins of dead Mules & Jackasses: what travellers have rather strongly written about the numbers in the Cordillera passes, is here actually verifyed. — Excepting the Vultur aura, which feeds on the Carcases, I saw neither bird, quadruped, reptile, or insect. On the coast mountains at about 2000 ft elevation, the bare sand was in places strewed over with an unattached greenish Lichen, in form like those which grow on old stumps: this in a few spots was sufficiently abundant to tinge the sand when seen from a little distance, of a yellowish color. I also saw another minute species of Lichen on the old bones. And where the first kind was lying, there were in the clefts of the rocks a *few* Cacti. These are supported by the dense clouds which generally rest on the land at this height. Excepting these, I saw no one|595| plant. — This is the first true desart I have ever seen; the effect on me was not impressive, I believe owing to having been weaned[1] to such a country whilst travelling from Coquimbo to Copiapò. — In common language, the Traversia between Guasco & the latter place is a frightful desart; however in truth few spots 200 y^{ds} square could be found without any vestige of vegetation. — This country is very remarkable by being in the greater part covered by a thick crust of Salt & saliferous Sandstone. The Salt is white, very hard, & compact, it occurs in water worn nodules, which project out of the soft sandstone. — The appearance of the mountains & valleys is that of the *last* remains of snow before all is thawed away: Many of the Strata contain Salt, this I suppose to have been washed out, & subsequently infiltering amongst the superficial sand is rehardened. The quantity is immense & it offers an incontestible proof of the dryness of the climate. At night I slept at the house of an[2] owner of one of the Saltpetre works. —

[1] The word 'weaned' is underlined in pencil, and a note in the margin, probably in Hensleigh Wedgwood's hand, runs: 'Wean *from* not *to* — a false metaphor'.
[2] Above the word 'an', '*the*?' is pencilled.

14th The country is here equally unproductive. They have a well (36 yards deep) from which some bitter Saltish water is procured, & firewood at twelve miles distance. Nearer to the main Cordillera there are some few little villages, such as Tarapaca, where having more water the inhabitants are enabled to irrigate a little land, & produce hay on which the mules & Jackasses employed in carrying the Saltpetre feed. — The owner|596| complained much of the heavy expences. The Nitrate of Soda, purified by solution in boiling water, is sold at the Ships side at 14 shillings the 100 pounds. — The mine consists of a thick (2–3 ft) hard layer of tolerably pure Salt, which is almost on the surface of the Land. The stratum follows the margin of a grand basin or plain which manifestly has once been either a lake or inland sea. — The present elevation is 3300 ft. — In my return I made a round by the famous Mineral of Guantajaya; the village entirely consists of the families of the miners; the place is utterly destitute, water is brought by animals from about 30 miles. — At present the mines produce scarcely anything; they have formerly been worked to a great extent; one having a depth of 400 yards. Masses, very many pounds weights, of Silver have been extracted, so pure as to require no process but running them down into bars. — We reached Iquique after sunset & I went on board, when the Beagle weighed her anchor for Lima. — I am very glad we have seen this place, I understand it a complete type of the greater part of the coast of Peru. —

19th In the night anchored in the outer part of the harbor of Callao. — Our passage was a short one owing to the steady trade wind; Rolling steadily onwards with our studding sails on each side, I was reminded of the Atlantic. — But there is a great difference in the interest of the two passages. |597| In the latter there is an ever varying & beautiful sky; the brilliant day is relieved by a cool refreshing evening & the cloudless sky is glorious. — The ocean teems with life, no one can watch the Flying-fish, Dolphin & Porpoises without pleasure. At night in the clear Heavens, the Europæan traveller views the new Constellations which foretell the new countries to which the good ship is onward driving. — Here in the Pacifick, although the water is never agitated by storms, it never is quiet, but feels through the unbroken continuity the violence which reigns in the South. Now, in the winter, a heavy dull bank of clouds intercepts during successive days even a glimpse of the sun. — The temperature is by no means warm; in approaching these low latitudes I did not experience that delicious mildness, which is known for a few days in the Spring of England, or in first entering the Tropics in the Atlantic.

20th During our whole stay the climate was far from pleasant; the ceaseless gloom which hangs over the country would render any landscape uninteresting. During 16 days I have only had one view of the Cordilleras behind Lima, which seen in stages through the openings of the clouds, bore a very grand aspect. — It is *proverbial* that rain never falls in this part of Peru; yet this is not correct, during nearly every day there is a thick drizzle or Scotch mist which is sufficient to make the streets muddy & ones clothese very damp. People are generally pleased to call this Peruvian dew. That much water does not fall is very manifest; the houses are covered with flat roofs, composed of hardened mud; |598| on the mole ship-loads of wheat are piled up & thus kept for months without any cover. Lastly, the country is quite sterile, excepting where irrigated. The *valley* of the Rimac, however, wears as green a clothing as those in central Chili. — I cannot say that I like what I have seen of Peru; in summer it is said that the Climate is much pleasanter; at all seasons of the year both inhabitants & foreigners suffer much from attacks of Ague. —[1]

No state in S. America, since the declaration of the Independence, has suffered more from Anarchy than Peru: at present there are four chiefs in arms for supreme government. If one should succeed in becoming very powerful, the others for a time coalesce against him, but afterwards are again disunited. The other day at the Anniversary of the Independence, high mass was performed, the President partaking of the Sacrament; during the "*Te Deum laudamus*" instead of each regiment displaying the Peruvian flag, a black one with death's head was unfurled. Imagine what a government, when such a scene could be ordered on such an occasion to be typical of their determination of fighting to death! — This state of affairs has happened very unfortunately for me, as I am precluded from making any excursions beyond the limits of the Towns. — The barren Isd of S. Lorenzo[2] which forms the harbor is nearly the only secure walk. — I climbed one day to the highest part, nearly 1200 ft high. This is within the limit of the region of Clouds at this season.|599| I there met with half a dozen different kinds of plants & an abundance of Cryptogamic vegetation; on the hills near Lima, at a little greater elevation, the ground is carpeted with moss & there are some beautiful yellow lilies called Amancaes. This shows a much greater humidity than in a corresponding situation at Iquique. Gradually travelling Northward, the climate becomes damper, & at Guyaquil there are luxuriant forests. —[3]

Callao is a most miserable filthy, ill built, small sea-port; the inhabitants both here & at Lima present every imaginable shade of mixture between Europœan, Negro & Indian blood. They appear a depraved, drunken set. The very atmosphere was loaded with foul smells; & that peculiar kind which can be perceived in nearly all towns within the Tropics was very strong. The Fortress which withstood L. Cochranes long siege, appears very imposing; the president is to-morrow going to dismantle it; he has not an officer to whom he could trust so important a charge. He himself obtained his present rank by being Governor & mutinying against the former president. — Callao being such as it is & Lima 7 miles distant, this is a disagreeable [place] to lie in a Ship; at present there are no means to take exercise. A short time since, M[r] Wilson the Consul general.—Lord E. Clinton & a Frenchman were riding & were attacked by a party of Soldier—robbers, who plundered them so completely, that they returned naked, excepting their drawers. — The robbers were actuated by warm Patriotism;|600| They waved the Peruvian banner & intermingled crys of "Viva la Patria"; "give me your jacket". "Libertad Libertad" with "Off with your trowsers". —

[1]Notes in margin: 'Common to coast of Pacifick—most mysterious—St Jago, Galapagos healthy—Not quantity of vegetation—Stagnant moisture'.
[2]It was of this island that CD wrote: 'I was much interested by finding embedded, together with pieces of sea-weed in the mass of shells, in the eighty-five foot bed, a bit of

cotton-thread, plaited rush, and the head of a stalk of Indian corn. This fact, coupled with another, which will be mentioned, proves I think the amount of eighty-five feet elevation since man inhabited this part of Peru.' See *Journal of Researches* pp. 451–2. Not only was this the earliest recorded finding of kernels of maize at such a site (see Duccio Bonavia in *Los Gavilanes*, Corporacion Financiera de Desarrollo S. A. Cofide, Lima, 1982, pp. 385–9), but the fact that CD does not mention the presence of ceramics suggests that the site was a preceramic one, and therefore that its elevation to 85 feet above sea-level had taken place within a period of the order of 2500 years.
[3] Note in margin: 'Change sudden'.

29th–August 3rd   I took a place in a coach which runs twice every day to Lima & spent five very pleasant days there. There is so much hospitality in these countries & the conversation of intelligent people in a new & foreign place cannot fail to be interesting. Moreover a residence of some years in contact with the polite & formal Spaniards certainly improves the manners of the English merchants. — I found the Consul General, Mr Wilson, most exceedingly obliging: having been Aid de Camp to Bolivar he has travelled over much of S. America & knows its inhabitants right well. —

Lima stands on a small plain formed during the gradual retreat of the sea; out of it rise barren steep hills like Islands. —[1] It is irrigated by two streams, the valleys of which rapidly contract & are hidden between the headlands of the first Cordilleras. The plain is divided into large green fields divided by straight mudwalls; there are very few trees excepting some willows & fruit trees. By the presence of an occasional cluster of Banana plants & Orange trees only does the landscape partake of a Tropical character. The city of Lima is now in a wretched state of decay; the streets are nearly unpaved & in all directions heaps of filth are piled up. — Amongst these the Gallinazoes, tame as Poultry,|601| are picking up bits of Carrion. There is little air of business; there are few Carriages, carts or even Cargo-Mules in the streets. — The houses have generally an irregular upper story, built on account of the Earthquakes of plastered wood-work; some of the old-houses now used by several families are immensely large & would rival in the suites of Apartments the most magnificent in London. Lima must indeed formerly have been a splendid, but small city; the extraordinary number of churches give to it, especially when seen from a short distance, a character quite distinct from the generality of towns.

There are two things in Lima, which all Travellers have discussed; the ladies "tapadas", or concealed in the saya y Manta, & fruit called Chilimoya.[2] To my mind the former is as beautiful as the latter is delicious. The close elastic gown fits the figure closely & obliges the ladies to walk with small steps which they do very elegantly & display

very white silk stockings & very pretty feet. — They wear a black silk
veil, which is fixed round the waist behind, is brought over the head, &
held by the hands before the face, allowing only one eye to remain
uncovered. — But then that one eye is so black & brilliant & has such
powers of motion & expression, that its effect is very powerful. —
Altogether the ladies are so metamorphised; that I at first felt as much
surprised, as if I had been introduced amongst a number of nice|602|
round mermaids, or any other such beautiful animal. And certainly
they are better worth looking at than all the churches & buildings in
Lima. — Secondly for the Chilimoya, which is a very delicious fruit, but
the flavour is about as difficult to describe, as it would be to a Blind man
some particular shade of colour; it is neither a nutritive fruit like the
Banana, or a crude fruit like the Apple, or refreshing fruit like the
Orange or Peach, but it is a very good & large fruit & that is all I have to
say about it. —

[1] Note in margin: 'What [illeg] 500 feet [illeg]'.
[2] See watercolour by Syms Covington labelled 'Walking Dress of the Females of Lima',
reproduced in Beagle Record p. 239, which corresponds exactly with CD's description of
the ladies concealing all but one eye beneath their mantilla. The chilimoya is Anana
cherimola, the Peruvian custard apple.

9th–September.   H.M.S. Blonde arrived with Capt. FitzRoy on board; he
subsequently during the Beagle stay resided in Lima. — The country
has continued in the same state of mis-rule; even the road between
Lima & Callao has been infested with gangs of mounted robbers. — In
consequence I have staid quietly on board. My occupation has been
writing up Geological notes about Chili. — if the time had not been
robbed either from England or the Pacifick it would have been pleasant;
the consciousness of this, gave a longing to proceed. — I paid Lima two
short visits; one day I went out with some Merchants, who have a few
dogs to hunt deer with. — Our sport was very poor; but I had an
opportunity of seeing the remains of one of the very numerous old
Indian villages, with its hill-like mound in the centre. — The ruins in this
plain of houses, enclosures, irrigating streams|603| & burial mounds,
give one a high idea of the ancient population. — When the Earthern
ware is considered; the woollen clothes, the utensils, of elegant forms
cut out of the hardest rocks, the tools of Copper & ornaments of the
precious metals, it is clear they were considerably advanced in civilization. — The burial mounds (called "Huacas") are really stupendous
works; in some places, however, a natural hill appears only to be
artificially encased & modelled. — Another & very different class of
ruins possesses some interest, namely those of old Callao, which was

destroyed by the great Earthquake of 1747. — The state of ruin is much more complete than that of Concepcion; quantities of shingle almost conceal the foundations of the walls. It is believed the Land at the time subsided: I found some curious geological facts, which are only explicable by a similar movement but at a very remote period, when the country stood at a less elevation by 100 ft & yet was inhabited by Indians.—[1] I do not think there is any place which the Beagle has visited, of which I have seen so little; so I will write no more. —

[1] In *Journal of Researches* p. 452, CD wrote: 'This bed with fossil earthenware occurring at about the same altitude with the terrace at San Lorenzo, confirms the supposed amount of elevation within the human period.'

September 6th   The little schooner "Constitution" in which Mr Sulivan surveyed North coast of Chili has been bought of the Capt. for Government.—[1] Mrs Usborne & Forsyth are left in her to survey the coast of Peru & afterwards return in a merchant vessel to England.

[1] FitzRoy wrote: 'Lieutenant Sulivan brought his little vessel safely to an anchor near the Beagle on the 30th [July], having accomplished his survey in a very satisfactory manner. So well did he speak of the Constitucion, as a handy craft and good sea boat, and so correctly did his own work in her appear to have been executed, that after some days' consideration I decided to buy her, and at once set on foot an examination of the coast of Peru, similar to that which Mr Sulivan had completed of the coast of Chile. Don Francisco Vascuñan had authorized the sale of his vessel at Callao: she was purchased by me for £400, and immediately fitted out afresh.' See *Narrative* 2: 482–3. But once again this generous and disinterested outlay of his own funds in furtherance of his mission was undertaken without obtaining prior permission from the Admiralty, and once again he was rewarded only with an official reprimand. The Minutes written across his letter to their Lordships informing them of his action refer to 'former papers forbidding him to hire a tender', and state: 'Inform Capt. FitzRoy that Lords highly disapprove of this proceeding, especially after the orders which he previously received on the subject.' Nevertheless the Hydrographer, Sir F. Beaufort, acknowledges that the subsidiary craft will materially assist the survey. See *Admiralty Records*, Record Office, ADM/1/3848.

7th   The Beagle sailed for the Galapagos:

15th   on the 15th she was employed in surveying the outer coast of Chatham Isd[1] the S. Eastern one of the Archipelago. |604|

[1] The Spanish name of Chatham Island now used is Isla San Cristobal.

16th   The next day we ran near Hoods Is^d[1] & there left a Whale boat. — In the evening the Yawl was also sent away on a surveying cruize of some length. — The weather, now & during the passage, has continued as on the coast of Peru, a steady, gentle breeze of wind & gloomy sky. — We landed for an hour on the NW end of Chatham Is^d. — These islands at a distance have a sloping uniform outline, excepting where broken by sundry paps & hillocks. — The whole is black Lava, *completely* covered by small leafless brushwood & low trees. — The fragments of Lava where most porous are reddish & like cinders; the stunted trees show

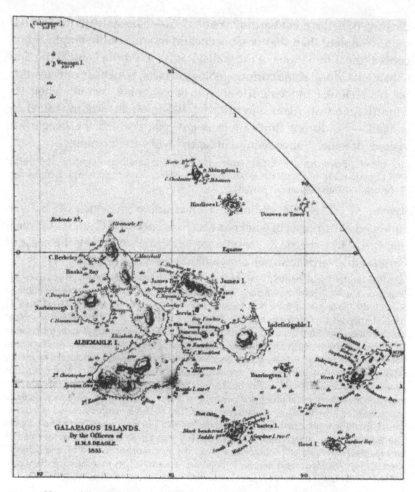

Chart of the Galapagos Islands (from part of a map in *Narrative* 1).

little signs of life. — The black rocks heated by the rays of the Vertical sun like a stove, give to the air a close & sultry feeling. The plants also smell unpleasantly. The country was compared to what we might imagine the cultivated parts of the Infernal regions to be. —[2]

This day, we now being only 40 miles from the Equator, has been the first warm one; up to this time all on board have worn cloth clothese; & although no one would complain of cold, still less would they of too much warmth. — The case would be very different if we were cruizing on the Atlantic side of the Continent.

[1] Now Isla Española.

[2] In a letter to Caroline Darwin from Lima, CD had written: 'I am very anxious for the

Galapagos Islands, — I think both the Geology & Zoology cannot fail to be very interesting. —' And also from Lima he wrote to W. D. Fox: 'I look forward to the Galapagos, with
more interest than any other part of the voyage. — They abound with active Volcanoes &
I should hope contain Tertiary strata. —' See *Correspondence* 1: 458 and 460. In the event,
he saw few examples of currently active volcanoes in the Galapagos, and it was the
zoology of the islands that proved to be of greatest interest.

17*th*   The Beagle was moved into St Stephens harbor. We found there an
American Whaler & we previously had seen two at Hoods Island. —
|605| The Bay swarmed with animals; Fish, Shark & Turtles were
popping their heads up in all parts. Fishing lines were soon put
overboard & great numbers of fine fish 2 & even 3 ft long were caught.
This sport makes all hands very merry; loud laughter & the heavy
flapping of the fish are heard on every side. — After dinner a party went
on shore to try to catch Tortoises, but were unsuccessful. — These
islands appear paradises for the whole family of Reptiles. Besides three
kinds of Turtles, the Tortoise is so abundant; that [a] single Ship's
company here caught from 500–800 in a short time. — The black Lava
rocks on the beach are frequented by large (2–3 ft) most disgusting,
clumsy Lizards. They are as black as the porous rocks over which they
crawl & seek their prey from the Sea. — Somebody calls them "imps of
darkness". — They assuredly well become the land they inhabit. —
When on shore I proceeded to botanize & obtained 10 different flowers;
but such insignificant, ugly little flowers, as would better become an
Arctic, than a Tropical country. — The birds are Strangers to Man &
think him as innocent as their countrymen the huge Tortoises.[1] Little
birds within 3 & four feet, quietly hopped about the Bushes & were not
frightened by stones being thrown at them. Mr King |606|killed one
with his hat & I pushed off a branch with the end of my gun a large
Hawk. —

[1]In Down House Notebook 1.17, CD wrote soon after arriving at the Galapagos: 'The
Thenca very tame & curious in these Islands. I certainly recognise S. America in
Ornithology. Would a botanist? 3/4 of plants in flower.' See *CD and the Voyage* p. 247.

18*th*   Again we moved our Anchorage & again after dinner took a long
walk. — We ascended the broken remains of a low but broard crater.
The Volcano had been sub-marine — the strata which dipped away on
all sides were composed of hard Sandstones composed of Volcanic
dust. A few leagues to the North a broken country was studded with
small black cones; the ancient chimneys for the subterranean melted
fluids. — The hunting party brought back 15 Tortoises: most of them
very heavy & large. One weighed   lbs. —[1]

[1]The space left for filling in the weight of the tortoise remains blank.

19th & 20th   During these two days surveyed the seaward coast of the Is^d & returned to an anchor where we had found the Whaler. — At one point there were little rills of water, & one small cascade. — The valleys in the neighbourhead were coloured a somewhat brighter green. — Upon first arriving I described the land as covered with leafless brushwood; & such certainly is the *appearance*. I believe however almost every plant or tree is now both in flower & its leaf. — But the most prevalent kinds are ornamented with but very few & these of a brown color.

21st   My servant & self were landed a few miles to the NE in order that I might examine the district mentioned above as resembling|607| chimney. The comparison would have been more exact if I had said the Iron furnaces near Wolverhampton. — From one point of view I counted 60 of these truncated hillocks, which are only from 50 to 100 ft above the plain of Lava. — The age of the various streams is distinctly marked by the presence & absence of Vegetation; in the latter & more modern nothing can be imagined more rough & horrid. — Such a surface has been aptly compared to a sea petrified in its most boisterous moments. No sea however presents such irregular undulations, — nor such deep & long chasms. The craters are all entirely inert; consisting indeed of nothing more than a ring of cinders. — There are large circular pits, from 30 to 80 ft deep; which might be mistaken for Craters, but are in reality formed by the subsidence of the roofs of great caverns, which probably were produced by a volume of gaz at the time when the Lava was liquid. — The scene was to me novel & full of interest; it is always delightful to behold anything which has been long familiar, but only by description. — In my walk I met two very large Tortoises (circumference of shell about 7 ft). One was eating a Cactus & then quietly walked away. — The other gave a deep & loud hiss & then drew back his head. — They were so heavy, I could scarcely lift|608| them off the ground. — Surrounded by the black Lava, the leafless shrubs & large Cacti, they appeared most old-fashioned antediluvian animals; or rather inhabitants of some other planet. —

22nd   We slept on the sand-beach, & in the morning after having collected many new plants, birds, shells & insects, we returned in the evening on board. — This day was glowing hot, & was the first when our closeness to the Equator was very sensible. —

23rd & 24th   Crossed over & came to an anchor at Charles Island. —[1] Here there is a settlement of only five to 6 years standing. An Englishman Mr

Lawson[2] is now acting as Governor. — By chance he came down to visit a Whaling Vessel & in the morning accompanied us to the Settlement. —

[1] Now Isla Florena.
[2] Mr Nicholas O. Lawson was an Englishman serving the Republic of the Equator, or Ecuador.

25th This is situated nearly in the centre of the Island, about 4 & ½ miles inland, & elevated perhaps 1000 ft above the sea. — The first part of the road passed through a thicket of nearly leafless underwood as in Chatham Is^d. — The dry Volcanic soil affording a congenial habitation only to the Lizard tribe. — The wood gradually becomes greener during the ascent. — Passing round the side of the highest hill; the body is cooled by the fine Southerly trade wind & the eye refreshed by a plain green as England in the Spring time. — Out of the wood extensive patches|609| have been cleared, in which sweet Potatoes (convolvulus Batata) & Plantains grow with luxuriance. — The houses are scattered over the cultivated ground & form what in Chili would be called a "Pueblo". — Since leaving Brazil we have not seen so Tropical a Landscape, but there is a great deficiency in the absence of the lofty, various & all-beautiful trees of that country. — It will not easily be imagined, how pleasant the change was from Peru & Northern Chili, in walking in the pathways to find *black mud* & on the trees to see mosses, ferns & Lichens & Parasitical plants adhæring. — Owing to an unusual quantity of rain at this time of year, I suspect we have seen the Island at its full advantage. — I suspect this the more from meeting with singularly few insects of any of the orders. — If such luxuriance is constant this scarcity of its universal concomitants is very remarkable. — The inhabitants are in number 200–300: nearly all are people of color & banished for Political crimes from the State of the Equator (Quito & Guyaquil &c) to which this Archipelago belongs. — It appears the people are far from contented; they complain, here as in Chiloe, of the deficiency of money: I presume there is some more essential want than that of mere Currency, namely want of sale of their produce. — This of course will gradually be ameliorated. —already on an average,|610| in the year 60–70 Whaling vessels call for provisions & refreshment. — The main evil under which these islands suffer is the scarcity of water. — In very few places streams reach the beach so as to afford facilities for the watering of Shipping. Every where the porous nature of the Volcanic rocks has a tendency to absorb without again throwing up the little water which falls in the course of the year. — At the Settlement there are several springs & small pools, three or four of which are said never to fail. — Generally the islands in the Pacifick are subject to years of drought & subsequent scarcity; I should be afraid this group will not afford an exception. — The inhabitants here lead a sort of Robinson Crusoe life; the houses are very simple, built of poles & thatched with grass. — Part of their time is employed in hunting the wild pigs & goats with which the woods abound; from the climate, agriculture requires but a small portion. — The main article however of animal food is the Terrapin or Tortoise: such numbers yet remain that it is calculated two days hunting will find food for the other five in the week. — Of course the numbers have been much reduced; not many years since the Ship's company of a Frigate brought down to the Beach in one day more than 200, —where|611| the settlement now is, around the Springs, they formerly swarmed. — Mr Lawson thinks there is yet left sufficient for 20 years: he has however sent a party to Jame's[1] Island to salt (there is a Salt mine there) the meat. — Some of the animals are there so very large, that upwards of 200 £b^s of meat have been procured from one. —Mr Lawson reccollect having seen a Terrapin which 6 men could scarcely lift & two could not turn over on its back. These immense creatures must be very old, in the year 1830 one was caught (which required 6 men to lift it into the boat) which had various dates carved on its shells; one was 1786. — The only reason why it was not at that time carried away must have been, that it was too big for two men to manage. — The Whalers always send away their men in pairs to hunt. —

[1]James is the name still in use, the Spanish alternatives being Santiago or San Salvador.

26th & 27th I industriously collected all the animals, plants, insects & reptiles from this Island. — It will be very interesting to find from future comparison to what district or "centre of creation" the organized beings of this archipelago must be attached. —[1]

I ascended the highest hill on the Is^d, 2000 ft. — it was covered in its upper part with coarse grass & Shrubs. — The remains of an old Crater were very evident; small as the whole island is, I counted 39 conical hills, in the summit of all of which there was a more or less perfect circular depression.|612| It is long since the Lava streams which form the lower parts of the Island flowed from any of these Craters: Hence we have a smoother surface, a more abundant soil, & more fertile vegetation. — It is probable that much of the Lava is of subaqueous origin. —

[1]In a letter to Henslow from Sydney written four months later, CD said: 'I last wrote to you from Lima, since which time I have done disgracefully little in Nat: History; or rather

I should say since the Galapagos Islands, where I worked hard. — Amongst other things, I collected every plant, which I could see in flower, & as it was the flowering season I hope my collection may be of some interest to you. — I shall be very curious to know whether the Flora belongs to America, or is peculiar. I paid also much attention to the Birds, which I suspect are very curious. —' See *Correspondence* 1: 485.

Later, when CD was completing his ornithological notes some time between mid-June and August 1836, he wrote: 'Thenca (Mimus Thenca). These birds are closely allied in appearance to the Thenca of Chile. They are lively, inquisitive, active, run fast, frequent houses to pick the meat of the tortoise which is hung up, — sing tolerably well, — are said to build a simple open nest, — are very tame, a character in common with other birds. I imagined, however, its note or cry was rather different from the Thenca of Chile—? Are very abundant over the whole Island; are chiefly tempted up into the high & damp parts by the houses & cleared ground.'

'I have specimens from four of the larger Islands; the specimens from Chatham & Albemarle Isd. appear to be the same, but the other two are different. In each Isd. each kind is exclusively found; habits of all are indistinguishable.'

'When I recollect the fact, that from the form of the body, shape of scales & general size, the Spaniards can at once pronounce from which Isd. any tortoise may have been brought:— when I see these Islands in sight of each other and possessed of but a scanty stock of animals, tenanted by these birds but slightly differing in structure & filling the same place in Nature, I must suspect they are only varieties. The only fact of a similar kind of which I am aware is the constant asserted difference between the wolf-like Fox of East & West Falkland Isds. — If there is the slightest foundation for these remarks, the Zoology of Archipelagoes will be well worth examining; for such facts would undermine the stability of species.' See Nora Barlow, 'Darwin's Ornithological Notes', *Bulletin of the British Museum (Natural History), Historical Series*, 2: 201–78, 1963; and F. J. Sulloway, 'Darwin's Conversion: The *Beagle* Voyage and Its Aftermath', *Journal of the History of Biology*, 15: 325–96, 1982.

The first stirrings of doubt about the immutability of species had evidently struck him by now.

28th   Steered towards the Southern end of Albermale Is[d],[1] which was surveyed.

[1] The correct spelling is Albemarle, and the island is now known as Isabela.

29th   Anchored at Noon in a small cove beneath the highest & boldest land which we have yet seen. — The Volcanic origin of all is but too plainly evident: Passed a point studded over with little truncated cones or Spiracles as some Author calls them; the Craters were very perfect & generally red-coloured within. — The whole had even a more *work-shop* appearance than that described at Chatham Is[d]. — A calm prevented us anchoring for the night. —

30th   The next day, a light breeze carried us over the calm sea, which lies between Narborough[1] & Albermale Is[d]. In the latter, high up, we saw a small jet of steam issuing from a Crater. — Narborough Isld presents a more rough & horrid aspect than any other; the Lavas are generally naked as when first poured forth. — When H.M.S. Blonde was here there was an active Volcano in that Island. — After sun-set, came to an

Albemarle and Charles Islands, by S. Bull after P. G. King (*Narrative* 2: 498).

anchor in Banks cove in Albermale Is[d] & which cove subsequently turned out to|613| be the Crater of an old Volcano.

[1] Now Isla Fernandina.

October 1st   Albermale I[s] is as it were the mainland of the Archipelago, it is about 75 miles long & several broard. —is composed of 6 or 7 great Volcanic Mounds from 2 to 3000 ft high, joined by low land formed of Lava & other Volcanic substances. — Since leaving the last Island, owing to the small quantity of water on board, only half allowance of water has been served out (ie ½ a Gallon for cooking & all purposes). — This under the line with a Vertical sun is a sad drawback to the few comforts which a Ship possesses. — From different accounts, we had hoped to have found water here. — To our disappointment the little pits in the Sandstone contained scarcely a Gallon & that not good. —it was however sufficient to draw together all the little birds in the country. — Doves & Finches[1] swarmed round its margin. — I was reminded of the manner in which I saw at Charles Isd a boy procuring dinner for his family. Sitting by the side of the Well with a long stick in his hand, as the doves came to drink he killed as many as he wanted & in half an hour collected them together & carried them to the house. —

To the South of the Cove I found a most beautiful Crater, elliptic in form, less than a mile in its longer axis & about 500 ft deep. — Its bottom was occupied by a lake, out of which a tiny Crater formed an Island. — The day|614| was overpoweringly hot; & the lake looked blue & clear. — I hurried down the cindery side, choked with dust, to my disgust on tasting the water found it Salt as brine. — This crater & some other neighbouring ones have only poured forth mud or Sandstone containing fragments of Volcanic rocks; but from the mountain behind, great bare streams have flowed, sometimes from the summit, or from small Craters on the side, expanding in their descent have at the base formed plains of Lava. — The little of the country I have yet seen in this vicinity is more arid & sterile than in the other Islands. — We here have another large Reptile in great numbers. —it is a great Lizard, from 10–15 lb in weight & 2–4 ft in length, is in structure closely allied to those imps of darkness which frequent the sea-shore. — This one inhabits burrows to which it hurrys when frightened with quick & clumsy gait. — They have a ridge & spines along the back; are colored an orange yellow, with the hinder part of back brick red. — They are hideous animals; but are considered good food: This day forty were collected. —

[1] This appears to be the only mention made by CD, either in the *Diary* or in his pocketbooks, of the family of finches that came to bear his name and to be most closely associated with the development of his ideas about speciation. However, the relative lack of interest in the Geospizidae displayed by CD when he was actually collecting birds in the Galapagos is consistent with the conclusion of Sulloway ('Darwin and his Finches: The Evolution of a Legend', *Journal of the History of Biology* **15**: 1–53, 1982) that it was not until the *Beagle*'s specimens were classified by John Gould early in 1837 that the true significance of their variability between the individual islands first became apparent to him.

By the time the *Journal of Researches* was published in 1839, CD no longer believed in the fixity of species, but the most radical of his ideas were still kept strictly to himself. He did not give a great deal away when he wrote: 'It has been mentioned, that the inhabitants can distinguish the tortoises, according to the islands whence they are brought. I was also informed that many of the islands possess trees and plants which do not occur on the others. For instance the berry-bearing tree, called Guyavita, which is common on James Island, certainly is not found on Charles Island, though appearing equally well fitted for it. Unfortunately, I was not aware of these facts till my collection was nearly completed:

it never occured to me, that the productions of islands only a few miles apart, and placed under the same physical conditions, would be dissimilar. I therefore did not attempt to make a series of specimens from the separate islands. It is the fate of every voyager, when he has just discovered what object in any place is more particularly worthy of his attention, to be hurried from it. In the case of the mocking-bird, I ascertained (and have brought home the specimens) that one species (*Orpheus trifasciatus*, Gould) is exclusively found in Charles Island; a second (*O. parvulus*) on Albemarle Island; and a third (*O. melanotus*) common to James and Chatham Islands. The last two species are closely allied, but the first would be considered by every naturalist as quite distinct. I examined many specimens in the different islands, and in each the respective kind was *alone* present. These birds agree in general plumage, structure, and habits; so that the different species replace each other in the economy of the different islands. These species are not characterized by the markings on the plumage alone, but likewise by the size and form of the bill, and other differences. I have stated, that in the thirteen species of ground-finches, a nearly perfect gradation may be traced, from a beak extraordinarily thick, to one so fine, that it may be compared to that of a warbler. I very much suspect, that certain members of the series are confined to different islands; therefore, if the collection had been made on any *one* island, it would not have presented so perfect a gradation. It is clear, that if several islands have each their peculiar species of the same genera, when these are placed together, they will have a wide range of character. But there is not space in this work, to enter on this curious subject.' See *Journal of Researches* pp. 474–5.

FitzRoy's ideas had also changed between the return of the *Beagle* and publication of the *Narrative*, since following his marriage he had become a firm believer in the absolute truth of the Bible. His view of the significance of the beaks of the finches differed somewhat from CD's, for he wrote: 'All the small birds that live on these lava-covered islands have short beaks, very thick at the base, like that of a bull-finch. This appears to be one of those admirable provisions of Infinite Wisdom by which each created thing is adapted to the place for which it was intended.' See *Narrative* 2: 503.

In the 1845 edition of the *Journal of Researches*, the theme of the gradation of the beaks of the ground finches was further expanded, and CD unwrapped his ideas just a little further: 'Seeing this gradation and diversity of structure in one small, intimately related group of birds, one might really fancy that from an original paucity of birds in this archipelago, one species had been taken and modified for different ends.' See *Journal of Researches*, 2nd edn, p. 380.

October 2ⁿᵈ   Sailed from this Crater Harbor: but were becalmed during the greater part of the day in the Straits which separates the two Islands:

3ʳᵈ   We then stood round the North end of Albermale Island. — The whole of this has the same sterile dry appearance; is studded with the small Craters which are appendages to the great Volcanic mounds, — & from which in very many places the black Lava has flowed, the configuration of the streams being like that of so much mud. — I should think it would be difficult to find in the intertropical latitudes a piece of land 75 miles long, so entirely useless to man or the larger animals. — From the evening of this day to the 8th was most unpleasant passed in struggling to get about 50 miles to Windward against a strong current.

8ᵗʰ   At last we reached Jame's Island, the rendezvous of Mʳ Sulivan. — Myself, Mʳ Bynoe & three men were landed with provisions, there to

wait till the ship returned from watering at Chatham Is^d. — We found
on the Isl^d a party of men sent by Mr Lawson from Charles Is^d to salt fish
& Tortoise meat (& procure oil from the latter). — Near to our Bivouac-
ing place, there was a miserable little Spring of Water. — We employed
these men to bring us sufficient for our daily consumption. — We
pitched our tents in a small valley a little way from the Beach. — The
little Bay was formed by two old Craters: in this island as in all the others
the mouths from which the Lavas have flowed are thickly studded over
the country.[1]

[1] Note in margin: 'Freshwater Cove of the Buccaniers'.

9th   Taking with us a guide we proceeded|616| into the interior & higher
parts of the Island, where there was a small party employed in hunting
the Tortoise. — Our walk was a long one. — At about six miles distance
& an elevation of perhaps 2000 ft the country begins to show a green
color. — Here there are a couple of hovels where the men reside. —
Lower down, the land is like that of Chatham Is^d, — very dry & the trees
nearly leafless. I noticed however that those of the same species
attained a much greater size here than in any other part. — The
Vegetation here deserved the title of a Wood: the trees were however
far from tall & their branches low & crooked.[1] About 2 miles from the
Hovels & probably at an additional 1000 ft elevation, the Springs are
situated. They are very trifling ones, but the water good & deliciously
cold. — They afford the only watering places as yet discovered in the
interior. — During the greater part of each day clouds hang over the
highest land: the vapor condensed by the trees drips down like rain.
Hence we have a brightly green & damp Vegetation & muddy soil. —
The contrast to the sight & sensation of the body is very doubtful after
the glaring dry country beneath. — The case is exactly similar to that
described in Charles Is^d. — So great a change with|617| so small a one of
elevation cannot fail to be striking. — On the 12th I paid a second visit to
the houses, bringing with me a blanket bag to sleep in. — I thus enjoyed
two days collecting in the fertile region. — Here were many plants,
especially Ferns; the tree Fern however is not present.[2] The tropical
character of the Vegetation is stamped by the commonest *tree* being
covered with compound flowers of the order of Syngynesia. — The
tortoise when it can procure it, drinks great quantities of water: Hence
these animals swarm in the neighbourhead of the Springs. — The
average size of the full-grown ones is nearly a yard long in its back shell:
they are so strong as easily to carry me, & too heavy to lift from the

ground. — In the pathway many are travelling to the water & others returning, having drunk their fill. — The effect is very comical in seeing these huge creatures with outstreched neck so deliberately pacing onwards. — I think they march at the rate 360 yards in an hour; perhaps four miles in the 24. — When they arrive at the Spring, they bury their heads *above* the eyes in the muddy water & greedily suck in great mouthfulls, *quite regardless* of lookers on. —

Wherever there is water, broard & well beaten roads lead from all sides to it, |618| these extend for distances of miles. — It is by this means that these watering places have been discovered by the fishermen. — In the low dry region there are but few Tortoises: they are replaced by infinite numbers of the large yellow herbivorous Lizard mentioned at Albermale Is^d. — The burrows of this animal are so very numerous; that we had difficulty in finding a spot to pitch the tents. — These lizards live entirely on vegetable productions; berrys, leaves, for which latter they frequently crawl up the trees, especially a Mimosa; never drinking water, they like much the succulent Cactus, & for a piece of it they will, like dogs, struggle [to] seize it from another. Their congeners the "imps of darkness" in like manner live entirely on sea weed. — I suspect such habits are nearly unique in the Saurian race.

In all these Isl^ds the dry parts reminded me of Fernando Noronha; perhaps the affinity is only in the similar circumstance of an arid Volcanic soil, a flowering leafless Vegetation in an Intertropical region, but without the beauty which generally accompanies such a position. —

During our residence of two days at the Hovels, we lived on the meat of the Tortoise fried in the transparent Oil which is procured from the fat. — |619| The Breast-plate with the meat attached to it is roasted as the Gauchos do the "Carne con cuero". It is then very good. — Young Tortoises make capital soup—otherwise the meat is but, to my taste, indifferent food. —[3]

[1] Note in margin: 'Saw some having circumference of 8 ft & several of 6 ft'.
[2] Note in margin: 'Not any Palm'.
[3] According to FitzRoy, several tortoises were eventually brought alive to England. He recorded that a hunting party brought 18 on board from Chatham Island on 18 September, and a further 30 on 12 October. 'The largest we killed was three feet in length from one end of the shell to the other: but the large ones are not so good to eat as those of about fifty pounds weight—which are excellent, and extremely wholesome food.' See *Narrative* 2: 504.

11th   The Mayór-domo took us in his boat to the Salina which is situated about 6 miles down the coast. — We crossed a bare & apparently recent stream of Lava which had flowed round an ancient but very perfect

Crater. — At the bottom of this Crater is a Lake, which is only 3 or 4 inches deep & lies on layers of pure & beautifully Crystallized Salt. The Lake is quite circular & fringed with bright breen succulent plants; the sides of Crater are steep & wooded; so that the whole has rather a pretty appearance. — A few years since in this quiet spot the crew of a Sealing vessel murdered their Captain. We saw the skull lying in the bushes. —

In rocky parts there were great numbers of a peculiar Cactus whose large oval leaves connected together formed branches rising from a cylindrical trunk. —[1] In places also a Mimosa was common; the shade from its foliage was very refreshing, after being exposed in the open wood to the burning Sun.

[1] Sketch in the margin:

12th–16th   We all were busily employed during these days in collecting all sorts of Specimens.|620| The little well from which our water was procured was very close to the Beach: a long Swell from the Northward having set in, the surf broke over & spoiled the fresh water. — We should have been distressed if an American Whaler had not very kindly given us three casks of water (& made us a present of a bucket of Onions). Several times during the Voyage Americans have showed themselves at least as obliging, if not more so, than any of our Countrymen would have been. Their liberality moreover has always been offered in the most hearty manner. If their prejudices against the English are as strong as our's against the Americans, they forget & smother them in an admirable manner. —

16th   The weather during nearly all the time has been cloudless & the sun very powerful; if by chance the trade wind fails for an hour the heat is very oppressive. During the two last days, the Thermometer within the Tents has stood for some hours at 93°. — In the open air, in the wind & sun, only 85°. — The sand was intensely hot, the Thermometer placed in a *brown* kind immediately rose to 137, & how much higher it would have done I do not know: for it was not graduated above this: — The *black* Sand felt far hotter, so that in *thick* boots it was very disagreeable to pass over it. —|621|

17th   In the afternoon the Beagle sent in her boats to take us on board. —

18th   Finished the survey of Albermale Isd; this East side of the Island is nearly black with recent uncovered Lavas. — The main hills must have immense Cauldron like Craters, — their height is considerable, above

4000 ft: yet from the outline being one uniform curve, the breadth of the mountain great, they do not appear lofty. —

19th During the night proceeded to Abingdon Is^d,[1] picked up Mr Chaffers in the Yawl in the morning & then steered for two small Is^{ds} which lie 100 miles to the North of the rest of the Group. —[2]

[1] Now Isla Pinta.
[2] Culpepper and Wenman Islands.

20th After having surveyed these the Ships head was put towards Otaheite & we commenced our long passage of 3200 miles. —

November 1st We are now travelling steadily onwards at the rate of 150 or 160 miles a day. — The trade wind night & day incessantly blows. — With studding sails set on each side we pleasantly cross the blue ocean. — Having now left the gloomy region which extends far from the coast of S. America, daily the sun shines brightly in the cloudless sky. —

9th This day we saw the first Island which can be truly said to belong to Polynesia. —[1] It is called Dog or Doubtful Isd. — The latter name expressing all which was known about it. — As may be seen in the Charts, it is an outlier on the East side of the large group of the Low Isds. This in its structure however is not truly one|622| of the Low Islands. — Its surface is raised considerably more than the average height from three to four ft. —otherwise in its level green surface, its apparently circular outline, in the detached rocks, or rather those which from the lowness of the sand beaches seem detached, there is a peculiar character, which may be understood by a drawing in Capt. Beecheys work. —[2] The insignificant patch of land bears no proportion & seems an intruder on the domain of the wide all-powerful ocean. — The proximity of this land was shown by the great & increased numbers of sea-birds, especially two species of Terns—To our great grief, the rainy season appears to have commenced; during the last four days, the sky has been very gloomy, with thunder, lightning, squalls of wind & heavy rain. From the extreme humidity of the Atmosphere the heat is rather oppressive. — The Thermometer in the Poop cabin remains constantly from 80°–83 degrees. — The air being thick & misty & the night dark, for the first time it has not been thought prudent to run on. So that we are now hove to, wasting the precious time till daylight comes & shows us the dangers of our course.

[1] According to FitzRoy 'we saw Honden Island, one of the low coral formations, only a few feet above water, yet thickly covered with cocoa-nut trees. Our observations corroborated the position assigned to it by Admiral Krusenstern, in his excellent chart

and memoir, the only documents of any use to us while traversing the archipelago of the Low Islands.' See *Narrative* **2:** 506.

[2] See F. W. Beechey, *Narrative of a voyage to the Pacific and Beering's Strait, to co-operate with the Polar expeditions: performed in His Majesty's Ship Blossom . . . in the years 1825, 26, 27, 28.* 2 parts. London, 1831.

13th By the evening of this day we succeeded in passing through the whole of the Archipelago, sometimes called the Dangerous or that of the Low Islands. At daylight & Noon we partially ascertained the figure & position of two Islands which are|623| not placed in the charts. — These & others have a very uninteresting appearance; a long brilliantly white beach is capped by a low bright line of green vegetation. This stripe on both hands rapidly appears to narrow in the distance & sinks beneath the horizon. — The width of dry land is very trifling: from the Mast-head it was possible to see at Noon Island across the smooth lagoon to the opposite side. — This great lake of water was about 10 miles wide.

Sunday 15th At daylight, Tahiti, an island which must for ever remain as classical to the Voyager in the South Sea, was in view. — At this distance the appearance was not very inviting; the luxuriant vegetation of the lower parts was not discernible & the centre as the clouds rolled past showed the wildest & most precipitous peaks which can be well imagined. — As soon as we got to an anchor in Matavai bay, we were surrounded by canoes. — This was our Sunday but their Monday; if the case had been reversed we should not have received a single visit, for the injunction not to launch a canoe on the Sabbath is rigidly obeyed. After dinner we landed to enjoy all the delights of the first impressions produced by a new country & that country the charming Tahiti. — Crowds of men, women & children were collected on the memorable point Venus[1] ready to receive us with laughing merry faces. — They marshalled us towards the house|624| of M[r] Wilson the missionary of the district, who met us on the road and gave us a very friendly reception.[2] After sitting a short time in the house we separated to walk about, but returned in the evening at tea-time.

The only ground cultivated or inhabited in this part of the Island is a stripe or points of low flat Alluvial soil accumulated at the base of the mountains & protected by the reef of coral, which encircles at [a] distance the entire land. The whole of this land is covered by a most beautiful orchard of Tropical plants. In the midst of bananas, orange, cocoa-nut & Bread fruit trees, spots are cleared where Yams, sweet potatoes, sugar cane & pineapples are cultivated. Even the brushwood is a fruit tree, namely the Guava, which from its abundance is noxious as a weed—In Brazil I have often admired the contrast of varied beauty in the banana, palm & orange trees; here we have in addition the Bread-fruit, conspicuous by its large, glossy & deeply digitated leaf. It is admirable to behold groves of a tree sending forth its branches with the force of an old oak in England, loaded with large nutritious fruit. |625| However little generally the utility explains the delight received from any fine prospect, in such cases as this it cannot fail largely to enter as an element in the feelings. — The little winding paths, cool from the surrounding shade, lead to the scattered houses. These have been too often described, for me to say anything about them: they are pleasant, airy abodes, but not quite so clean as I had been led to expect.

In nothing have I been so much pleased as with the inhabitants. — There is a mildness in the expression of their faces, which at once banishes the idea of a savage, — & an intelligence which shows they are advancing in civilization. — No doubt their dress is incongruous, as yet no settled costume having taken the place of the ancient one. — But even in its present state it is far from being so ridiculous as described by travellers of a few years standing. — Those who can afford it, wear a white shirt & sometimes a jacket, with a wrapper of coloured cotton round their middles, thus making a short petticoat like the Chilipa of the Gaucho. — This appears so general with the chiefs, that probably it|626| will become the settled fashion. They do not, even to the Queen, wear shoes or stockings, & only the chiefs a straw hat on their heads. — The common people when working, have the whole of the upper part of their bodies uncovered; & it is then that a Tahitian is seen to advantage. — In my opinion, they are the finest men I have ever beheld;—very tall, broad-shouldered, athletic, with their limbs well proportioned. It has been remarked that but little habit makes a darker tint of the skin more pleasing & natural to the eye of an Europæan than his own color. — To see a white man bathing along side a Tahitian, was like comparing a plant bleached by the gardeners art, to the same growing in the open fields. — Most of the men are tattooed, the ornaments so gracefully follow the curvature of the body that they really have a very elegant & pleasing effect.[3] One common figure varying only in its detail branches somewhat like palm leaves (the similarity is not closer than between the capital of a Corinthian column & a tuft of Acanthus) from the line of the back bone & embraces each side. — The simile is a fanciful one, but I thought the body of a man was thus ornamented like the trunk of a noble tree by a delicate creeper. —

Many of the older people have their feet|627| covered with small figures, placed in order so as to resemble a sock. — This fashion is however partly gone by & has been succeeded by others. Here, although each man must for ever abide by the whim which reigned in his early days, yet fashion is far from immutable. An old man has his age for ever stamped on his body & he cannot assume the air of a young dandy. — The women are also tattooed much in the same manner as the men & very commonly on their fingers. — An unbecoming fashion in another respect is now almost universal; it is cutting the hair, or rather shaving it from the upper part of the head in a circular manner so as only to leave an outer ring of hair. — The Missionaries have tried to persuade the people to change this habit, but it is the fashion & that is answer enough at Tahiti as well as Paris. — I was much disappointed in the personal appearance of the women; they are far inferior in every respect to the men. The custom of wearing a flower in the back of the head or through a small hole in each ear is pretty. The flower is generally either white or scarlet & like the Camelia Japonica. — The women also wear a sort of crown of woven cocoa nut leaves, as a shade to their eyes. — They are in greater want of some becoming costume even than the men.|628|

Hospitality is here universal. — I entered many of their houses & everywhere received a merry pleasant welcome. — All the men understand a little English, that is they know the names of common things; with the aid of this & signs a lame sort of conversation could be carried on. — After thus wandering about each his own way we returned to Mr Wilson's. In going afterwards to the boat we were interrupted by a very pretty scene, numbers of children were playing on the beach, & had lighted bonfires which illuminated the placid sea & surrounding trees: others in circles were singing Tahitian verses. — we seated ourselves on the sand & joined the circle. The songs were impromptu & I believe relating to our arrival; one little girl sang a line which the rest took up in parts, forming a very pretty chorus. — the air was singular & their voices melodious. The whole scene made us unequivocally aware that we were seated on the shores of an Island in the South Sea. —

[1] Site for the observation of the transit of Venus on 3 June 1769 by Cook and Banks in HMS *Endeavour*.

[2] Followed by two deleted sentences: 'Neither the person or manners of Mr Wilson tend to give any idea of a high or devoted character, but rather of a goodnatured quiet trader. I fully believe however from all which I heard & saw, that this exterior hides a great deal of most unpretending excellent merit.'

[3] Rough sketch in margin.

Tuesday 17ᵗʰ   This day is reckoned in the log book as Tuesday 17ᵗʰ instead of Monday 16ᵗʰ, owing to our, so far successful, chase of the sun. — Before breakfast the ship was hemmed in by a flotilla of canoes, & when the natives were allowed to come on board, I suppose the number could not have|629| been less than 200 on our decks. It was the opinion of every one, that it would have been difficult to have picked out an equal number from any other nation who would have given so little trouble. — Every body brought something for sale; shells were the main article of trade. — The Tahitians now fully understand the value of money & prefer it to old clothes or other articles. — The various coins of English & Spanish denomination puzzle the inhabitants & they never seem to think the small Silver quite secure until changed into dollars. — Some of the *chiefs* have accumulated considerable sums of money; one not long since offered 800 $ for a small vessel & frequently they purchase horses & whale-boats at from 50–100 $.

After breakfast I went on shore & ascended the slope of the nearest part of the mountains to an elevation between two and three thousand feet. — The form of the land is rather singular & may be understood by explaining its hypothetical origin. I believe a group of the interior mountains stood as a smaller island in the sea, & around their steep flanks streams of Lavas & beds of sediment were accumulated in a conical mass under water. This after having been raised was cut by numerous profound ravines, which all diverge from the common centre; the intervening ridges thus belonging to one slope & being flat-topped. — Having crossed the|630| narrow girt of inhabited land, I followed the line of one of these ridges; having on each hand very steep & smooth sided valleys. — The vegetation is singular, consisting almost exclusively of small dwarf fern, mingled higher up with coarse grass. — The appearance was not very dissimilar from that of some of the hills in North Wales; and this so close above the orchard of Tropical plants on the coast was very surprising. At the highest point which I reached trees again appeared. — The wood here was very pretty. — tree ferns having replaced the Cocoa Nut. — It must not however be supposed that these woods at all equalled the forests of Brazil. — In an island, that vast number of productions which characterize a continent cannot be expected to occur. —

From this point, there was a good view of the distant island of Eimeo, dependant on the same Sovereign with Tahiti. — On the lofty & broken pinnacles white massive clouds were piled, which formed an island in the blue sky, as Eimeo itself in the blue ocean. The island is completely encircled by a reef, with the exception of one small gateway; at this distance a narrow but well defined line of brilliant white where the waves first encountered the wall of coral, was alone visible; Within this line was included the smooth glassy water of the|631| lagoon, out of which the mountains rose abruptly. — The effect was very pleasing & might be aptly compared to a framed engraving, where the frame represents the breakers, the marginal paper the lagoon, & the drawing the Island itself. — When I descended in the evening from the mountain, a man whom I had pleased with a trifling gift met me bringing with him hot roasted Bananas, a pineapple & Cocoa Nuts. — I do not know anything more delicious than the milk of a young Cocoa Nut, after walking under a burning sun. — The pineapples here are also of such excellence as to be better than those reared in England & this I believe to be the last & highest compliment which can be paid to a fruit or indeed anything else. — Before going on board I went to Mr Wilson, who interpreted to my friend who had paid me so adroit an attention, that I wanted him & some one other man to accompany me on a short excursion into the mountains. —

18th In the morning I came on shore early bringing with me some provisions in a bag & two blankets for myself & servant. — These were lashed to each end of a pole & thus carried by my Tahitian companions. From custom a man will walk a whole day with fifty pounds|632| at each end of such a stick. — I had before told my guides to provide themselves with food & clothing; for the latter however they said their skins were sufficient & for the former that there was plenty of food in the mountains. The line of march was the valley of Tia-auru, in which the river that enters the sea by point Venus flows. This is one of the principal streams in the Island & its source lies at the base of the loftiest mountains which attain the elevation of about 7000 ft. — The whole Island may be considered as one group of mountains, so that the only way to penetrate the interior is to follow up the valleys. Our road at first lay through the wood which bordered each side of the river; the glimpses of the lofty central peaks, seen up the avenue of the valley with here & there a waving Cocoa Nut tree on one side, were extremely picturesque. — The valley soon began to narrow & the sides to grow higher & more precipitous. — After having walked for three or four hours, the width of the ravine scarcely exceeded that of the bed of the stream; on each hand the walls were nearly vertical, yet from the soft nature of the volcanic strata trees & a rank vegetation sprung from every projecting ledge. These precipices must have been some thousand ft high; the|633| whole formed a mountain gorge far more magnificent than anything I had ever beheld. — Till the midday sun stood vertically over the ravine, the air had felt cool & damp, but now it became very sultry. — Shaded by a ledge or rock beneath a façade of columnar Lava we ate our dinner. — My guide before this had procured a dish of small fish & fresh-water prawns. — They carried with them a small net stretched on a hoop; where the water was deep in eddies, they dived & like otters by their eyesight followed the fish into holes & corners & thus secured them. The Tahitians have the dexterity of Amphibious animals in the water; an anecdote mentioned by Ellis shows how much at home they feel in that element. — When a horse was landing for Pomarre in 1817, the slings broke, & it fell into the water; immediately the natives jumped overboard & by their crys & vain efforts at assistance, almost drowned the animal. As soon as it reached the shore, the whole population took to flight & tried to hide themselves from the man-carrying pig, as they christened the horse. —

A littler higher up, the river divided itself into three little streams; the two Northern ones were impracticable from a succession of waterfalls which|634| descended from the jagged summit of the highest mountain; the other to all appearance was equally inaccessible, but we managed to ascend in that direction by the most extraordinary road which I ever beheld. — The sides of the valley were here quite precipitous, but as generally happens small ledges projected which were thickly covered by wild bananas, liliaceous plants & other luxuriant productions of the Tropics. — The Tahitians by climbing amongst these ledges hunting for fruit had discovered a track by which the whole precipice could be scaled. — The first leaving the bottom of the valley was very dangerous; a face of naked rock had to be passed by the aid of ropes which we brought with us. — How any person discovered that this formidable spot was the only point where the side of the mountain could be attempted, I cannot imagine. — We then cautiously followed one of the ledges, till we came to the stream already alluded to. — This ledge formed a flat spot, above which a beautiful cascade of some hundred ft poured down its waters, & beneath it another high one emptied them into the main stream. — From this cool shady recess we made a circuit to avoid the overhanging cascade. As before we followed little project-ing ledges, the apparent danger|635| being partly hidden by the thick-ness of the vegetation. In passing from one ledge to another there was a vertical wall of rock:—one of the Tahitians, a fine active man, placed the trunk of a tree against this, swarmed up it & then by the aid of crevices reached the summit. — He fixed the ropes to a projecting point & lowered them for us & then hauled up the dog & luggage. — Beneath the ledge on which the dead tree was reared the precipice must have been five or six hundred feet deep; if the abyss had not been partly concealed by the overhanging ferns & lilies, my head would have turned giddy & nothing should have induced me to have attempted it. — We continued to ascend sometimes by ledges & sometimes by knife edge ridges, having on each hand profound ravines. — In the Cordilleras I have seen mountains on a far greater scale, but for abruptness no part was at all comparable to this. In the evening we reached a flat little spot on the banks of the same stream which I have mentioned as descending by a chain of beautiful waterfalls. Here we bivouacked for the night. — On each side of the ravine there were great beds of the Feyè or mountain Banana, covered with ripe fruit. — Many of these plants grew to a height from twenty to twenty five feet high|636| & from three to four in circumference. By the aid of strips of bark for twine, the stems of the bamboos & the large leaf of the banana, the Tahitians in a few minutes built an excellent house; & with the withered leaves made a soft bed. —

They then proceeded to make a fire & cook our evening meal. A light was procured by rubbing a blunt pointed stick in a groove, as if with the intention of deepening it, until by friction the dust became ignited. — A peculiarly white & very light wood is alone used for this purpose;[1] it is the same which serves for poles to carry any burthen & for the floating out-rigger to steady the canoe. — The fire was produced in a few seconds; to a person however, who does not understand the art, it requires the greatest exertion, as I found before I at last to my great pride succeeded in igniting the dust. The Gaucho in the Pampas uses a different method; taking an elastic stick of about eighteen inches long, he presses one end on his breast & the other pointed one in a hole in a piece of wood, & then rapidly turns the curved part, like a carpenter does a Centre-bit. — The Tahitians, having made a small fire of sticks, placed a score of stones about the size of a cricket ball on the burning wood. In about ten minutes time, the sticks were consumed & the stones hot. They had previously folded up in small parcels made of leaves, pieces of beef, fish, ripe & unripe Bananas, & the tops of the wild Arum. — These green parcels were laid in a layer between two|637| of the hot stones & the whole then covered up by earth so that no smoke or steam escaped. — In about a quarter of an hour the whole was most deliciously cooked; the choice green parcels were laid on a cloth of

Banana leaves; with a Cocoa nut shell we drank the cool water of the running stream & thus enjoyed our rustic meal. —

I could not look on the surrounding plants without wonder. On every side were forests of Banana, & the fruit which served for food in many ways, lay in heaps decaying on the ground. — In front of us there was an extensive brake of wild Sugar Cane. — The banks of the stream were shaded by the dark green knotted stem of the Ava, so famous in former days for its powerful intoxicating effects; I chewed a piece & found that it had an acrid & unpleasant taste which would induce any one at once to pronounce it poisonous. — Thanks be to the Missionaries this plant now thrives in these deep ravines innocuous to every one. — In the close neighbourhood I saw the wild Arum, the roots of which when well baked are good to eat & the young leaves better than spinach: — there was the wild Yam & a liliaceous plant called Tì, which grows in abundance, & has a soft brown root in shape & size like a huge log of wood. This served us for dessert, for it is as sweet as treacle & with a pleasant taste. |638| There were moreover several other wild fruits & useful vegetables. The little stream, besides its cool water, produces also eels & cray-fish. — I did indeed admire this scene, when I compared it with an uncultivated one in the temperate zone. — I felt the force of the observation that man, at least savage man, with his reasoning powers only partly developed, is the child of the Tropics. — [2]

As the evening drew to a close, I strolled alongside the stream beneath the gloomy shade of the Bananas. — My walk was soon brought to a close by coming to a Waterfall of two or three hundred feet high; — and above this was another. — I mention all these waterfalls in this one brook, to give an idea of the general inclination of the land. — In the little recess where the water fell, it did not appear that a breath of wind ever entered. — The leaves of the Bananas, damp with spray, showed one unbroken edge, instead of as commonly happens, being split into a thousand shreds. — Suspended, as it were, on the side of the mountain, there were glimpses into the depth of the neighbouring valleys; & the highest pinnacles of the central mountains towering up within sixty degrees of the Zenith, hid half the|639| evening sky. Thus seated it was a sublime spectacle to watch the shades of night gradually obscuring the highest points.

Before we laid ourselves down to sleep, the elder Tahitian fell on his knees & with closed eyes repeated a long prayer in his native tongue. He prayed as a Christian should do, with fitting reverence, & without fear of ridicule or ostentation of piety. — In a like manner, neither of the men would taste food without saying before hand a short grace. — Those Travellers who hint that a Tahitian prays only when the eyes of the missionary are fixed on him, should have slept with us that night on the mountain side. — Rigidly to scrutinize how far a man, born under idolatry, understands the full motive & effect of prayer, does not appear to me a very charitable employment. During the night it rained heavily, but the good thatch of Banana leaves kept us dry. —

[1] In *Journal of Researches* p. 488 this is identified as *Hibiscus tiliaceus*.
[2] Followed by deleted sentence: 'One cannot however say that one is more natural than the other; if an animal exerts its instinct to procure food, the law of nature clearly points out that man should exert his reason & cultivate the ground.'

19th   At Daylight, after their morning prayer, my friends prepared an excellent breakfast in the same manner as in the evening. — They themselves certainly partook of it largely; indeed I never saw any men eat anything nearly so much in quantity. They did not, however, over eat themselves, that is their activity was anything but impaired. — I should suppose such capacious stomachs must be the result of a large part of their diet consisting of fruits & vegetables |640| which do not contain in a given bulk very much nutriment. — Unwittingly I was the means of my companions breaking one of their own laws & resolutions. — I took with me a flask of spirits, which they could not resolve to refuse, but as often as they drank a little, they put their fingers before their mouths & uttered the word "Missionary". — About two years ago, although the use of the Ava was prevented, drunkedness from the introduction of spirits became very prevalent. The Missionaries prevailed on a few good men, who saw their country rapidly going to ruin, to join with them in a Temperance Society. — From good sense & shame all the chiefs & Queen were thus at last united. — Immediately a law was passed that no spirits should be allowed to be introduced into the island & that he who sold & he who bought the forbidden article should be punished by a fine. — With remarkable justice a certain period was allowed for stock in hand to be sold before the law came in effect. — On that day a general search was made in which even the houses of the Missionaries were not exempted, & all the Ava (as the natives call all ardent spirits) was poured out on the ground. — When one reflects on the effect of intemperance on the aboriginals of the two Americas, I think it will be acknowledged that every |641| well wisher of Tahiti owes no common debt of gratitude to the Missionaries. —

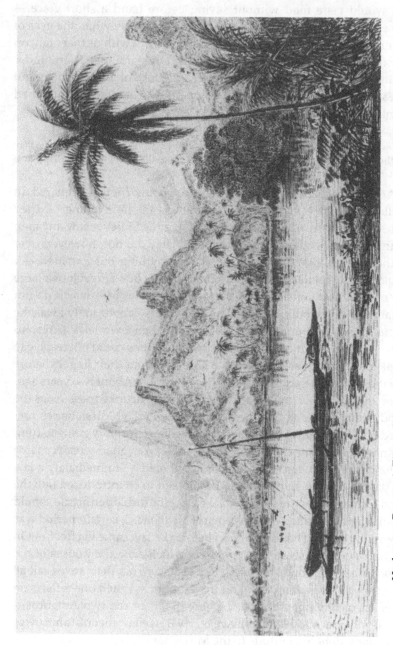

Harbour at Papetoai, Eimeo (Moorea) Island, by T. Landseer after C. Martens (*Narrative* **2**: 509).

After breakfast we proceeded on our journey: as my object was merely to see a little of the interior scenery, we returned by another track, which descended into the main valley lower down. For some distance we wound along the side of the mountain which formed the valley; the track was extraordinarily intricate; in the less precipitous parts it passed through very extensive groves of the wild Banana. — The Tahitians with their naked tattooed bodies, their heads ornamented with flowers, & seen in the dark shade of the woods, would have formed a fine picture of Man inhabiting some primeval forest. — In our descent we followed the line of ridges; these were exceedingly narrow, & for considerable lengths steep as the inclination of a ladder, but all clothed by Vegetation. The extreme care necessary in poising each step, rendered the walking fatiguing. — I am not weary of expressing my astonishment at these ravines & precipices. — The mountains may be almost described as merely rent by so many crevices. — When viewing the surrounding country from the knife edged ridges, the point of support was so small that the effect was nearly the same as would, I imagine, be observed from a balloon.|642|. In this descent we only had need of using the ropes once, at the point where we entered the main valley. — Proceeding downwards we slept under the same ledge of rocks where we had before dined. — The night was fine, but from the depth and narrowness of the gorge profoundly dark. — Before actually seeing this country, I had difficulty in understanding two facts mentioned by Ellis. Namely, that after the murderous battles, the survivors on the conquered side retired into the mountains, where a handful of men could resist a multitude. — Certainly half a dozen men at the spot where the Indians reared the old tree could easily have repelled thousands. — Secondly that after the introduction of Christianity, there were wild men who lived in the mountains, & whose retreats were unknown to the more civilized inhabitants. —

20th In the morning we started by times & reached Matavai at Noon. — On the road we met a large party of noble athletic men going for the wild Bananas. I found the Ship, on account of difficulty in watering, had moved four miles to the harbor of Papawa, to which place I immediately walked. — This is a very pretty spot; the cove is so surrounded by reefs that the water is smooth as in a lake. — The cultivated ground, with all its beautiful productions & the cottages, reach close down to the water's edge.|643|

21st The Beagle returned to her old quarters at Matavai. —in the evening I took a pleasant ramble on shore. —

Sunday 22nd   The harbor of Papiete, which may be considered as the capital of the Island, is about seven miles distant from Matavai, to which the Beagle had returned. — The Queen resides there, it is the seat of Government & chief resort of shipping. — Capt. FitzRoy took a party there in the morning to hear divine service in the Tahitian language & afterwards in our own. —Mr Pritchard the leading Missionary in the Island performed service. —[1] Mr Pritchard was regularly educated in the mission college; he appears a sensible agreeable gentleman & good man. — I have already mentioned with respect Mr Wilson. The third Missionary whom we have seen is Mr Nott, who has resided 40 years on the Island. — His occupations are chiefly literary, & has now finished his great task of translating the whole bible. — He bears universally a high & respectable character. — The characters of this class of men have been so frequently attacked, that I have mentioned my opinion on the three who reside in this vicinity. We met at Mr Pritchards house three young ladies, daughters of Missionaries, who were here on a visit. Their appearance & manners showed that they had been properly educated.[2] It was curious to see those who could not be distinguished in appearance from our countrywomen, speaking the|644| Tahitian language with even greater fluency than English. — Even the characters of these quiet young women have not escaped the bitter attacks of the enemies of the Missionaries. — Before reaching these islands, from the varying accounts I had read I felt great interest to form from my own eyes a judgment of their moral state, although such judgment would necessarily be imperfect. — The first impression on any subject very much depends on ones previously acquired ideas. — Mine were drawn from Ellis' Polynesian Researches,[3] an admirable & most interesting work, but naturally looking at everything under a favourable point of view; from Beechey,[4] neutral; & Kotzebue,[5] strongly adverse. — He who compares these three accounts will I think form a tolerably accurate conception of the present state of Tahiti. —

One of my impressions, which I took from the two last authorities, was decidedly incorrect: viz that the Tahitians had become a gloomy race & lived in fear of the Missionaries. Of the latter feeling I saw no trace, without indeed fear & respect are confounded under one name. — Instead of discontent being a common feeling, it would be difficult in Europe in a crowd to pick out half so many merry, happy faces. — The prohibition of the Flute & dancing is inveighed against as wrong & foolish. — the more than Presbyterian manner of keeping the Sabbath is looked on in a similar view. I will not pretend to offer any opinion on this subject against men who have resided|645| as many years as I have days in the Island. On the whole it is my opinion that the state of morality & religion is highly creditable. — There are many who attack even more acrimoniously than Kotzebue, both the Missionaries, their system & the effect produced. — Such reasoners never compare the present to the former state only twenty years before; nor even to that of Europe in this day, but to the high standard of Gospel perfection. — They expect the Missionaries to effect what the very Apostles failed to do. — By as much as things fall short of this high scale, blame is attached to the Missionaries, instead of credit for what has been effected. They forget or will not remember that human sacrifices & the power of an idolatrous priesthood, — a system of profligacy unparalleled in the world, & consequent infanticide as part of that system, — bloody wars where the conquerors spared neither women or children have been abolished; that dishonesty, intemperance & licentiousness have been greatly reduced by the introduction of Christianity. — It is base ingratitude in a Voyager to forget these things; at the point of Shipwreck on some unknown coast he will most devoutly pray that the lesson of the Missionary may have extended thus far.

In their morality, the virtue of the women is said to be most open to exception; but before|646| they are blamed too severely, it will be well distinctly to call to mind the scenes described by Capt. Cook & Mr Banks in which the grandmothers & mothers of the present race played a part. Those who are most severe should consider how much of the morality in Europe is owing to the system early impressed by mothers on their daughters & how much in each individual case to the precepts of religion. — But it is useless to argue against such men; I believe that disappointed at not finding the field of licentiousness quite so open as formerly, they will not give credit to a morality which they do not wish to practice, or to a religion which they undervalue, if not despise. —[6]

The Tahitian service was a very interesting spectacle. — The Chapel is a large airy framework of wood; it was filled to excess by tidy clean people of all ages & sexes. — I was rather disappointed in the apparent degree of attention; but I believe my expectations were raised too high. — Anyhow the appearance was quite equal to that in a country Church in England. — The singing of the hymns was decidedly very pretty; the language however from the pulpit, although fluently

delivered had not much euphony: a constant repetition of sounds like
"*tata, ta, mata mai*", rendered it monotomous. — After English service, a
party returned on foot to Matavai;—it was a pleasant walk|647| either
along the sea-beach or under the shade of the many beautiful trees.

[1,2] The intervening passage is marked in pencil for deletion.

[3] See William Ellis, *Polynesian researches, during a residence of nearly six years on the South Sea Islands* . . . 2 vols. London, 1829.

[4] See note 2 for 9 November, p. 365.

[5] See Otto von Kotzbehue, *A voyage of discovery, into the South Sea and Beering's Straits* . . . Translated by H. E. Lloyd. 3 vols. London, 1821.

[6] A paper entitled 'A letter containing remarks on the moral state of Tahiti, New Zealand &c' was published by FitzRoy and CD in *South African Christian Recorder* 2: 221–38. The first part, written in the first person except for three introductory paragraphs, and signed by FitzRoy alone, contained excerpts from CD's journal. See *The Collected Papers of Charles Darwin*, edited by P. H. Barrett, Chicago, 1977, Vol. 1, pp. 19–38.

23rd   I hired a canoe & men to take me on the reef. — These canoes are
from their extreme narrowness comical little boats. — They would
immediately be upset if it was not for a floating log of very light wood
joined to the canoe by two long transverse poles. — We paddled for
some time about the reef admiring the pretty branching Corals. — It is
my opinion, that besides the avowed ignorance concerning the tiny
architects of each individual species, little is yet known, in spite of the
much which has been written, of the structure & origin of the Coral
Islands & reefs. —[1]

[1] Although from his reading, and from his thinking about the consequences of the elevation of the South American continent, CD's theory about the formation of coral reefs had probably already come to him by this time, this was the first reef that he actually saw. See Appendix V in *Correspondence* 1: 567–71.

24th   About two years ago a small vessel under English colors was
plundered by the inhabitants of the Low Islands, which were then
under the dominion of the Queen of Tahiti. — It is believed they were
instigated to this act by some indiscreet laws issued by Her Majesty. —
The British Government demanded compensation; this was acceeded
to & a sum nearly equal to 3000 dollars was agreed to be paid on the first
of last September. — The Commodore at Lima ordered Capt. FitzRoy to
enquire concerning this debt & to demand satisfaction if not paid. —
Capt. FitzRoy asked for an interview with the Queen:— For this
purpose a Parliament was held where all the principal chiefs of the
Island & the Queen were assembled. |648| I will not attempt to describe
what took place, as so interesting an account has been given by Capt.
FitzRoy.[1] The money had not been paid:—perhaps the alledged reasons
were rather equivocating; otherwise I cannot sufficiently express our

general surprise at the extreme good sense; reasoning powers, modera-
tion, candor & prompt resolution which were displayed on all sides. —
I believe every one in our party left the meeting with a very different
opinion of the Tahitians from what he entertained when entering. —
The chiefs & people resolved to subscribe & complete the sum which
was wanting. — Capt. FitzRoy urged that it was hard that their private
property should be sacrificed for the crimes of distant Islanders. They
replied that they were grateful for his consideration, but that Pomarre
was their queen, & they were determined to help her in this her
difficulty. — This resolution & its prompt execution, for a book was
opened early the next morning, is an uncommon instance of loyalty.
After the main discussion was ended, several of the chiefs took the
opportunity of asking Capt. FitzRoy many intelligent questions concerning international customs & laws. These related to the treatment of
ships & foreigners. On some points, as soon as their decision was
made, the law was issued verbally on the spot. — This Tahitian parlia-
ment lasted for several|649| hours and when it was over Capt. FitzRoy
invited the Queen to pay the Beagle a visit. — We all dined with M^r
Pritchard, & after it was dark pulled back to the ship. —

[1]CD originally wrote 'as it will probably be published', and evidently revised this
sentence after reading FitzRoy's account of the Parliament.

25th   Capt. FitzRoy & myself breakfasted with Mr Wilson & afterwards
the Beagle got under weigh: from light airs we did not get into Papiete
till the evening. Four boats were sent on shore for Her Majesty. The
Ship was dressed with flags & the yards manned on her coming on
board. — With her came most of the chiefs: the behaviour of all was very
proper; they begged for nothing & appeared much gratified by the
presents which were given them. — The Queen is an awkquard large
woman, without any beauty, gracefulness or dignity of manners. — She
appears to have only one royal attribute, viz a perfect immoveability of
expression (& that generally rather a sulky one) under all circumstances. —Sky rockets & the Seamens songs appeared to give most
amusement. — The Queen remarked that one song, a very noisy comic
one, certainly could not be *"Hymeni"*. — The Royal party did not leave
us till past midnight: they all appeared well contented with their visit. —

26th   Capt. FitzRoy & myself went on shore. The object was to wait for a
deed signed by the Queen & two principal chiefs stating how much of
the required sum was paid & their|650| determination immediately to
collect the remainder. —[1] In the course of the day we paid two visits to

the Queen in her house, which I must need say is a most paltry place. —
In the evening dined with Mr Pritchard, whom I regret I had not time to
become better ackquainted with, & afterwards we went on board the
Ship, which was under weigh waiting for us in the offing.[2] In the
evening with a gentle land breeze a course was steered for New
Zealand, & as the Sun set we took a farewell look at the mountains of
Tahiti, — the island to which every traveller has offered up his tribute of
admiration. —

[1,2] Intervening passage marked in pencil to be deleted.

December 3rd   After several days of light winds, we passed near to the
island Whytootacke; We here saw a union of the two prevailing kinds
of structure united. A hilly irregular mass was surrounded by a well
defined circle of reefs, which in great part have been converted into low
narrow strips of land, which as Cook calls them are half drowned,
consisting merely of sand & Corall rocks heaped up on the dead part of
a former reef. — The inhabitants made a smoke to attract our attention. —

19th   In the evening we saw in the distance New Zealand. — We may now
consider ourselves as nearly without the limits of the Pacific Ocean. It is
necessary to sail over this great sea to understand its immensity.|651|
Moving quickly onwards for weeks together, we meet with nothing but
the same blue profoundly deep ocean. Even within the Archipelagoes
the Islands are mere specks & far distant one from the other. Accustomed to look at Maps drawn on a small scale, where dots, shading, &
names are crowded together, we do not judge rightly how infinitely
small the proportion of dry land is to the water of this great sea. — The
Meridian of the Antipodes is likewise passed; every league, thanks to
our good fortune, which we now travel onwards, is one league nearer
to England. These Antipodes call to mind old recollections of childish
doubt & wonder. Even but the other day, I looked forward to this airy
barrier as a definite point in our voyage homewards; now I find it & all
such resting places for the imagination are like shadows which a man
moving onwards cannot catch. — A gale of wind, which lasted for some
days, has just lately given us time & inclination to measure the future
stages in our long voyage of half the world, & wish most earnestly for
its termination. —

20th   During the night the Ship stood off & on under easy sail; in the
morning it was found that a current had carried her 20 miles to Leeward
of the Bay of Islands. — The country is irregular in form & hilly but not

high; the whole is scattered over with wood. — The coast in parts is|652| formed by high reddish cliffs more or less broken down. — We may compare this coast to that of East T. del Fuego, Chiloe, or the Indian territory South of Concepcion, or I do not doubt to a score of other places, where no very marked feature occurs. —

21st   Early in the morning we entered the Bay of Islands, being becalmed near the mouth for some hours; we did not reach the anchorage till the middle of the day. — The country is hilly but with a smooth outline; & it is deeply intersected by numerous arms extending from the Bay. The surface appears from a distance as if clothed with coarse pasture, but this in truth is nothing but fern. On the more distant hills, as well as in patches in some of the valleys, there is a good deal of Woodland. The general tint however of the landscape is not a very bright green, but resembles the country a short distance to the Southward of Concepcion in Chili. — In several parts of the Bay, close down to the waters edge, little villages of square tidy looking houses were scattered. Three Whaling Ships were lying at anchor; but with the exception of these & of a few canoes now & then crossing from one shore to the other an air of extreme quietness reigned over the whole district. — Only one single canoe came alongside; this & the whole scene afforded a remarkable & not very pleasing contrast to our joyful boisterous welcome at Tahiti. — |653|

In the afternoon we went on shore to one of the larger groups of houses, which yet hardly deserves the title of a village. It's name is Pahia; it is the residence of the Missionaries, & with the exception of their servants & labourers there are no native residents. — In the vicinity of the Bay of Islands, the number of Englishmen including their families amounts to between two & three hundred; all the cottages, many of which are whitewashed, & look as I have said very neat, are the property of Englishmen. The hovels of the natives are so diminutive & paltry that they can scarcely be perceived from any distance. — At Pahia it was quite pleasing to behold in the platforms before the houses so many English flowers; there were roses of several kinds, honeysuckle, jessamine, stocks & whole hedges of sweet briar.

22nd   In the morning I went out walking: I soon found that the country is very impracticable; the hills are all thickly covered by tall fern, together with a low bush which grows like a cypress; & very little ground in this neighbourhood has been cleared or cultivated. I then tried the sea beach, but proceeding towards either hand, my walk was soon stopped short by creeks & deep streams of fresh water. — The communication between the inhabitants of different parts of the Bay, is as in Chiloe almost entirely kept up by boats. |654| I was surprised to find that almost every hill which I ascended had been at some former time more or less fortified. The summits were cut into steps or successive terraces and they had frequently been protected by deep trenches. I afterwards observed that the principal hills inland, in a like manner showed an artificial outline. These are the Pas, so frequently mentioned by Capt Cook under the name of "hippah"; the difference being owing to the prefixed article. — That the Pas had formerly been used was evident from the piles of shells & the pits, in which, as I was informed, sweet potatoes were kept buried as reserved provisions. As there was no water on these hills, the defenders could never have anticipated a long siege, but only a hurried attack for plunder, under which circumstances the successive terraces would afford good protection. — The general introduction of fire-arms has changed the whole system of warfare; an exposed situation on the summit of a hill would now be worse than useless. — The Pas in consequence is at the present day always built on a level piece of ground. —it consists of a double stockade of thick & tall posts, placed in a zigzag line so that every part can be flanked. Within the stockade a mound of earth is thrown up, behind which the defenders can rest in|655| safety, or use their fire-arms over the top. On the level of the ground, little archways sometimes pass through this breastwork, by means of which the defenders can crawl out to the stockade to reconnoitre their enemies. — The Rev.ᵈ W. Williams, who gave me this account, added that in one Pas he had noticed on the inside of the mound of earth projecting spurs or buttresses: on asking the chief the use of these, he replied, that if two or three men should be shot, their neighbours would not see their bodies & so be discouraged. — These Pas are considered by the New Zealanders as a very perfect means of defence. The attacking force is never so well disciplined as to rush in a body to the stockade, cut it down & effect their entry. — When a tribe goes to war, the chief cannot order one party to go here & another there, but every man fights in the manner which best pleases himself, & to individuals to approach a stockade defended by firearms must appear certain death. I should think in no part of the world a more war-like race of inhabitants could be found than the New Zealanders. Their conduct in first seeing a Ship, as described by Capt. Cook, strongly illustrates this. The act of throwing volleys of stones at so great & novel an object as a ship & their defiance of "come on shore

& we will kill & eat you all" shows uncommon boldness. — This warlike spirit is evident in many|656| of their customs & smallest actions. If a New Zealander is struck, although but in joke, the blow must be returned; of this I saw an instance with one of our officers.

At the present day, from the progressive civilization, there is much less warfare. When Europæans first traded here, muskets & ammunition far exceeded in value any other article; now they are in little request & are indeed often offered for sale. — Amongst some of the Southern tribes there is, however, yet much hostility; I heard a characteristic anecdote of what took place there some time ago. — A Missionary found a chief & his tribe in full preparation for war; their muskets clean & bright & their ammunition ready. — He reasoned long on the inutility of the war & the little cause which had been given; the chief was much shaken in his resolution & seemed in doubt. — But at length it occurred to him that a barrel of his gunpowder was in a bad state & would not keep much longer; this was brought forward as an unanswerable argument for the necessity of immediately declaring war, —the idea of allowing so much gunpowder to spoil was not tolerable & it settled the point.[1]

I was told by the|657| Missionaries that in the life of Shongi, the chief who visited England, the love of War was the one & lasting spring of every action. The tribe in which he was a principal chief had at one time been much oppressed by another from the Thames river. A solemn resolution was agreed on, that when their boys should grow up into men & they should be powerful enough, they would never forget or forgive these injuries. To fulfil this appears to have been Shongi's chief motive for going to England; when there it was his sole object; presents were only valued which could be converted into arms; of the arts, those alone were interesting which were concerned with the manufactory of arms. — When at Sydney Shongi by a strange coincidence met at the house of Mr Marsden the hostile chief of the Thames: — their conduct was civil to each other. But Shongi told him that when again in New Zealand he would never cease to carry war into his country. — The challenge was accepted; & Shongi on his return fulfilled the threat to the utmost letter; the tribe on the Thames river was utterly overthrown, & the chief to whom the challenge had been given was himself killed. — Shongi, although concealing such deep feelings of hatred & revenge, is described to have been a goodnatured sort of person. —

In the evening of this day I went with Capt. FitzRoy & Mr Baker, one of the Missionaries, to pay a visit|658| to Kororarika. This is the largest village & will one day no doubt increase into the chief town. Besides a considerable native population there are many English residents. — These latter are of the most worthless character; & amongst them are many run away convicts from New South Wales. There are many spirit shops, & the whole population is addicted to drunkenness & all kinds of vice. As this is the capital, a person would be inclined to form his opinion of the New Zealanders from what he here saw; but in this case his estimate of their character would be too low. — This little village is the very strong-hold of vice; although many tribes, in other parts, have embraced Christianity, here the greater part are yet remain in Heathenism. In such places the Missionaries are held in little esteem; but they complain far more of the conduct of their countrymen than of the natives. It is strange, but I here heard these worthy men say that the only protection which they need & on which they rely is from the native Chiefs against Englishmen!

We wandered about the village & saw & conversed with many of the people, both men, women & children. Looking at the New Zealander, one naturally compares him with the Tahitian; both belonging to the same family of mankind. The comparison however tells heavily against the New Zealander. He may perhaps be superior in energy, but in every other respect his character is of a much lower order. One glance at their|659| respective expressions, brings conviction to the mind that one is a savage, the other a civilized man. It would be in vain to seek in the whole of New Zealand a person with the face & mien of the old Tahitian chief Utamme. No doubt the extraordinary manner in which tattooing is here practised gives a disagreeable expression to their countenances. The complicated but symmetrical figures covering the whole face, puzzle & mislead an unaccustomed eye; it is moreover probable that the deep incisions, by destroying the play of the superficial muscles, would give an air of rigid inflexibility. — But besides all this, there is a twinkling in the eye which cannot indicate anything but cunning & ferocity. — Their figures are tall & bulky, but in elegance are not comparable with those of the working classes in Tahiti; this I believe was the opinion of all on board, though we had expected otherwise from having read Mr Earles[2] work. — Both their persons & houses are filthily dirty & offensive; the idea of washing either their persons or clothes never seems to have entered their heads. I saw a chief, who was wearing [a] shirt black & matted with filth; when asked how it came to be so dirty, he replied with surprise "Do not you see it is an old one?" — Some of the men have shirts, but the common dress is one or two large|660| blankets generally black with dirt, which are thrown over

their shoulders in a very inconvenient and awkward fashion. A few of
the principal chiefs have decent suits of English clothes, but these are
only worn on great occasions.

Considering the number of foreigners residing in New Zealand & the
amount of commerce carried on there, the state of government of the
country is most remarkable. It is however incorrect to use the term
government, where absolutely no such thing exists. The land is divided
by well determined boundaries between the various tribes, which are
totally independent of each other. The individuals in each tribe consist
of free men, & slaves taken in war; the land is common to all the
freeborn, that is each may occupy & till any part that is vacant; in a sale
therefore of land, every such person must receive part payment. —
Amongst the free men, there will always be some one who from riches,
from talents, or from descent from some noted character, will take the
lead, & in this respect he may be considered as the chief. — But if the
united tribe should be asked who was their chief, no one would be
acknowledged. Without doubt in many cases the individuals thus
obtain great influence, but as far as I|661| understand their power is not
legitimate. Even the authority of a master over his slave, or parent over
his children, appears to be regulated by no kind of ordinary custom.
Proper laws are of course quite unknown; certain lines of action are
generally considered right & others wrong. —if such customs are
infringed upon, the injured person or his tribe if they have power, seek
retribution; if not they treasure up the recollection of it, till the day of
revenge arrives. — If the state in which the Fuegians live should be fixed
on as zero in the scale of governments, I am afraid the New Zealand
would rank but a few degrees higher, while Tahiti, even as when first
discovered, would occupy a respectable position.[3]

[1] Followed by a deleted sentence: 'I should imagine that formerly the different tribes could
hardly have ever been at peace with each other; but only that there were cessations of
hostilities. —'
[2] See Augustus Earle, *A narrative of a nine months' residence in New Zealand in 1827* . . .
London, 1832. In a letter to Caroline Darwin written on 27 December, CD said: 'We are
quite indignant with Earle's book, beside extreme injustice it shows ingratitude. — Those
very missionaries, who are accused of coldness, I know without doubt that they always
treated him with far more civility, than his open licentiousness could have given reason
to expect. —' See *Correspondence* 1: 472.
[3] Followed by a deleted passage: 'Continuing our ramble about the village, Mr Baker took
us to see a chapel which was building. I presume the Missionaries have fixed on this spot,
where there are so few Christians, in order to attack vice in her very Citadel; certainly at
present the old adage is true, "the nearer the Church, the further from Heaven".
Altogether the village of Kororarika is a disgusting scene; & I am glad it is not necessary
to take this as a specimen of New Zealand.'

23rd   At a place called Waimate, about fifteen miles from the Bay of

Islands & midway between the Eastern & Western coasts, the Missionaries|662| have purchased some land for agricultural purposes. I had been introduced to the Rev.[d] W Williams, who, upon my expressing a wish, invited me to pay him a visit there. — M[r] Busby the British Resident offered to take me in his boat up a creek, where I should see a pretty waterfall & which would also shorten my walk. — He likewise procured for me a guide: upon asking a neighbouring chief to recommend a man, the chief himself offered to go; but his ignorance for the value of money was so complete, that he at first asked how many *pounds* I would give him, but afterwards was well contented with two dollars. When I showed the chief a very small bundle which I wanted carried, it became absolutely necessary to take a slave for that purpose; — such feelings are beginning to wear away, but formerly a leading man would have died sooner than undergone the indignity of carrying the smallest burthen. — My companion was a light active man, dressed in a dirty blanket, & with his face completely tattooed; he had formerly been a great warrior. He appeared to be on very cordial terms with M[r] Busby; but at various times they had quarrelled violently. M[r] Busby remarked that a little quiet irony would frequently silence one of these natives in his most blustering moments. This chief has come & harangued M[r] Busby in a hectoring|663| manner, saying, "A great chief, a great man, a friend of mine, has come to pay me a visit, you must give him something to eat, some fine presents &c." M[r] Busby has allowed him to finish his discourse & then has quietly replied by some such answer as "What else shall your slave do for you?" The man would then instantly with a very comical expression cease his braggadocio. —

Some time ago M[r] Busby suffered a far more serious attack; a chief & a party of men tried to break into his house in the middle of the night, & not finding this so easy, commenced a brisk firing with their muskets. M[r] Busby was slightly wounded, but the party was at length driven away. Shortly afterwards it was discovered who was the aggressor, & a general meeting of the chiefs was convened to consider the case. — It was considered by the New Zealanders as very atrocious, in as much as it was a night attack, & that Mrs Busby was lying ill in the house after her confinement: this circumstance, much to their honour, being considered in all cases as a protection. The chiefs agreed to confiscate the land to the King of England: — The whole proceeding, however, in thus trying & punishing a chief was entirely without precedent. The aggressor moreover lost|664| caste in the estimation of his equals; & this was considered by the British as of more consequence than the confiscation. —

As the boat was shoving off, a second chief stepped in her, who only wanted the amusement of the passage up & down the creek. I never saw a more horrid & ferocious expression than this man had: it immediately struck me I had seen his likeness; it will be found in Retzch's outlines of Schiller's ballad,[1] where two men are pushing Robert into the burning iron furnace; it is the man who has his arm on Robert's breast. Physiognomy here spoke the truth; this chief had been a notorious murderer & was to boot an arrant coward. — At the point where the boat landed, Mr Busby accompanied me a few hundred yards on the road; I could not help admiring the cool impudence of the hoary old villain, whom we left lying in the boat, when he shouted to Mr Busby, "Do not you stay long, I shall be tired of waiting here". —

We now commenced our walk; the road lay along a well beaten path, bordered on each side by the tall fern which covers the whole country. After travelling some miles, we came to a little country village where a few hovels were collected together & some patches of ground cultivated for potato crops. The introduction of the potato had been|665| of the most essential benefit to the island; it is now much more used than any native vegetable. New Zealand is favourable by one great natural advantage, namely that the inhabitants can never perish from famine. The whole country abounds with fern, & the roots of this, if not very palatable, yet contain much nutriment: — A native can always subsist on them & on the shell fish, which is very abundant on all parts of the sea shore. The villages are chiefly conspicuous by the platforms which are raised on four posts, ten or twelve feet above the ground & on which the produce of the fields is kept secure from all accidents. — On coming near to one of the huts, I was much amused by seeing in due form the ceremony of rubbing, or as it would be more properly called, pressing noses. The women on our first approach began uttering something in a most dolorous plaintive voice, they then squatted themselves down & held up their faces; my companions standing over them placed the bridges of their own noses at right angles to theirs, & commenced pressing; this lasted rather longer than a cordial shake of the hand would with us; as we vary the force of the grasp of the hand in shaking, so do they in pressing. During the process they utter comfortable little grunts, very much in the same|666| manner as two pigs do when rubbing against each other. I noticed that the slave would press noses with any one he met, indifferently either before or after his master, the Chief. —[2] Although amongst savages the chief has absolute power of life & death over his slave, yet there is generally an entire absence of ceremony between them. Mr Burchell[3] has remarked the same thing in Southern Africa with the rude Bachapins. Where civilization has arrived at a certain point, as among the Tahitians, complex formalities are soon instituted between the different grades of life. For instance in the above island every one was formerly obliged to uncover themselves as low as the waist in presence of the king.

The ceremony of pressing noses having been completed with all present, we seated ourselves in a circle in front of one of the houses & rested there half an hour. — All the native hovels which I have seen, have nearly the same form & dimensions & all agree in being filthily dirty. They resemble a cow shed with one end open; but having a partition a little way within, with a square hole in it, which cuts off a part & makes a small gloomy chamber. When the weather is cold the inhabitants sleep there & likewise keep all their property. They eat, however, & pass their time in the open part in front.

My guides having finished their pipes, we continued our walk. The path led through the same undulating country, the whole uniformly clothed as before with fern. On our right hand we had a serpentine river, the banks of which were fringed with trees & here & there on the hill sides there were clumps of wood. — The whole scene, in spite of its green color, bore rather a desolate aspect; the sight of so much fern impresses the mind with an idea of useless sterility; this, however, is not the case, for wherever the fern grows thick & breast high, the land|667| by tillage becomes productive. I have heard it asserted, & I think with much probability, that all this extensive open country was once covered by forests, & that it had been cleared ages past by the aid of fire. — It is said that frequently by digging in the barest spots, lumps of that kind of rosiñ which flows from the Kauri pine, are found. — The natives had an evident motive in thus clearing the country, for in such parts the fern, formerly so staple an article of food, best flourishes. The almost entire absence of associated grasses which forms so remarkable a feature in the vegetation of this Island, may perhaps be accounted for by the open parts being the work of man, while Nature had designed the country for forest land. —[4] The soil is volcanic; in several parts we passed over slaggy and vesicular lavas & the form of a crater was clearly to be distinguished in several of the neighbouring hills. — Although the scenery is nowhere beautiful & only occasionally pretty I enjoyed my walk; I should have enjoyed it more if my companion, the chief, had not possessed extraordinary conversational powers. I only knew three words, good—bad—& yes: with these I answered all his remarks, without of course having understood one word he said. This was quite sufficient. I was a good listener, —an agreeable person, —& he never ceased talking to me.|668|

At length we reached Waimate; after having passed over so many miles of an uninhabited useless country, the sudden appearance of an English farm house & its well dressed fields, placed there as if by an enchanter's wand, was exceedingly pleasing. —Mr Williams not being at home, I received in Mr Davies' house a cordial & pleasant welcome. — After drinking tea with his family party, we took a stroll about the farm. — At Waimate there are three large houses, where the Missionary gentlemen Mrs Williams, Davies & Clarke reside; near to these are the huts of the native labourers. — On an adjoining slope fine crops of barley & wheat in full ear, & others of potatoes & of clover were standing; but I cannot attempt to describe all I saw; there were large gardens, with every fruit & vegetable which England produces & many belonging to a warmer clime. — I may instance asparagus, kidney beans, cucumbers, rhubarb, apples & pears, figs, peaches, apricots, grapes, olives, gooseberries, currants, hops, gorse for fences, & English oaks! & many different kinds of flowers. Around the farm yard were stables, a threshing barn with its winnowing machine, a blacksmiths forge & on the ground ploughshares & other tools; in the middle was that happy mixture of pigs & poultry which may be seen so comfortably lying together in every English farm yard.|669| At the distance of a few hundred yards, where the water of a little rill has been dammed up into a pool, a large & substantial water-mill had been erected. All this is very surprising when it is considered that five years ago nothing but the fern here flourished. Moreover native workmanship taught by the Missionaries has effected this change:—the lesson of the Missionary is the enchanter's wand. The house has been built, the windows framed, the fields ploughed, even the trees grafted by the New Zealander. At the mill a New Zealander may be seen powdered white with flour, like his brother miller in England. — When I looked at this whole scene I thought it admirable. — It was not that England was vividly brought before my mind; yet as the evening drew to a close, the domestic sounds, the fields of corn, the distant country with its trees now appearing like pasture land, all might well be mistaken for such. — Nor was it the triumphant feeling at seeing what Englishmen could effect: but a thing of far more consequence;—the object for which this labor had been bestowed, —the moral effect on the native inhabitant of New Zealand. —⁵|670|

The whole Missionary system appears to me very different from that of Tahiti; much more attention is there paid to religious instruction & to the direct improvement of the mind; here more to the arts of civilization. I do not doubt in both cases the same object is in view: — judging from the success alone I should rather lean to the Tahiti side; probably however each system is best adapted to the country where it is followed. The mind of a Tahitian is certainly one of a higher order, & on the other hand the New Zealander, not being able to pluck from the tree that shades his house the breadfruit & banana, would naturally turn his attention with more readiness to the Arts. — When comparing the state of New Zealand to Tahiti it must always be remembered that from the respective forms of government, the Missionaries have here to labor at a task many times more difficult. — The Reviewer of Mr Earle's travels in the Quarterly Journal, by pointing out a more advantageous line of conduct for the Missionaries, evidently considers that too much attention has been paid to religious instruction in proportion to other subjects. This opinion being so very different from the one at which I arrived, any third person hearing the two sides would probably conclude that the|671| Missionaries had been the best judges & had chosen the right path.

Several young men were employed about the farm, who had been brought up by the Missionaries, having been redeemed by them from slavery. They were dressed in a shirt & jacket & had a respectable appearance. Judging from one trifling anecdote I should think they must be honest; when walking in the fields, a young labourer came up to Mr Davies & gave him a knife & gimlet, saying he had found them on the road & did not know to whom they belonged!— These young men & boys appeared very merry & good-humoured; in the evening I saw a party of them playing cricket; when I thought of the Austerity of which the Missionaries have been accused, I was amused at seeing one of their sons taking an active part in the game.— A more decided & pleasing change was manifest in the young women who acted as servants within the houses; their clean tidy & healthy appearance, like that of dairy maids in England, formed a wonderful contrast with the women of the filthy hovels in Kororarika.— The wives of the Missionaries tried to persuade them not to be tattooed; but a famous operator having arrived from the South they said, "We really must just have a few lines on our lips; else when we grow old our lips will shrivel & we shall be so very ugly".— Tattooing is not generally nearly|672| so much practised as formerly; but as it is a badge of distinction between the Chief & the

Slave, it will not probably very soon be disused. So soon does any train of ideas become habitual, that the Missionaries told me that even in their eyes a plain face looks mean & not like that of a New Zealand gentleman.

Late in the evening I went to Mr Williams' house where I passed the night. — I found there a very large party of children, collected together for Christmas day, & who were sitting round a table at tea. I never saw a nicer or more merry group: — & to think that this was in the centre of the land of cannibalism, murder & all atrocious crimes! The cordiality & the happiness so plainly pictured in the faces of the little circle is, I believe from what I could see, equally felt by the older persons of the Mission.

¹Schiller's *Ballad of Fridolin.*
²The passage from here to the end of the paragraph is written on a separate page, marked to be inserted.
³See W. J. Burchell, *Travels in the interior of Southern Africa.* London, 1822–4.
⁴This passage was almost unaltered in the first edition of *Journal of Researches* p. 506, but in the 1845 edition, p. 424, the words 'by the open parts being the work of man, while Nature had designed the country for forest land' were changed to 'by the land having been aboriginally covered with forest-trees.'
⁵Followed by a deleted sentence: 'Much could not be expected in so short a time; but the neighbouring people appeared to me rather cleaner & certainly with better expressions of countenance, that those at the Bay of Islands.'

24th In the morning prayers were read in the native tongue to the whole family: after breakfast I rambled about the gardens & farm. — This was market day when the natives of the surrounding hamlets bring their stock of potatoes, Indian corn or pigs, to exchange for blankets, tobacco & sometimes (from the persuasions of the Missionaries) for soap. Mr Davies' eldest son, who manages a farm of his own, is the man of business in the|673| market. The children of the Missionaries, who came whilst young to the Island, understood the language better than their parents, & can get anything more easily done by the natives. Mr Williams & Davies walked with me to part of a neighbouring forest to show me the famous Kauri pine. I measured one of these noble trees & found it to be thirty one feet in circumference; there was another close by which I did not see, thirty-three, & I have heard of one no less than forty feet. — The trunks are also very remarkable by their smoothness, cylindrical figure, absence of branches, & having nearly the same girth for a length from sixty even to ninety feet. The crown of this tree, where it is irregularly branched is small & out of proportion to the trunk; & the foliage is again diminutive as compared to the branches. The forest in this part was almost composed of the Kauri; amongst which the great

ones from the parallelism of their sides stood up like gigantic columns
of wood. — The timber of this tree is the most valuable product of the
island; besides this, quantities of a resin oozes from the bark, which is
collected & sold at a penny a pound to the North Americans, but its use
is kept secret. —

On the outskirts of the wood I saw plenty of the New Zealand hemp
plant growing in the swamps; this is the second most valuable export. —
This plant resembles (but not botanically) the common iris; the
under|674| surface of the leaf is lined by a layer of strong silky fibres; the
upper green vegetable matter being scraped off with a broken shell, the
hemp remains in the hand of the workwoman. In the forest besides the
Kauri there are some fine timber trees: I saw numbers of beautiful
Tree-Ferns & heard of Palms. Some of the New Zealand forests must be
impenetrable to a very extraordinary degree; Mr Matthews gave me an
account of one which although only thirty four miles wide & separating
two inhabited districts, like the central forest of Chiloe, had never been
passed. He & another Missionary each with a party of about fifty men,
undertook to open a road; but it cost them more than a fortnight's
labor! — In the woods I saw very few birds; with respect to animals it is
most remarkable that so large an island, extending over nearly a
thousand miles in latitude, & in many parts one hundred & fifty broad,
with varied stations, a fine climate & land of all heights from 14,000 feet
downwards, should not possess one indigenous animal with the
exception of a small rat. — It is moreover said that the introduction of
the common Norway kind has entirely annihilated the New Zealand
species in the short space of two years, from the Northern extremity of
the island. In many places I noticed several sorts of weeds, which like
the rats I was forced to own as countrymen. A leek, however, which has
overrun whole districts & will be very troublesome, was imported lately
as a favour by a French vessel. — The common dock is widely dissemi-
nated & will|675| I am afraid for ever remain a proof of the rascality of
an Englishman who sold the seeds for those of the tobacco plant.

On returning from our pleasant walk to the houses, I dined with M^r
Williams; & then a horse being lent me, I returned to the Bay of
Islands. — I took leave of the Missionaries, with thankfulness for their
kind welcome & upright characters. I think it would be difficult to find
a body of men better adapted for the high office which they fulfil. —

Christmas day. — In a few more days the fourth year of our absence from
England will be completed. Our first Christmas day was spent at
Plymouth; the second at St Martins Cove near Cape Horn; the third at

Port Desire in Patagonia; the fourth at anchor in the peninsula of Tres Montes; this fifth here, & the next I trust in Providence again in England. —

We attended Divine Service in the Chapel of Pahia; part of the Service was read in English & part in the New Zealand language.[1]|676|

As far as I was able to understand, the greatest proportion of the population in this northern part of the island profess Christianity. It is curious that even the religion of those who do not, is altered & is now partly Christian, partly Heathen. — Moreover, so excellent is the Christian faith, that the outward conduct of the believers is said most decidedly to have been improved by its doctrines, which are to a certain extent generally known. — It is however beyond doubt that much immorality still exists; that there are very many who would not hesitate to commit the heavy crime of killing a slave for a trifling offence; polygamy is still common, indeed I believe general. — We did not hear of any recent act of cannibalism; but Mr Stokes found on a small Island burnt human bones strewed round an old fire-place; these remnants, however, of some quiet banquet might have been lying there for several years. — Notwithstanding the above facts it is probable that the moral state of the people will rapidly improve. — Mr Busby mentioned one pleasing anecdote, as a proof of the sincerity of some at least of those who profess Christianity; one of his young men left him, who had been accustomed to read prayers to the rest of the servants. Some weeks afterwards, happening to pass late in the evening by an outhouse, he saw & heard one of his men reading with difficulty by the light of the fire, the Bible to the|677| rest; after this the party knelt & prayed; in their prayers they mentioned Mr Busby & his family & the Missionaries, each separately in his respective district. — Mr Busby then went in & told them how glad he was to see how they were employed: — they replied they had done so ever since the first young man had gone, & so should continue. —

[1] Followed by a deleted passage: 'This appears a clumsy method, for the one half could not fail to be tedious to every individual. The number of New Zealanders, who attended, was not large; the singing, although aided by a small organ was inferior to that of Tahiti. —'

26th   Mr Busby offered to take Mr Sulivan & myself in his boat some miles up the river Cowa-Cowa, & then to walk on to the village of Waiomio, where there are some curious rocks. — Following one of the arms of the Bay, we enjoyed a pleasant row, passing through pretty scenery till we came to a village beyond which the boat could not proceed. — The chief

& a party of men volunteered to walk on with us to Waiomio, a distance only of four miles. This chief is at present rather notorious, from having hung one of his wives & a slave for adultery. When remonstrated with by one of the Missionaries he said he thought he was following the English method. Old Shongi who happened to be in England at the time of the Queen's trial, expressed great disapprobation at the whole proceedings; he said he had five wives, & he would sooner cut off all their heads than suffer so much trouble about one. — Leaving this village we crossed over to another one seated on a|678| hill side at a little distance. The daughter of the chief of this place, who yet followed heathen customs, had died five days before; the hovel in which she had expired was burnt to the ground; her body being enclosed between two small canoes, was placed upright in the ground & protected by an enclosure bearing wooden images of their gods, & the whole was painted bright red, so as to be conspicuous from afar. Her gown was fastened to the coffin, & her hair being cut off was cast at its foot. The relatives of the family had torn the flesh of their arms, bodies & faces, so as to be covered with clotted blood; & the old women looked most filthy, disgusting objects: On the following day some of the officers visited this place, & again found the women howling & cutting themselves.

We continued our walk & soon reached Waiomio; here there are some singular masses of limestone resembling in their forms ruined castles. — These rocks have long served for burial places, & hence are sacred. One of the young men cried out "Let us be brave", & run on ahead; but when within a hundred yards, the whole party stopped short; they allowed us however with perfect indifference to examine the whole place. — At this village we found several old men; we rested here some hours, during which time there was a long discussion with|679| Mr Busby, concerning the right of sale of certain lands. An old man who appeared a perfect genealogist, illustrated the successive possessors by bits of stick driven in the ground. — Before leaving, a little basket full of roasted sweet potatoes was given to each of our party, & we all, according to the custom, carried them away to eat on the road. I noticed that amongst the women employed in cooking there was one Slave; it must be humiliating to a man thus to be employed in what is only considered as woman's work: in a like manner, slaves do not go to war; but this perhaps can hardly be considered as a hardship. — I heard of one poor wretch, who during hostilities ran away to the opposite party; being met by two men he was immediately seized; but they not agreeing to whom he should belong, each stood over him with a stone hatchet & seemed determined at least that the other should not take him alive. — the poor man, almost dead with fright, was only saved by the address of a Chief's wife. — We then enjoyed a pleasant walk back to the boat, but did not reach the ship till late in the evening.

27*th*–29*th*   Chiefly employed in writing letters, & in collecting some specimens.

30*th*   In the afternoon we stood out of the Bay of Islands on our course to Sydney. I believe we were all glad to leave New Zealand; it|680| is not a pleasant place; amongst the natives there is absent that charming simplicity which is found at Tahiti; & of the English the greater part are the very refuse of Society. Neither is the country itself attractive. — I look back but to one bright spot & that is Waimate with its Christian inhabitants. —

NB (February) I must confess, that after having visited Sydney, my admiration of the Missionary establishment is considerably diminished; I looked at New Zealand in its position as near the Antipodes of England, & not as being within a few hundred miles of a great & highly civilized Colony. It makes much difference to the Beholder, whether he comes from the West or the East. —[1]

[1] This note, dated February in brackets and preceding the January entries, suggests that once again CD has been catching up with his writing some weeks after the events described.

*January 12*th   Early in the morning, a light air carried us towards the entrance of Port Jackson: instead of beholding a verdant country scattered over with fine houses, a straight line of yellowish cliff brought to our mind the coast of Patagonia. A solitary lighthouse, built of white stone, alone told us we were near to a great & populous city. — Having entered the harbor, it appeared fine & spacious; but the level country, showing on the cliff-formed shores bare & horizontal strata of sandstone, was covered by woods of thin scrubby trees that bespoke useless sterility. — Proceeding further inland, parts of the country improved; beautiful Villas & nice Cottages were|681| here & there scattered along the beach; and in the distance large stone houses, two or three stories high, & Windmills standing on the edge of a bank, pointed out to us the neighbourhead of the Capital of Australian civilization.

At last we anchored within Sydney Cove; we found the little basin, containing many large ships & surrounded by Warehouses. — In the evening I walked through the town & returned full of admiration at the whole scene. — It is a most magnificent testimony to the power of the British nation: here, in a less promising country, scores of years have effected many times more than centuries in South America. — My first feeling was to congratulate myself that I was born an Englishman: — Upon seeing more of the town on other days, perhaps it fell a little in my estimation; but yet it is a good town; the streets are regular, broad, clean & kept in excellent order; the houses are of a good size & the Shops well furnished. — It may be faithfully compared to the large suburbs which stretch out from London & a few other great towns: — but not even near London or Birmingham is there an aspect of such rapid growth; the number of large houses just finished & others building is truly surprising; nevertheless every one complains of the high rents & difficulty in procuring a house. — In the streets|682| gigs, phaetons & carriages with livery servants are driving about; of the latter many are extremely well equipped. Coming from S. America, where in the towns every man of property is known, no one thing surprised me more, than not readily being able to ascertain to whom this or that carriage belonged. — Many of the older residents say that formerly they knew every face in the Colony, but now that in a morning's ride, it is a chance if they know one. — Sydney has a population of twenty-three thousand, & is as I have said rapidly increasing; it must contain much wealth; it appears a man of business can hardly fail to make a large fortune; I saw on all sides fine houses, one built by the profits from steam-vessels, another from building, & so on. An auctioneer who was a convict, it is said intends to return home & will take with him 100,000 pounds. — Another who is always driving about in his carriage, has an income so large that scarcely anybody ventures to guess at it, the least assigned being fifteen thousand a year. — But the two crowning facts are, first that the public revenue has increased 60,000£ during this last year, & secondly that less than an acre of land within the town of Sydney sold for 8000 pounds sterling. —[1]|683|

I hired a man & two horses to take me to Bathurst, a village about one hundred & twenty miles in the interior, & the centre of a great pastoral district; by this means I hoped to get a general idea of the appearance of the country. —

[1] Followed by a deleted sentence: 'There is one advantage which the town enjoys in the number of pleasant walks in the Botanic Garden & Government domain; there are no fine trees, but the walks wind about the Shrubberies & are to me infinitely more pleasing than the formal Alamedas of S. America.'

16th   In the morning of the 16th I set out on my excursion; the first stage took us through Paramatta, a small country town, but second to Sydney in importance. — The roads were excellent & made on the Macadam principle, whinstone being brought from the distance of several miles for this purpose; nor had turnpikes been forgotten. — The road appeared much frequented by all sorts of carriages. — I met two Stage Coaches. — In all these respects there was a most close resemblance to England; perhaps the number of Ale-houses was here in excess. The Iron gangs, or parties of convicts, who have committed some trifling offence in this country, appeared the least like England: they were dressed in yellow & grey clothes, & were working in irons under the charge of sentrys with loaded arms. — I believe one chief cause of the early prosperity in these Colonies is government thus being able by means of forced labour to open at once good roads throughout the country. —

I slept at night at a very comfortable Inn at Emu ferry, which is thirty-five miles from Sydney|684| & near the ascent of the Blue Mountains. — This line of road is the most frequented & has longest been inhabited of any in the Colony. — The whole land is enclosed with high railings, for the farmers have not been able to rear hedges. — There are many substantial houses & good cottages scattered about; but although considerable pieces of the land are under cultivation, the greater part yet remains as when first discovered. — Making allowances for the cleared parts, the country here resembles all that I saw during the ten succeeding days. — The extreme uniformity in the character of the Vegetation, is the most remarkable feature in the landscape of the greater part of New S. Wales. — Everywhere we have an open woodland, the ground being partially covered with a most thin pasture. The trees nearly all belong to one family;[1] & have the surface of their leaves placed in a vertical instead of as in Europe a nearly horizontal position; This fact & their scantiness makes the woods light & shadowless; although under the scorching sun of the summer this is a loss of comfort, it is of importance to the farmer, as it allows grass to grow where it otherwise could not. — The greater number of the trees, with the exception of some of the Blue|685| Gums, do not attain a large size; but they grow tall & tolerably straight & stand well apart. It is singular that the bark of some kinds annually falls, or hangs dead in long shreds, which swing about with the wind; & hence the woods appear desolate & untidy. — Nowhere is there an appearance of verdure or fertility, but rather that of arid sterility: — I cannot imagine a more complete contrast in every respect, than the forest of Valdivia or Chiloe, with the woods of Australia. —

Although this country flourishes so remarkably, the appearance of infertility is to a certain degree real; the soil without doubt is good, but there is so great a deficiency in rain & running water, that it cannot produce much. — The Agricultural crops & indeed often those in gardens, are estimated to fail once in three years; & it has even thus happened on successive years: — hence the Colony cannot supply itself with the bread & vegetables which its inhabitants consume. — It is essentially pastoral, & chiefly so for sheep & not the larger quadrupeds: the alluvial land near Emu ferry is some of the best cultivated which I have seen; & certainly the scenery on the banks of the Nepean, bounded to the West by the Blue Mountains, was pleasing even to the eye of a person thinking of England. —

At Sunset by good fortune a party of a score of the Aboriginal Blacks passed by, each carrying in their accustomed manner a bundle of spears & other|686| weapons. — By giving a leading young man a shilling they were easily detained & they threw their spears for my amusement. — They were all partly clothed & several could speak a little English; their countenances were good-humoured & pleasant & they appeared far from such utterly degraded beings as usually represented. — In their own arts they are admirable; a cap being fixed at thirty yards distance, they transfixed it with the spear delivered by the throwing stick, with the rapidity of an arrow from the bow of a practised Archer; in tracking animals & men they show most wonderful sagacity & I heard many of their remarks, which manifested considerable acuteness. — They will not however cultivate the ground, or even take the trouble of keeping flocks of sheep which have been offered them; or build houses & remain stationary. — Never the less, they appear to me to stand some few degrees higher in civilization, or more correctly a few lower in barbarism, than the Fuegians. —

It is very curious thus to see in the midst of a civilized people, a set of harmless savages wandering about without knowing where they will sleep, & gaining their livelihood by hunting in the woods. — Their numbers have rapidly decreased; during my whole ride with the exception of some boys brought up in the houses, I saw only one other party. — These were rather more numerous & not so well clothed. — I should have mentioned|687| that in addition to their state of independence of the Whites, the different tribes go to war. In an engagement which took place lately the parties, very singularly chose the centre of the village of Bathurst as the place of engagement; the conquered party took refuge in the Barracks. — The decrease in numbers must be owing to the drinking of Spirits, the Europæan diseases, even the milder ones of which such as the Measles are very destructive, & the gradual extinction of the wild animals. It is said that from the wandering life of these people, great numbers of their children die in very early infancy. When the difficulty in procuring food is increased, of course the population must be repressed in a manner almost instantaneous as compared to what takes place in civilized life, where the father may add to his labor without destroying his offspring.

[1] Followed by deleted words: 'the foliage is scanty & of a rather peculiar light green tint; it is not periodically shed'.

17th Early in the morning we passed the Nepean in a ferry boat. The river, although at this spot both broad & deep, has a very small body of running water. Having crossed a low piece of land on the opposite side we reached the slope of the Blue Mountains. The ascent is not steep, the road having been cut with much care on the side of the Sandstone cliffs; at no great elevation we come to a tolerably level plain, which almost imperceptibly rises to the Westward, till at last its height exceeds three thousand feet. By the term Blue Mountains, & hearing of their absolute elevation, I had expected to see a bold|688| chain crossing the country; instead of this a sloping plain presents merely an inconsiderable front to the low country. — From this first slope, the view of the extensive woodland towards the coast was interesting, & the trees grew bold & lofty; but when once on the sandstone platform, the scenery became exceedingly monotomous. On each side the road was bordered by a scrubby wood of small trees of the never-failing Eucalyptus family; with the exception of two or three small Inns there were no houses or cultivated land. The road was likewise solitary, the most frequent object being a bullock-waggon piled up with bales of Wool. —

In the middle of the day we baited our horses at a little Inn, called the Weather-board. The country here is elevated 2800 feet above the sea. About a mile & a half from this place there is a view exceedingly well worth visiting; following down a little valley & its tiny rill of water, an immense gulf is suddenly & without any preparation seen through the trees which border the pathway at the depth of perhaps 1500 ft. Walking a few yards farther, one stands on the brink of a vast precipice, & below is the grand bay or gulf, for I know not what other name to give it, thickly covered with forest. The point of view is situated as it were at the head of the bay, for the line of cliff diverges away on each side, showing headland behind headland, as on a|689| bold Sea coast. These cliffs are composed of horizontal strata of whitish Sandstone; & so absolutely vertical are they, that in many places a person standing on the edge & throwing a stone can see it strike the trees in the abyss below: so unbroken is the line, that it is said to be necessary to go round a distance of sixteen miles in order to reach the foot of the waterfall made by this little stream. — In front & about five miles distant another line of cliff extends, thus having the appearance of completely encircling the valley; hence the name of Bay is justified as applied to this grand amphitheatrical depression. — If we imagine that a winding harbor with its deep water surrounded by bold cliff shores was laid dry, & that a forest sprung up on the sandy bottom, we should then have the appearance & structure which is here exhibited. The class of view was to me quite novel & extremely magnificent. In the evening we reached the Blackheath; the Sandstone plateau has here attained the elevation of 3411 ft, & is covered as before, with one monotomous wood. — On the road, there were occasional glimpses of a profound valley, of the same character as the one described; but from the steepness & depth of its sides, the bottom was scarcely ever to be seen. — The Blackheath is a very comfortable inn, kept by an old Soldier; it reminded me of the small inns in North Wales. I was surprised to|690| find that here, at the distance of more than seventy miles from Sydney, fifteen beds could be made up for travellers. —

18*th*   Very early in the morning I walked about three miles to see Govett's Leap; a view of a similar, but even perhaps more stupendous character than that of the Weatherboard. So early in the day the gulf was filled with a thin blue haze, which, although destroying the general effect, added to the apparent depth of the forest below, from the country on which we stood. M*r* Martens who was formerly in the Beagle & now resides in Sydney, has made striking & beautiful pictures from these two views. —

   A short time after leaving the Blackheath, we descended from the sandstone platform by the pass of Mount Victoria. To effect this pass, an enormous quantity of stone has been cut through; the design & its manner of execution would have been worthy of a line of road in England, even that of Holyhead. — We now entered upon a country less elevated by nearly a thousand feet & consisting of granite: with the change of rock the vegetation improved; the trees were both finer & stood further apart, & the pasture between them was a little greener & more plentiful. —

At Hassan's walls I left the high road & made a short detour to a farm called Walerawang; to the superintendent of which I had a letter of introduction from the owner|691| in Sydney. Mr Browne had the kindness to ask me to stay the ensuing day, which I had much pleasure in doing. This place offers an example of one of the large farming or rather sheep grazing establishments of the Colony; cattle & horses are however in this case rather more numerous than usual, owing to some of the valleys being swampy & producing a coarser pasture. The sheep were 15,000 in number, of which the greater part were feeding under the care of different shepherds on unoccupied ground, at the distance of more than a hundred miles beyond the limits of the Colony. Mr Browne had just finished this day the last of the shearing of seven thousand sheep; the rest being sheared in another place. — I believe the value of the average produce of wool from 15,000 sheep would be more than 5000£ sterling. Two or three flat pieces of ground near the house were cleared & cultivated with corn, which the harvest men were now reaping. No more wheat is sown than sufficient for the annual support of the labourers; the general number of assigned convict servants being here about forty; but at present there were rather more. Although the farm is well stocked with every requisite, there was an apparent absence of comfort; & not even one woman resided here. — The Sunset of a fine day will generally cast an air of happy contentment|692| on any scene; but here at this retired farmhouse the brightest tints on the surrounding woods could not make me forget that forty hardened profligate men were ceasing from their daily labours, like the Slaves from Africa, yet without their just claim for compassion.

19th   Early on the next morning Mr Archer, the joint superintendent, had the kindness to take me out Kangaroo hunting. We continued riding the greater part of the day; but had very bad sport, not seeing a Kangaroo or even a wild dog. — The Grey-hounds pursued a Kangaroo Rat into a hollow tree out of which we dragged it: it is an animal as big as a rabbit, but with the figure of a Kangaroo. A few years since this country abounded with wild animals; now the Emu is banished to a long distance & the Kangaroo is become scarce; to both the English Greyhound is utterly destructive; it may be long before these animals are altogether exterminated, but their doom is fixed. The Natives are always anxious to borrow the dogs from the farmhouses; their use, offal when an animal is killed, & milk from the cows, are the peace offerings of the Settlers, who push further & further inland. — The thoughtless Aboriginal, blinded by these trifling advantages, is delighted at the approach of the White Man, who seems predestined to inherit the country of his children.

Although having bad sport, we enjoyed a pleasant ride;|693| The woodland is generally so open that a person on horseback can gallop through it; it is traversed by a few flat bottomed valleys, which are green & free from trees; in such spots the scenery was like that of a Park & pretty. — In the whole country I scarcely saw a place without the marks of fire; whether these had been more or less recent, whether the stumps were more or less black, was the greatest change which varied the monotony so wearisome to the traveller's eye. In these woods there are not many birds; I saw, however, some large flocks of the white Cockatoo feeding in a Corn field; & a few most beautiful parrots; crows, like our jackdaws, were not uncommon & another bird something like the magpie. The English have not been very particular in giving names to the productions of Australia; trees of one family (Casuarina) are called Oaks, for no one reason that I can discover without it is that there is no one point of resemblance; animals are called tigers & hyenas, simply because they are Carnivorous, & so on in many other cases. In the dusk of the evening I took a stroll along a chain of ponds, which in this dry country represent the course of a river, & had the good fortune to see several of the famous Platypus or Ornithorhyncus paradoxicus. They were diving & playing about the surface of the water; but showed very little of their bodies,|694| so that they might easily have been mistaken for many water rats. Mr Browne shot one; certainly it is a most extraordinary animal; the stuffed specimens do not give at all a good idea of the recent appearance of the head & beak; the latter becoming hard & contracted. —

A little time before this, I had been lying on a sunny bank & was reflecting on the strange character of the Animals of this country as compared to the rest of the World. An unbeliever in everything beyond his own reason, might exclaim "Surely two distinct Creators must have been [at] work; their object however has been the same & certainly the end in each case is complete". — Whilst thus thinking, I observed the conical pitfall of a Lion-Ant:— A fly fell in & immediately disappeared; then came a large but unwary Ant; his struggles to escape being very violent, the little jets of sand described by Kirby (Vol. I. p. 425)[1] were promptly directed against him. — (*Note in margin:* NB The pitfall was not above half the size of the one described by Kirby.) His fate however was better than that of the poor fly's:— Without a doubt this predacious Larva belongs to the same genus, but to a different species from the

Europæan one. — Now what would the Disbeliever say to this? Would any two workmen ever hit on so beautiful, so simple & yet so artificial a contrivance? It cannot be thought so. — The one hand has surely worked throughout the universe.[2] A Geologist perhaps would suggest, that the periods of Creation have|695| been distinct & remote the one from the other; that the Creator rested in his labor. —[3]

[1] See William Kirby and William Spence, *An introduction to entomology; or, elements of the natural history of insects.* 4 vols. London, 1815–26.
[2] Altered in pencil from 'I cannot think so. — The one hand has worked over the whole world. —'
[3] In *Journal of Researches* pp. 526–7, the lion-ant episode appears much as it does here, except that 'sceptic' is substituted for 'Disbeliever', and the final sentence beginning 'A Geologist perhaps would . . .' is omitted. In the 1845 edition (p. 442) the episode is relegated to a footnote, and the last five sentences are omitted.

20th   A long days ride to Bathurst; before joining the high road we followed a mere path through the forest; the country with the exception of a few squatters huts was very solitary. A "squatter" is a freed or "ticket of leave" man, who builds a hut with bark in unoccupied ground, buys or steals a few animals, sells spirits without a license, receives stolen goods & so at last becomes rich & turns farmer: he is the horror of all his honest neighbours. — A "crawler" is an assigned convict, who runs away & lives how he can by labor or petty theft. — The "Bush Ranger" is an open villain, who subsists by highway robbery & plunder; generally he is desperate & will sooner be killed than taken alive. — In the country it is necessary to understand these three names, for they are in perpetual use. —

This day we had an instance of the sirocco-like wind of Australia; which comes from the parched deserts of the interior. While riding, I was not fully aware, as always happens, how exceedingly high the temperature was. — Clouds of dust were travelling in every part, & the wind felt like that which has passed over a fire. — I afterwards heard the thermometer out of doors stood at 119° & in a room in a closed house 96°. — In the afternoon we came in|696| view of the downs of Bathurst. These undulating but nearly level plains are very remarkable in this country by being absolutely destitute of a single tree: they are only covered by a very thin, brown pasture. We rode some miles across this kind of country, & then reached the township of Bathurst, seated in [the] middle of what may be described as a very broad valley, or narrow plain. I had a letter of introduction to the commandant of the troops, & with him I staid the ensuing day. —

21st   Bathurst has a singular & not very inviting appearance; groups of small houses, & a few large ones, are scattered rather thickly over two or three miles of a bare country which is divided into numerous fields by lines of rails. A good many gentlemen live in the neighbourhood & some possess very comfortable houses. A hideous little red brick Church stands by itself on a hill & there are barracks & government buildings. — I was told not to form too bad an opinion of the country by judging of it on the road side, nor too good a one from Bathurst; in this latter respect I did not feel myself in the least danger of being prejudiced. It must be confessed that the season had been one of great drought, & that the country did not wear a favourable aspect; although I understand two or three months ago it was|697| incomparably worse. The secret of the rapidly growing prosperity of Bathurst is that the pasture, which appears to the stranger's eye wretched, is for sheep grazing excellent. The town stands on the banks of the Macquarie: this is one of the rivers whose waters flow into the vast unknown interior. The North & South line of watershed which divides the inland streams from those of the coast has an elevation of about 3000 ft., (Bathurst is 2200) & runs at a distance of about eighty or a hundred miles from the seaside. — The Macquarie figures in the maps as a respectable river, & is the largest of those draining this part of the inland slope:—yet to my surprise I found it a mere chain of ponds, separated from each other by spaces almost dry; generally a little water does flow, & sometimes there are high & impetuous floods. Very scanty as the supply of water is in all this district, it becomes, further in the interior, still scarcer. —

The Officers all seemed very weary of this place & I am not surprised at it: it must be to them a place of exile: Last year there had been plenty of Quail to shoot, but this year they have not appeared; this resource exhausted, the last tie which bound them to existence, seemed on the point of being dissolved. — Capt. Chetwode had attempted gardening; but to see the poor|698| parched herbs was quite heart-breaking. Yesterday's hot wind had alone cut off many scores of young apples, peaches & grapes. —

22nd I commenced my return, taking a new road called Lockyer's line, in which the country is rather more hilly & picturesque. At noon we baited at a farm house; the owner had only come out two years before, but he appeared to be going on very well; he had two pretty daughters, who, I suspect, would not remain long on his hands. — This was a long day's ride & the house where I wished to sleep was some way off the road & not easy to find. — I met on this, & indeed on all other occasions, a very general & ready civility amongst the lower orders; when one considers what they are & what they have been, this is rather surprising. — The farm where I passed the night, was owned by two young Englishmen, who had only lately come out & were beginning a settlers life; the total want of almost every comfort was not very attractive; but future prosperity was certain & not far distant.

23rd The next day we passed through large tracts of country in flames; volumes of smoke sweeping across the road. — Before noon we joined our former track and ascended Mount Victoria: I slept at the Weatherboard, & before dark took another walk to the grand Amphitheatre.

24th In the morning I did not feel well, & I thought it more prudent|699| not to set out. — The ensuing day was one of steady drizzling rain; all was still, excepting the dropping from the eaves; the horizon of the undulating Woodland was lost in thin mist; the air was cold & comfortless—it was a day for tedious reflection. —

26th Escaped from my prison; Having crossed the wearisome Sandstone plain, descended to Emu ferry. A few miles further on I met Capt. King who took me to his house at Dunheved. I spent a very pleasant afternoon walking about the farm & talking over the Natural History of T. del Fuego.

27th Accompanied by Capt. King rode to Paramatta. Close to the town, his brother in law Mr Mac Arthur lives[1] & we went there to lunch. The house would be considered a very superior one, even in England. — There was a large party, I think about 18 in the Dining room. — It sounded strange in my ears to hear very nice looking young ladies exclaim, "Oh we are Australian, & know nothing about England". — In the afternoon I left this most English-like house & rode by myself into Sydney. —

[1] Hannibal Hawkins Macarthur had just built a Greek Revival mansion called Vianaco on his property at Vineyard.

28th & 29th Before we came to the Colony, the things about which I felt most interest were the state of Society amongst the higher & Convict classes, & the degree of attraction to emigrate. Of course after so very short a visit, one's opinion is worth little more than a conjecture; but it is as difficult not to form some opinion, as it is to form a correct judgment. —|700| On the whole, from what I heard more than from what I saw, I was disappointed in the state of Society. — The whole community is rancorously divided into parties on almost every subject. Amongst those who from their station of life, ought to rank with the best, many live in such open profligacy, that respectable people cannot associate with them. There is much jealousy between the children of the rich emancipist & the free settlers; the former being pleased to consider honest men as interlopers. The whole population poor & rich are bent on acquiring wealth; the subject of wool & sheep grazing amongst the higher orders is of preponderant interest. The very low ebb of literature is strongly marked by the emptiness of the booksellers shops; these are inferior to the shops of the smaller country towns of England. — There are some very serious drawbacks to the comforts of families, the chief of these is perhaps being surrounded by convict servants. How disgusting to be waited on by a man, who the day before was by your representation flogged for some trifling misdemeanour? The female servants are of course much worse; hence children acquire the use of the vilest expressions, & fortunately if not equally vile ideas.[1] On the other hand, the capital of a person will without trouble produce him treble interest as compared to England: & with care he is sure|701| to grow rich. The luxuries of life are in abundance, & very little dearer, as most articles of food are cheaper, than in England. The climate is splendid & most healthy, but to my mind its charms are lost by the uninviting aspect of the country. Settlers possess one great advantage in making use of their sons, when very young men from sixteen to twenty years of age, in taking charge of remote farming stations; this however must happen at the expence of their boys associating entirely with convict servants. — I am not aware that the tone of Society has yet assumed any peculiar character; but with such habits & without intellectual pursuits, it can hardly fail to deteriorate. —[2] The balance of my opinion is such, that nothing but rather severe necessity should compel me to emigrate. —

The rapid prosperity of this colony is to me, not understanding such subjects,[3] very puzzling. — The two main exports are Wool & Whale Oil, to both of which productions there is a limit. The country is totally unfit for Canals; therefore there is a not very distant line beyond which the land carriage of wool will not repay the expence of shearing & tending sheep: The pasture everywhere is so thin that already settlers have pushed far into the interior; moreover very far|702| inland the country appears to become extremely poor. — I have before said agriculture can never succeed on a very extended scale. So that, as far as I can see, Australia must ultimately depend upon being the centre of commerce for the Southern Hemisphere; & perhaps on her future Manufactories: from the habitable country extending along the coast, & from her English extraction she is sure to be a maritime nation: possessing coal, she always has the moving power at hand. — I formerly imagined that Australia would rise into as grand & powerful a country as N. America, now it appears to me, as far as I can understand such subjects, that such future power & grandeur is very problematical. —

With respect to the state of the convicts, I had still fewer opportunities of judging than on the other points. The first question is whether their state is at all one of punishment; no one will maintain that it is a very severe one. But this, I suppose, is of little consequence as long as it continues to be an object of dread to Criminals at home. The corporeal wants of the convicts are tolerably well supplied; their prospect of future liberty & comfort is not distant & on good conduct certain. A "ticket of leave", which makes a man, as long as he keeps clear of suspicion as well as crime, free within a certain district, is given upon good conduct after years proportional to the length of the sentence: — for life, eight years is the time of probation; for seven years, four, &c. — Yet, with all this, & overlooking the previous imprisonment & wretched passage out, I believe the years of assignment are passed with discontent & unhappiness: as an intelligent man remarked to me, they|703| know no pleasure beyond sensuality, and in this they are not gratified. The enormous bribe which government possesses in offering free pardons, & the deep horror of the secluded penal settlements, destroy confidence between the convicts & so prevents crime. — As to a sense of shame, such a feeling does not appear to be known; of this I witnessed some singular proofs.[1] — It is a curious fact, but I was universally told that the character of the convict population is that of arrant cowardice, — although not unfrequently some become desperate & quite indifferent of their lives, yet that a plan requiring cool or continued courage was seldom put into execution. — The worse feature in the whole case is, that although there exists what may be called a legal reform, or that very little which the law can touch is committed, yet that any moral reform should take place appears to be quite out of the question. — I was assured by well informed people that a man who should try to improve could not, while living with the other assigned servants; — his life would be one of intolerable misery & persecution. — Nor must the contamination of the Convict ships & prisons both here & in England be forgotten. — On the whole, as a place of punishment, its object is scarcely gained; as a real system of reform, it has failed as[2] perhaps would every other plan.[3] —

[1] Followed by deleted sentence: 'I heard of one instance where the dear little innocent must have perfectly astounded its Mama. —'
[2] Followed by deleted words: '& became like that of the people of the United States'.
[3] 'Political Economy' deleted.

30th The Beagle made sail for Hobart Town: Capt. King & some other people accompanied us a little way out of Harbour.— Philip King remains behind & leaves the Service.— |704|

February 5th After a six days passage, of which the first part was fine & the latter very cold & squally, we entered the mouth of Storm Bay: the weather justified this awful name.— This Bay should rather be called a deep Estuary, which receives at its head the waters of the Derwent.— Near its mouth there are extensive basaltic platforms, the sides of which show fine façades of columns; higher up the land becomes mountainous, & is all covered by a light wood.— The bases of these mountains, following the edges of the bay, are cleared & cultivated; the bright yellow fields of corn & dark green ones of potato crops appear very luxuriant. Late in the evening we came to an anchor in the snug cove on the shores of which stands the capital of Tasmania, as Van Diemen's land is now called.— The first aspect of the place was very inferior to that of Sydney; the latter might be called a city, this only a town.—

In the morning I walked on shore.— The streets are fine & broad; but the houses rather scattered: the shops appeared good: The town stands at the base of M. Wellington, a mountain 3100 ft high, but of very little picturesque beauty: from this source however it receives a good supply of water, a thing much wanted in Sydney.— Round the cove there are some fine warehouses; & on one side a small Fort.— Coming from the Spanish Settlements, where such magnificent|705| care has generally been paid to the fortifications, the means of defence in these colonies appeared very contemptible.— Comparing this town to Sydney, I was chiefly struck with the comparative fewness of the large houses, either built or building.— I should think this must indicate that fewer people are gaining large fortunes. The growth however of small houses has been most abundant; & the vast number of little red brick dwellings, scattered on the hill behind the town, sadly destroys its picturesque appearance.— In London I saw a Panorama of a Hobart town; the scenery was very magnificent, but unfortunately there is no resemblance to it in nature.— The inhabitants for this year are 13,826; in the whole of Tasmania 36,505.— The Aboriginal blacks are all removed & kept (in reality as prisoners) in a Promontory, the neck of which is guarded. I believe it was not possible to avoid this cruel step; although without doubt the misconduct of the Whites first led to the Necessity.—

7th–10th During these days I took some long pleasant walks examining the geology of the country.— The climate here is damper than in New

S. Wales & hence the land is more fertile. Agriculture here flourishes; the cultivated fields looked very well & the gardens abounded with the most luxuriant vegetables & fruit trees. Some of the farm houses, situated in retired spots, had a very tempting appearance. The general aspect of the Vegetation is similar to that of Australia; perhaps it is a little more|706| green & cheerful & the pasture between the trees rather more abundant. — One long walk which I took was on the opposite side of the Bay; I crossed in a Steam boat, two of which are constantly plying backwards & forwards. — The machinery of one [of] these vessels was entirely manufactured in this Colony, which from its very foundation only numbers three & thirty years!

11th I ascended Mount Wellington. I made the attempt the day before, but from the thickness of the wood failed. — I took with me this time a guide, but he was a stupid fellow & led me up by the South or wet side. Here the vegetation was very luxuriant & from the number of dead trees & branches, the labor of ascent was almost as great as in T. del Fuego or Chiloe. — It cost us five & a half hours of hard climbing before we reached the summit. — In many parts the gum trees grew to a great size & the whole composed a most noble forest. — In some of the dampest ravines, tree-ferns flourished in an extraordinary manner;—I saw one which must have been about twenty five feet high to the base of the fronds, & was in girth exactly six feet:—the foliage of these trees forming so many most elegant parasols created a shade gloomy like that of the first hour of night. — The summit of the mountain is broard & flat & is composed of huge angular masses of naked greenstone; its elevation is 3100 ft above the level of the Sea. — The day was splendidly clear & we enjoyed a most extensive view. — To the Northward the country appeared a mass of wooded mountains|707| of about the same elevation & tame outline as the one on which we stood. To the South the intricate outline of the broken land & water forming many bays was mapped with clearness before us. — After staying some hours on the summit we found a better way to descend, but did not reach the Beagle till eight oclock, after a severe day's work.

$12^{th}-15^{th}$ I had been introduced [to] Mr Frankland, the Surveyor General, & during these days I was much in his Society. — He took me two very pleasant rides & I passed at his house the most agreeable evening since leaving England. There appears to be a good deal of Society here: I heard of a Fancy Ball, at which 113 were present in costumes! I suspect also the Society is much pleasanter than that of Sydney. — They enjoy an advantage in there being no wealthy Convicts. — If I was obliged to emigrate I certainly should prefer this place: the climate & aspect of the country almost alone would determine me. — The Colony moreover is well governed; in this convict population, there certainly is not *more* , if not *less* , crimes, than in England.

16th   The weather has been cloudy, which has prolonged our stay beyond what was expected. — I went this day in a Stage Coach to New Norfolk. This flourishing village contains 1822 inhabitants. It is distant 22 miles from Hobart town; the line of road follows the Derwent. — We passed very many nice farms & much Corn land. — Returned in the evening by the same Coach.

17th   The Beagle stood out with a fair wind on her passage to K. George's Sound. The Gun-room officers|708| gave a passage to England to Mr Duff of the 21st Reg:

March 6th   In the evening came to an anchor in the mouth of the inner harbor of King Georges Sound. Our passage has been a tolerable one; & what is surprising, we had not a single encounter with a gale of wind. — Yet to me, from the long Westerly swell, the time has passed with no little misery. We staid there eight days & I do not remember since leaving England having passed a more dull, uninteresting time. The country viewed from an eminence, appears a woody plain, with here & there rounded & partly bare hills of granite. — One day I went out with a party in hopes of seeing a Kangaroo hunt, & so walked over a good many miles of country. — Every where we found the soil sandy & very poor; it either supported a coarse vegetation of thin low brushwood & wiry grass, or a forest of stunted trees. — The scenery resembled the elevated sandstone platform of the Blue Mountains: the Casuarina (a tree which somewhat resembles a Scotch fir) is however in greater proportion as the eucalyptus is rather less. In the open parts there are great numbers of the grass-tree, a plant which in appearance has some affinity with the palm, but instead of the crown of noble leaves, it can boast merely of a tuft of coarse grass. The general bright green color of the brushwood & other plants viewed from a distance seems to bespeak fertility; a single walk will however quite dispel such an illusion; & if he thinks like me, he will|709| never wish to walk again in so uninviting a country.

The settlement consist [of] from 30–40 small white washed cottages, which are scattered on the side of a bank & along a white sea beach. — There are a very few small gardens; with these exceptions all the land remains in the state of Nature & hence the town has an uncomfortable appearance. — At the distance of a mile over the hill, Sir R. Spencer has a small & nice farm, & which is the only cultivated ground in the district. The inhabitants live on salted meat & of course have no fresh meat or vegetables to sell; they do not even take the trouble to catch the fish with which the bay abounds: indeed I cannot make out what they are or intend doing. — I understand & believe it is true, that thirty miles inland there is excellent land for all purposes; this is already granted into allotments & will soon be under cultivation. The settlement of King George's Sound will ultimately be the Sea port of this inland district. — Certainly I have formed a very low opinion of the place; it must however be remembered that only from two to three years have elapsed since its effectual colonization, & for this great allowances must be made. Whether, however, it will ever be able to compete with the Colonies which possess the cheap labor of convicts, time alone will show. — They possess here some advantages, the climate is very pleasant, & more rain falls than in the Eastern colonies. I judge of this from the fact that all|710| the broad flat bottomed valleys which are covered over with the rush-like grasses & brushwood, are in winter so swampy as scarcely to be passable. — The second grand advantage is the good disposition of the aboriginal blacks; it is not easy to imagine a more truly good natured & good humoured expression than their faces show: Moreover they are quite willing to work & to make themselves very useful; in this respect they are very different from those in the other Australian colonies. — In their habits, manners, instruments & general appearance they resemble the natives of New S. Wales. — Like them, they are very remarkable by the extreme slightness of their limbs, especially their legs; yet without, as it would appear, muscles to move their legs, they will carry a burthen for a longer time than most white men. — Their faces are very ugly, the beard is curly & not at all deficient, the skin of the whole body is very hairy & their persons most abominably filthy. Although true Savages, it is impossible not to feel an inclination to like such quiet good-natured men. —

During the two first days after our arrival, there happened to be a large tribe called the White Coccatoo men, who come from a distance paying the town a visit. — Both these men & the K. George's Sound men were asked to hold a "Corrobery" or dancing party near one of the Residents houses. — They were tempted with the offer of some tubs of boiled|711| rice or sugar. As soon as it grew dark they lighted small fires & commenced their toilet, which consisted in painting themselves in spots & lines with a white colour. — As soon as all was ready, large fires were kept blazing, round which the women & children were collected as spectators. — The Cockatoo and King George's men formed two distinct parties & danced generally in answer to each other. The dancing consisted in the whole set running either sideways or in Indian file into an open space & stamping the ground as they marched all together & with great force. — Their heavy footsteps were accompanied by a kind of grunt, & by beating their clubs & weapons, & various other gesticulations, such as extending their arms & wriggling their bodies. It was a most rude barbarous scene, & to our ideas without any sort of meaning; but we observed that the women & children watched the whole proceeding with the greatest pleasure. — Perhaps these dances originally represented some scenes such as wars & victories; there was one called the Emu dance in which each man extended his arm in a bent manner, so as to imitate the movement of the neck of one of those birds. In another dance, one man took off all the motions of a Kangaroo grazing in the woods, whilst a second crawled up & pretended to spear him. — When both tribes mingled in one dance, the ground trembled with the heaviness of their steps & the air resounded with their wild crys. — Every one appeared in high|712| spirits; & the group of nearly naked figures viewed by the light of the blazing fires, all moving in hideous harmony, formed a perfect representation of a festival amongst the lowest barbarians. — [1] In T. del Fuego we have beheld many curious scenes in savage life, but I think never one where the natives were in such high spirits & so perfectly at their ease. — After the dancing was over, the whole party formed a great circle on the ground & the boiled rice & sugar was distributed to the delight of all. [2]

[1] Followed by a deleted sentence: 'I imagine from what I have read that similar scenes may be seen amongst the same coloured people, who inhabit the Southern extremity of Africa.'
[2] FitzRoy wrote: 'Mr Darwin ensured the compliance of all the savages by providing an immense mess of boiled rice, with sugar, for their entertainment.' See *Narrative* 2: 626.

8ᵗʰ   One day I accompanied Capt. FitzRoy to Bald head; this is the spot mentioned by so many navigators, where some have imagined they have seen Coral & other trees petrified in the position in which they grew. — According [to] our view of the case, the rocks have been formed by the wind heaping up Calcareous sand, which by the percolation of rain water has been consolidated & during this process enclosed trees, roots & land shells. — In time the wood would decay & as this took place, lime was washed into the cylindrical cavities & became hard like stalactites. — The weather is now again in parts wearing away these soft rocks & hence the harder casts of roots & branches stand out in exact imitation of a dead shrubbery. — The day was to me very interesting, as I had never before heard of such a case. — |713|

14th Our departure was delayed by strong winds & cloudy weather until this day. Since leaving England I do not think we have visited any one place so very dull & uninteresting as K. George's Sound. Farewell Australia, you are a rising infant & doubtless some day will reign a great princess in the South; but you are too great & ambitious for affection, yet not great enough for respect; I leave your shores without sorrow or regret.

April 1st We arrived in view of the Southern Keeling or Cocos Is^d. Our passage would have been a very good one, if during the last five days when close to our journey's end, the weather had not become thick & tempestuous. Much rain fell, & the heat & damp together were very oppressive: in the Poop cabin the thermometer however only stood at 81° or 82°. Keeling Is^d is one of the low circular Coral reefs, on the greater part of which matter has accumulated & formed strips of dry land. Within the chain of Is^{ds} there is an extensive shallow lake or lagoon. The reef is broken on the Northern side & there lies the entrance to the anchorage. The general appearance of the land at a distance is precisely similar to what I have mentioned at the Low Is^{ds} of the Pacifick. On entering the Lagoon the scene is very curious & rather pretty, its beauty is however solely derived from the brilliancy of the surrounding colors. The shoal, clear & still water of the lagoon, resting in its |714| greater part on white sand, is when illuminated by a vertical sun of a most vivid green. This brilliant expanse, which is several miles wide, is on all sides divided either from the dark heaving water of the ocean by a line of breakers, or from the blue vault of Heaven by the strip of land, crowned at an equal height by the tops of the Cocoa nut trees. As in the sky here & there a white cloud affords a pleasing contrast, so in the lagoon dark bands of living Coral are seen through the emerald green water. — Looking at any one & especially a smaller Islet, it is impossible not to admire the great elegant manner in which the young & full grown Cocoa-nut trees, without destroying each others sym-metry, mingle together into one wood: the beach of glittering white Calcareous sand, forms the border to these fairy spots.

When the ship was in the channel at the entrance, Mr Liesk, an English resident, came off in his boat. The history of the inhabitants of this place, is, in as few words as possible, as follows. About nine years ago a Mr Hare, a very worthless character, brought from the E. Indian

Archipelago a number of Malay slaves which now including children amount to more than a hundred. Shortly afterwards Capt. Ross, who had before visited these Isds in his merchant ship, arrived from England bringing with him his family & goods for Settlement. — Along with him came Mr Liesk, who had been a Mate in the same|715| ship. The Malay slaves soon ran away from the Is^d on which Mr Hare was settled & joined Capt. Ross's party: Mr Hare upon this was ultimately obliged to leave these Islands.[1] The Malays are now nominally in a state of freedom, & certainly so as far as respects their personal treatment; but in most other points they are considered as slaves. From the discontented state of the people, the repeated removals & perhaps from a little mismanagement, things are not very prosperous. The Island has no quadruped excepting pigs, & no vegetables in any quantity excepting Cocoa nuts. On this tree depends the prosperity of the Isl^d. — The only export is Cocoa nut oil. At this present time Capt. Ross has taken, in a small schooner which was built here, a cargo of this oil & that of the nuts to Singapore. He will bring back rice & goods for the Malays. — On the Cocoa nuts, the Pigs, which are loaded with fat, almost entirely subsist, as likewise do the poultry & ducks. Even a huge land-crab is furnished by nature with a curious instinct & form of legs to open & feed on the same fruit. There is no want of animal food at these Islands, for turtle & fish abound in the lagoon. — The situation of this Isl^d & its facilities for shipping must one day make it of some consequence, & then its natural|716| advantages will be more fully developed. The ship came to an anchor in the evening, but on the following morning was warped nearer to Direction or Rat Is^d. —

[1] Followed by a deleted sentence: 'Capt. Ross then occupied a more convenient place (which is now called Water Is^d) where all the inhabitants are collected.'

2nd I went on shore. The strip of dry land is only a few hundred yards wide; on the lagoon side we have the white beach, the radiation from which in such a climate is very oppressive; & on the outer coast a solid broad flat of coral rock, which serves to break the violence of the open ocean. Excepting near the lagoon where there is some sand, the land is entirely composed of rounded fragments of coral. In such a loose, dry, stony soil, nothing but the climate of the intertropical regions could produce a vigorous vegetation. Besides the Cocoa nut which is so numerous as at first to appear the only tree, there are five or six other kinds. One called the Cabbage tree, grows to a great bulk in proportion to its height, & has an irregular figure; its wood being very soft. Besides these trees the number of native plants is exceedingly limited; I suppose it does not exceed a dozen. Yet the woods, from the dead branches of the trees, & the arms of the Cocoa nuts is a thick jungle. — There are no true land birds; a snipe & land-rail are the only two "waders", the rest are all birds of the sea. Insects are very few in number; I must except some spiders & a small ant, which swarms in countless numbers in every spot & place. These strips of|717| land are raised only to the height to which during gales of wind the surf can throw loose fragments; their protection is due to the outward & lateral increase of the reef, which must break off the sea. The aspect & constitution of these Islets at once calls up the idea that the land & the ocean are here struggling for the mastery: although terra firma has obtained a footing, the denizens of the other element think their claim at least equal. In every part one meets Hermit-Crabs of more than one species,[1] The large claw or pincers of some of these crabs are most beautifully adapted, when drawn back to form an operculum to the shell, which is nearly as perfect as the proper one which the living molluscous animal formerly possessed. I was assured, and as far as my observation went, it was confirmed, that there are certain kinds of these hermits which always use certain kinds of old shells. —carrying on their backs the houses they have stolen from the neighbouring beach. Overhead, the trees are occupied by numbers of gannets, frigate birds & terns; from the many nests & smell of the air, this might be called a sea rookery; but how great the contrast with a rookery in the fresh budding woods of England! The gannets, sitting on their rude nests look at an intruder with a stupid yet angry air. The noddies, as their name expresses, are silly little creatures. But there is one charming bird, it is a small and snow white tern, which smoothly hovers at the distance of an arm's length from ones head, its large black eye scanning with quiet curiosity your expression. Little imagination is required to fancy that so light & delicate a body must be tenanted by some wandering fairy spirit.|718|

[1] The passage from 'The large claw & pincers . . .' to '. . . kinds of old shells' is written on a separate unnumbered page and marked to be inserted here.

Sunday 3rd   After service I accompanied Capt. FitzRoy to the Settlement. We found on a point thickly scattered over with tall Cocoa nut trees, the town. Capt Ross & Mr Liesk live in a large barn-like house open at both ends & lined with mats made of the woven bark: the houses of the Malays are arranged along the shore of the lagoon. The whole place bore rather a desolate air, because there were no gardens to show the signs of care & cultivation. The natives come from different islands of the East Indian Archipelago, but all speak the same language; we saw inhabitants of Borneo, Celebes, Java & Sumatra. In color of the skin they resemble the Tahitians, nor widely differ from them in form of features: some of the women, however, showed a good deal of the Chinese character. I liked both their general expression & the sound of their voices. They appeared poor & their houses were destitute of furniture; but it was evident from the plumpness of the little children, that cocoa nuts & turtle afford no bad sustenance. On this island the wells occur from which ships obtain water; at first sight it appears not a little remarkable that the fresh water regularly ebbs & flows with the usual tide. We must believe that the compressed sand & porous Coral rock act like a sponge, & that|719| the rain water which falls on the ground, being specifically lighter than the salt, merely floats on its surface & is subject to the same movements. There can be no actual attraction between salt & fresh water, & the spongy texture must tend to prevent all mixture from slight movements; on the other hand, where the land solely consists of loose fragments, a well being dug, salt or brackish water enters, of which fact we saw an instance.

After dinner we staid to see a half superstitious scene, acted by the Malay women. They dress a large wooden spoon in garments — carry it to the grave of a dead man — & then at the full of the moon they pretend it becomes inspired & will dance & jump about. After the proper preparations the spoon held by two women became convulsed & danced in good time to the song of the surrounding children & women. It was a most foolish spectacle, but Mr Liesk maintained that many of the Malays believed in its spiritual movements. The dance did not commence till the moon had risen & it was well worth remaining to behold her bright globe so quietly shining through the long arms of the Cocoa nuts, as they waved in the evening breeze. These scenes of the Tropics are in themselves so delicious, that they almost equal those dearer ones to which we are bound by each best feeling of the mind. —
|720|

4th   I was employed all the day in examining the very interesting yet simple structure & origin of these islands. The water being unusually smooth, I waded in as far as the living mounds of coral on which the swell of the open sea breaks. In some of the gullies & hollows, there were beautiful green & other colored fishes, & the forms & tints of many of the Zoophites were admirable. It is excusable to grow enthusiastic over the infinite numbers of organic beings with which the sea of the tropics, so prodigal of life, teems; yet I must confess I think those

Naturalists who have described in well known words the submarine grottoes, decked with a thousand beauties, have indulged in rather exuberant language. —

6th I accompanied Capt. FitzRoy to an island near the head of the Lagoon; the channel was exceedingly intricate, winding through fields of delicately branched Corals. We saw several turtle & two boats were then employed in catching them. — The method is rather curious; the water is so clear & shallow that although at first the turtle dives away with much rapidity, yet a canoe or a boat under sail will after no very long chase overtake it; a man standing ready in the bows at this moment dashes through the|721| water upon its back. Then clinging with both hands by the shell of the neck, he is carried away till the turtle becomes exhausted & is secured. It was quite an interesting chase to see the two boats doubling about, & the men dashing into the water till at last their prey was seized.

When we arrived at the head of the lagoon we crossed the island, & found a great surf breaking on the windward coast. I can hardly explain the cause, but there is to my mind a considerable degree of grandeur in the view of the outer shores of these Lagoon Islands. There is a simplicity in the barrier-like beach, the margin of green bushes & tall Cocoa nuts, the solid flat of Coral rock, strewed with occasional great fragments, & the line of furious breakers all rounding away towards either hand. The ocean throwing its waters over the broard reef appears an invincible all-powerful enemy, yet we see it resisted & even conquered by means which would have been judged most weak & inefficient. The little sketch of Whit sunday Is^d in Capt. Beechey's voyage, gives as accurate an idea of the scene as can be well imagined. We did not return on board till late in the evening, as we staid sometime in the lagoon, looking at the Coral fields & collecting specimens of the giant Chama. —|722|

7th–11th During these days nearly every one was employed in parts of the examination of the Island; but the winds being very strong rendered the most important part, the deep sea sounding, scarcely practicable. I visited Horsburgh & West Is^d. — In the latter the vegetation is perhaps more luxuriant than in any other part. Generally the Cocoa nut trees grow separate, but here the young ones flourish beneath their tall parents & formed with their long & curved fronds the most shady arbors. Those alone who have tried it, can tell how delicious it is to be seated in such shade & there drink the cool pleasant fluid of the Cocoa nut which close by hangs in great bunches. In this Is^d there is a large bay, or little lagoon, composed of the finest white sand; it is quite level & is only covered by the tide at high water. From this large bay smaller creeks penetrate the surrounding woods; thus to see a field of glittering sand representing water, & around the border of which the Cocoa nut trees extend their tall waving trunks, formed a singular & very pretty view.

12th In the morning we stood out of the Lagoon. I am glad we have visited these Islands; such formations surely rank high amongst the wonderful objects of this world. It|723| is not a wonder which at first strikes the eye of the body, but rather after reflection, the eye of reason. We feel surprised when travellers relate accounts of the vast piles & extent of some ancient ruins; but how insignificant are the greatest of them, when compared to the matter here accumulated by various small animals. Throughout the whole group of Islands, every single atom, even from the most minute particle to large fragments of rocks, bear the stamp of once having been subjected to the power of organic arrangement. Capt. FitzRoy at the distance of but little more than a mile from the shore sounded with a line 7200 feet long, & found no bottom.[1]

Hence we must consider this Isl^d as the summit of a lofty mountain; to how great a depth or thickness the work of the Coral animal extends is quite uncertain. If the opinion that the rock-making Polypi continue to build upwards, as the foundation of the Isl^d from volcanic agency, after intervals gradually subsides, is granted to be true; then probably the Coral limestone must be of great thickness. We see certain Is^{ds} in the Pacifick, such as Tahiti & Eimeo, mentioned in this journal, which are encircled by a Coral reef separated from the shore by channels & basins of still water. Various causes tend to check the growth of the most efficient kinds of Corals in these situations. Hence if we imagine such an Island, after long successive intervals|724| to subside a few feet, in a manner similar, but with a movement opposite to the continent of S. America; the coral would be continued upwards, rising from the foundation of the encircling reef. In time the central land would sink beneath the level of the sea & disappear, but the coral would have completed its circular wall. Should we not then have a Lagoon Island? — Under this view, we must look at a Lagoon Is^d as a monument raised by myriads of tiny architects, to mark the spot where a former land lies buried in the depths of the ocean. —[2]

The Beagle stood over to the Northern Is^d, distant about 12 miles. This likewise is a small Lagoon Is^d, but its centre is nearly filled up: the entrance is not deep enough even for a boat to enter. — The plan being completed; in the evening a course was taken for the Isle of France. —

[1] In *Journal of Researches* pp. 554–69, a detailed account follows of CD's theory of the formation of coral reefs. His ideas on this subject appear to have been first thought out while he was on the west coast of South America, and although he had seen coral reefs in Tahiti, the Cocos (Keeling) Islands were the first isolated coral islands that he visited. Writing to Caroline Darwin two weeks later he said: 'We then proceeded to the Keeling Is^{ds}:— These are low lagoon Is^{ds}: about 500 miles from the coast of Sumatra. — I am very glad we called there, as it has been our only opportunity of seeing one of those wonderful productions of the Coral polypi. — The subject of Coral formation has for the last half year, been a point of particular interest to me. I hope to be able to put some of the facts in a more simple & connected point of view, than that in which they have hitherto been considered. The idea of a lagoon Island, 30 miles in diameter being based on a submarine crater of equal dimensions, has alway appeared to me a monstrous hypothesis.' See *Correspondence* 1: 567–71.

[2] A footnote, probably in Hensleigh Wedgwood's hand (see *Correspondence* 1: 530), says: 'Good, but the 1st pt not quite clear.'

29th   In the morning we passed round the northern extremity of the Isle of France or Mauritius. From this point of view the aspect of the island equalled the expectations raised by the many well known descriptions of its beautiful scenery. The sloping plain of the Pamplemousses, scattered over with houses & coloured bright green from the large fields of sugar cane, composed the foreground. The brilliancy of the green|725| was the more remarkable because it is a colour which generally is only conspicuous from a very short distance. Towards the centre of the island groups of wooded mountains arose out of the highly cultivated plain, their summits, as so commonly happens with ancient volcanic rocks, being jagged by the sharpest points. Masses of white clouds were collected around these pinnacles, as if merely for the sake of pleasing the stranger's eye. The whole island, with its sloping border & central mountains, was adorned with an air of perfect elegance; — the scenery, If I may use such an expression, appeared to the senses harmonious. —

Shortly after midday we came to an anchor at Port Louis. —

30th   I spent the greater part of the next day in walking about the town & visiting different people. The town is of considerable size, & is said to contain 20,000 inhabitants; the streets are very clean & regular. Although the island has been so many years under the English government, the general character of the place is quite French. Englishmen speak to their servants in French, & the shops are all French; indeed I should think that Calais or Boulogne was much more Anglefied. There is a very pretty little theatre, in which operas are excellently performed, & are much preferred by the|726| inhabitants[1] to common plays. We were also surprised at seeing large booksellers shops with well stored shelves:—music & reading bespeak our approach to the old world of civilization, for in truth both Australia & America may be considered as New Worlds. — One of the most interesting spectacles in Port Louis is the number of men of various races which may be met with in the streets. Convicts from India are banished here for life; of them at present there are about 800 who are employed in various public works. Before seeing these people I had no idea that the inhabitants of India were such noble looking men; their skin is extremely dark,[2] and many of the older men had large moustachios & beards of a snow white colour; this, together with the fire of their expressions, gave to them an aspect quite imposing. The greater number have been banished for murder & the worst crimes; others for causes which can scarcely be considered as moral faults, such as for not obeying, from superstitious motives, the English Government & laws. I saw one man of high cast, who had been banished because he would not bear witness against his neighbour who had committed|727| some offence; this poor man was also remarkable as being a confirmed opium eater, of which fact his emaciated body & strange drowsy expression bore witness. These convicts are generally quiet & well conducted; from their outward conduct, their cleanliness, & faithful observance of their strange religious enactments, it was impossible to look at these men with the same eyes as at our wretched convicts in New S. Wales.— Besides such prisoners, large numbers of free people are yearly imported from India; for the planters feared that the negroes, when emancipated, would not work: from these causes the Indian population is very considerable. With respect to the negroes, they appeared a very inferior race of men to those of Brazil, & as I believe, of the W. Indies: they come from Madagascar & the Zanzibar coast. The great act of emancipation caused no excitement amongst these people; it seems a general opinion that at first when free, nothing will tempt them to undergo much labor. I was however surprised to find how little the few people with whom I conversed seemed to care about the subject. Feeling confident in a resource in the countless population of India, the result of the emancipation was here much less regarded than in the West Indies. —|728|

[1] Followed by the deleted words in brackets 'praise be to their tastes'.
[2] Followed by the deleted words: 'certainly deeper than an intermediate shade between an American & a Negro'.

Sunday May 1ˢᵗ   I took a quiet walk along the sea coast to the north of the town; the plain is there quite uncultivated, consisting of a field of black lava smoothed over with coarse grass & bushes, the greater part of which are mimosas. Capt. FitzRoy before arriving here said he expected the island would have a character intermediate between the Galapagos & Tahiti. This is a very exact comparison, but it will convey a definite idea to a very few excepting those on board the Beagle. It is a very pleasant country, but it has not the charms of Tahiti or the grandeur of a Brazilian landscape.

The next day I ascended La Pouce, a mountain so called from a thumb like projection, which rises close behind the town to a height of 2600 feet. M. Lesson in the voyage of the Coquille has stated that the central plain of the Island appeared like the basin of a grand crater, & that La Pouce & the other mountains once formed parts of a connected wall; thus it likewise appeared to me.[1] From our elevated position, we enjoyed an excellent view over this great mass of volcanic matter: the country on this side of the island appears pretty well cultivated, the whole being divided into fields & studded with farmhouses. I am, however, assured that of the whole land not more than a half is yet in a productive state; if such|729| is the case & considering the present great export of sugar, at some period this island when thickly peopled, will be of very great value. Since England took possession, which is only twenty five years ago, the export of sugar is said to have increased in the proportion of seventy five to one. — One great cause of this prosperity is due to the excellent roads & means of communication throughout the island. At the present day in the neighbouring island of Bourbon under the French Government, the roads are in the same miserable order as they were only a few years past in this place. The Macadamizing art has perhaps been of greater advantage to the colonies, even than to the parent country. Although the French residents must have largely profited by the increased prosperity of their island, yet the English government is far from popular. It seems unfortunate that among the higher order of French & English there appears to exist scarcely any intercourse.

[1] See Louis Isidore Duperrey, *Voyage autour du monde . . . sur la corvette . . . La Coquille 1822–5. Zoologie* par MM. Lesson et Garnot. Paris, 1826–30.

3ʳᵈ   In the evening Capt. Lloyd, the surveyor general so well known from his survey across the Isthmus of Panama, invited Mʳ Stokes & myself to his country house, which is situated on the edge of Wilheim plains & about six miles from the port. We staid at this delightful place two days;

being elevated nearly 800 ft above the sea, the air is pleasantly cool & fresh; & on every side there|730| are delightful walks. Close by a grand ravine extends which is about 500 ft deep, & worn through the slightly inclined streams of lava that have flowed from the central platform.

5th Capt. Lloyd took us to the Rivière Noire which is several miles to the southward, in order that I might examine some rocks of elevated coral. We passed through pleasant gardens & fine fields of sugar cane growing amidst huge blocks of lava. The roads were bordered by hedges of mimosa, & near many of the houses there were avenues of the Mango. Some of the views, where the peaked hills & the cultivated farms were seen together, were exceedingly picturesque, & we were constantly tempted to exclaim, "how pleasant it would be to pass one's life in such quiet abodes". — Capt. Lloyd possessed an elephant; he sent it half way on the road, that we might enjoy a ride in true Indian fashion. I should think, as is commonly said to be the case, that the motion must be fatiguing for a long journey. The circumstance which surprised me most was the perfectly noiseless step: the whole ride on so wonderful an animal was extremely interesting. This elephant is the only one at present on the island; but it is said that others will be sent for. —

9th In the evening we sailed from Port Louis on our way to the C. of Good Hope; since|731| leaving England I have not spent so idle & dissipated a time. I dined out almost every day in the week: all would have been very delightful, if it had been possible to have banished the remembrance of England. Pleasant as the society appeared to us, it was manifest even during our short visit that no small portion of jealousy, envy & hatred was common here, as in most other small societies. — Alas, there does not exist a terrestrial paradise where such feelings have not found an entrance!

31st In the evening came to an anchor in Simon's Bay. — In the early part of the passage we passed in sight of the south end of Madagascar; we subsequently made the coast of Africa at Natal, & from that part coasted along a considerable length of the southern shores. We lost a week near Cape Lagullas by contrary winds & a severe gale. — The little town of Simon's Bay offers but a cheerless aspect to the stranger. About a couple of hundred square whitewashed houses, with very few gardens & scarcely a single tree, are scattered along the beach at the foot of a lofty, steep, bare wall of horizontally stratified sandstone. —

June 1st There being nothing worth seeing here, I procured a gig & set
out for the Cape town, which is 22 miles distant. Both of these towns are
situated within the heads, but at opposite extremities of a range of
mountains, which|732| is joined to the mainland by a low sandy flat.
The road skirted the base of these mountains: for the first 14 miles the
country is very desert; & with the exception of the pleasure which the
sight of an entirely new vegetation never fails to communicate, there
was very little of interest. The view however of the mountains on the
opposite side of the flat, brightened by the declining sun, was fine.
Within seven miles of Cape town, in the neighbourhood of Wynberg, a
great improvement was visible. In this vicinity are situated all the
country houses of the more wealthy residents of the Capital. The
numerous woods of young Scotch firs & stunted oak trees form the
chief attraction of this locality; there is indeed a great charm in shade &
retirement after the unconcealed bleakness of a country like this. —[1]
The houses & plantations are backed by a grand wall of mountains
which gives to the scene a degree of uncommon beauty. I arrived late in
the evening in Cape Town, & had a good deal of difficulty in finding
quarters: in the|733| morning several ships from India had arrived at
this great inn on the great highway of nations, & they had disgorged on
shore a host of passengers, all longing to enjoy the delights of a
temperate climate. There is only one good hotel, so that all strangers
live in boarding houses—a very uncomfortable fashion to which I was
obliged to conform, although I was fortunate in my quarters. —

[1]Followed by a deleted sentence: 'Poor in dimensions as the trees generally were, &
planted in straight lines & clipped according to the old Dutch fashion, yet the appearance
of a turnpike road strewed with decaying oak leaves & smelling like a wood in the autumn
of England, was quite delightful.'

2nd In the morning I walked to a neighbouring hill to look at the town. It
is laid out with the rectangular precision of a Spanish city; the streets are
in good order & macadamized, & some of them have rows of trees on
each side; the houses are all white-washed & look clean. In several
trifling particulars the town has a foreign air; but daily it is becoming
more English. There is scarcely a resident in the town, excepting among
the lowest order, who does not speak some English; in this facility in
becoming Anglefied, there appears to exist a wide difference between
this colony & that of Mauritius. This however does not arise from the
popularity of the English, for the Dutch, as the French at Mauritius,
although having profited to an immense degree by the English government, yet thoroughly dislike our whole nation. In the country universally there is one price for a Dutchman, & another & much higher one,

for an Englishman; nevertheless some few of the Dutchmen have lately sent their sons|734| to England to learn a proper system of agriculture.[1]

All the fragments of the civilized world, which we have visited in the southern hemisphere, all appear to be flourishing; little embryo Englands are hatching in all parts. The Cape Colony, although possessing but a moderately fertile country, appears in a very prosperous condition. In one respect it suffers like New South Wales, namely in the absence of water communication, and in the interior being separated from the coast by a high chain of mountains. This country does not possess coal, & timber, excepting at a considerable distance, is quite deficient. Hides, tallow & wine, are the chief export, & latterly a considerable quantity of corn. The farmers are beginning also to pay attention to sheep grazing,|735| a hint taken from Australia. It is no small triumph to Van Diemens Land, that live sheep have been exported from a colony of thirty three years standing to this one, founded in 1651. —

In Cape town it is said the present number of inhabitants is about 15,000, and in the whole colony, including coloured people, 200,000. Many different nations are here mingled together; the Europæans consist of Dutch, French & English, & scattered people from other parts. The Malays, descendants of slaves brought from the East Indian archipelago, form a large body; they appear a fine set of men; they can always be distinguished by conical hats, like the roof of a circular thatched cottage, or by a red handkerchief on their heads. — The number of negroes is not very great, & the Hottentots, the ill treated aboriginals of the country, are, I should think, in a still smaller proportion. The first object in Cape town which strikes the eye of a stranger, is the number of bullock waggons; several times I saw eighteen & heard of twenty four oxen being all yoked together in one team;[2] Besides these, in all parts waggons with four, six, & eight horses in hand, go trotting about the streets. — I have as yet not mentioned the well known Table mountain; this great mass of horizontally stratified sandstone rises quite close behind the town to a height of|736| 3500 feet; the upper part forms an absolute wall, often reaching into the region of the clouds. I should think so high a mountain, not forming part of a platform & yet being composed of horizontal strata, must be a rare phenomenon; it certainly gives the landscape a very peculiar, & from some points of view, a grand character. —

[1] Followed by a deleted passage: 'One young man, who had just returned from Norfolk, proposed to his father to drain a large shallow lake; the Father with difficulty consented to so strange an idea, as to convert a lake into a cornfield. The plan, as might be expected,

succeeded well, but after the three first years the ground was so overrun with weeds, that the old gentleman perceived with joy, that the new fashion seemed likely to fail. The son however soon closed the drain, flooded the land, & so killed the weeds; the old gentleman was amazed at these unheard of expedients, but the next year reaped a fine crop of corn.'
[2]Followed by deleted words: 'the line looks as long as if all the cows in a field had been caught & tied together for sport.'

4th   I set out on a short excursion to see the neighbouring country, but I saw so very little worth seeing, that I have scarcely anything to say. I hired a couple of horses & a young Hottentot groom to accompany me as a guide. he spoke English very well, & was most tidily drest; he wore a long coat, beaver hat, & white gloves! The Hottentots, or Hodmadods as old Dampier calls them, to my eye look like partially bleached negroes; they are of small stature, & have most singularly formed heads & faces. The temple & cheek bones project so much, that the whole face is hidden from a person standing in the same side position, in which he would be enabled to see part of the features of a Europæan. Their hair is very short & curly.

Our first days ride was to the village of the Paarl, situated between thirty & forty miles to the NE of the Cape town. After leaving the neighbourhood of the town, where white houses stand as if picked out of a street & then by chance dropped down in the open country, we had to cross a wide level sandy flat totally unfit for cultivation. In the hopes of finding some hard materials, the|737| sands have been bored along the whole line of road to the depth of forty feet. Leaving the flat, we crossed a low undulating country thinly clothed with a slight, green vegetation. This is not the flowering season, but even at the present time, there were some very pretty oxalis's & mesembryanthemums, & on the sandy spots, fine tufts of heaths. Even at this short distance from the coast, there were several very pretty little birds. — If a person could not find amusement in observing the animals & plants, there was very little else during the whole day to interest him: only here & there we passed a solitary white farm house.

Directly after arriving at the Paarl, I ascended a singular group of rounded granite hills close behind the village. I enjoyed a fine view from the summit; directly in front extended the line of mountains which I had to cross on the following morning. Their colors were grey, or partly rusty red, their outlines irregular but far from picturesque. The general tint of the lower country was a pale brownish green & the whole entirely destitute of wood-land. In the naked state of the mountains, seen also through a very clear atmosphere, I was reminded of Northern Chili, but the rocks there possess at least a brilliant colouring.

Immediately beneath the hill on which I was standing the long village of the|738| Paarl extended, all the houses were very tidy & comfortable & white-washed; there was not a single hovel. Each house had its garden & a few trees planted in straight rows, & there were many considerable vineyards being at this time of year destitute of leaves. The whole village possessed an air of quiet & respectable comfort. —

5th After riding about three hours, we came near to the French Hoeck pass. This is so called from a number of emigrant protestant Frenchmen having originally settled in a flat valley at the foot of the mountain: it is one of the prettiest places I saw in the colony. — The pass is a considerable work, an inclined road having been cut along the steep side of the mountain: it forms one of the principal roads from the low land of the coast to the mountains & great plains of the interior. We reached the foot of the mountains on the opposite or SE side of the pass a little after noon; here at the Toll-bar we found comfortable lodgings for the night. The surrounding mountains were destitute of trees & even of brushwood, but they supported a scattered vegetation of rather a brighter green than usual, the quantity however of white siliceous sandstone which every where protruded itself uncovered, gave to the country a bleak & desolate aspect. — |739|

6th My intention was to return by Sir Lowry Cole's pass, over the same chain of mountains but a little further to the South. Following unfrequented paths we crossed over an irregular hilly country to the other line of road. During the whole long day I met scarcely a single person, & saw but few inhabited spots or any number of cattle. A few Raebucks were grazing on the sides of the hills, & some large dirty white Vultures like the Condor of America slowly wheeled over the place where probably some dead animal was lying. — There was not even a tree to break the monotomous uniformity of the sandstone hills. I never saw a much less interesting country. — At night we slept at the house of an English farmer. —

7th At an early hour the next day we descended by Sir Lowry's pass, which like the last has been cut at much expence along the side of a steep mountain. From the summit there was a noble view of the whole of False Bay & of the Table mountain; & immediately below, of the cultivated country of Hottentot Holland. — The flat, covered with sand dunes, did not appear from the height of the tedious length, which we found it to be before reaching in the evening Cape town. —

8th–15th During these days I became acquainted with several very pleasant people:|740| With Dr A. Smith, who has lately returned from his most interesting expedition to beyond the Tropic, I took some long geological rambles. — I dined out several days, — with Mr Maclear (the astronomer), with Colonel Bell, and with Sir J. Herschel; this last was the most memorable event which, for a long period, I have had the good fortune to enjoy. —[1]

[1] In his *Autobiography* p. 107, CD wrote: 'I felt a high reverence for Sir J. Herschel, and was delighted to dine with him at his charming house at the C. of Good Hope and afterwards at his London house. I saw him, also, on a few other occasions. He never talked much, but every word which he uttered was worth listening to. He was very shy and he often had a distressed expression. Lady Caroline Bell, at whose house I dined at the C. of Good Hope, admired Herschel much, but said that he always came into a room as if he knew that his hands were dirty, and that he knew that his wife knew that they were dirty.' See also CD's letter to Henslow of 9 July, *Correspondence* 1: 500.

16th Returned to Simon's bay; the bad weather having set in caused our stay to be rather longer here than usual. —

17th Took a long walk with Mr Sulivan to examine several interesting features in the geology of the surrounding mountains. —

18th In the afternoon put to sea; our usual ill fortune followed us; first with a gale of wind, & then with scarcely any wind at all.

29th The Beagle crossed the Tropic of Capricorn for the sixth & last time. — We were surprised & grieved by finding light northerly breezes, within limits generally occupied by a strong trade wind. —

July 8th In the morning arrived off St Helena. This island, the forbidding aspect of which has been so often described, rises like a huge castle from the ocean. A great wall, built of successive streams of black lava, forms around its whole circuit, a bold coast. — Near to the town, as if in aid of the natural defence, small forts & guns are everywhere built up & mingled with the rugged rocks. The town extends up a flat|741| & very narrow valley; the houses look respectable & from among them a few green trees arise. When approaching the anchorage, there is one striking view; an irregular castle perched on the summit of a lofty hill & surrounded by a few scattered fir trees, boldly projects against the sky. — It is called High Knoll hill. —

9th–13th I obtained lodgings in a cottage within stone's throw of Napoleon's tomb.[1] I confess this latter fact possessed with me but little inducement. The one step between the sublime & the ridiculous has on this subject been too often passed. Besides, a tomb situated close by cottages & a frequented road does not create feelings in unison with the imagined resting place of so great a spirit. — With respect to the house in which Napoleon died, its state is scandalous, to see the filthy & deserted rooms, scored with the names of visitors, to my mind was like beholding some ancient ruin wantonly disfigured. — During the four days I staid in this central position, from morning to night I wandered over the Is^d & examined its geological history. The house was situated at an elevation of about 2000 ft; here the weather was cold & very boisterous, with constant showers of rain;—every now & then the whole scene was veiled by thick clouds.

Near to the coast the rough lava is entirely destitute of vegetation; in the central & higher parts a different series of rocks have, from extreme decomposition, produced a clayey soil which is stained in broad ^{2} bands|742| of many colours, such as purple, red, white & yellow. — At this season, the land moistened by constant showers produces a singularly bright green pasture; this lower & lower down gradually fades away & at last disappears. — In latitude 16° & at the trifling elevation of 1500 ft, it is surprising to behold a vegetation possessing a decided English character. But such is the case; the hills are crowned with irregular plantations of scotch firs; the sloping banks are thickly scattered over with thickets of gorze, covered with its bright yellow flowers; along the course of the rivulets weeping willows are common, & the hedges are formed of the blackberry, producing its well known fruit. When we consider the proportional numbers of indigenous plants being 52, to 424 imported species, of which latter so many come from England, we see the cause of this resemblance in character.[3] These numerous species, which have been so recently introduced, can hardly have failed to have destroyed some of the native kinds. I believe there is not any account extant of the vegetation at the period when the island was covered with trees; such would have formed a most curious comparison with its present sterile condition and limited Flora. It is not improbable that even at the present day similar changes may be in progress. — Many English plants appear to flourish here better than in their native country; some also from the opposite quarter of Australia succeed remarkably well, & it is only on the highest & steep mountain crests where the native Flora is predominant. The English, or rather the Welsh character of the scenery, is kept up by the numerous cottages & small white houses, some buried at the bottom of the deepest valleys & others stuck up on the lofty ridges. — Certainly some of the views are very striking; I may instance that of Sir W. Doverton's house, where the bold peak called Lott is seen over a dark wood of firs, the whole being backed by the red, waterworn mountains of the|743| Southern shore. —

But a glowing tropical style of landscape would have afforded a finer contrast than the homely English scenery, with the wile arid rocks of the coast.

On viewing the Is^d from an eminence, the first remark which occurs is on the infinite number of roads & likewise of forts. The public expenses, if one forgets its character as a prison, seems out of all proportion to the extent or value of the Island. So little level or useful land is there, that it seems surprising how so many people (about 5000) can subsist. The lower orders, or the emancipated slaves, are I believe extremely poor; they complain of want of work; a fact which is also shewn by the cheap labour. — From the reduction in number of public servants owing to the island being given up by the East Indian Company & consequent emigration of many of the richer people, the poverty probably will increase. — The chief food of the working class is rice with a little salt meat; as these articles must be purchased the low wages tell heavily: — the fine times, as my old guide called them, when "Bony" was here, can never again return. — Now that the people are blessed with freedom, a right which I believe they fully value, it seems probable their numbers will quickly increase: if so, what is to become of the little state of St Helena? —

My guide was an elderly man, who had been a goatherd when a boy, & knew every step amongst the|744| rocks. He was of a race many times mixed, & although with a dusky skin, he had not the disagreeable expression of a Mulatto: he was a very civil, quiet old man, & this appears the character of the greater part of the lower class. — It was strange to my ears to hear a man nearly white, & respectably dressed, talking with indifference of the times when he was a slave. — With my companion, who carried our dinners & a horn of water, which latter is quite necessary, as all in the lower valleys is saline, I every day took long walks. Beyond the limits of the elevated & central green circle, the wild valleys are quite desolate & untenanted. Here to the geologist, there are scenes of interest, which shew the successive changes & complicated violence, which have in past times happened. According to my views, St Helena has existed as an Is^d from a very remote period, but that originally like most Volcanic Isds it has been raised in mass from beneath the waters.[4] St Helena, situated so remote from any continent, in the midst of a great ocean & possessing an unique Flora, this little world, within itself excites our curiosity. — Birds & insects, as might be expected, are very few in number, indeed I believe all the birds have been introduced within late years. — Partridges & pheasant are tolerably abundant; the Is^d is far too English not to be subject to strict game laws. I was told of a more unjust sacrifice to|745| such ordinances, than I ever heard of even in England: the poor people formerly used to burn a plant which grows on the coast rocks, & export soda;—a peremptory order came out to prohibit this practice, giving as a reason, that the Partridges would have no where to build!—

In my walks, I passed more than once over the grassy plain bounded by deep valleys, on which stands Longwood.— Viewed from a short distance, it appears like a respectable gentleman's country seat. In front there are a few cultivated fields, & beyond them at some distance the hill of coloured rocks called the Flagstaff, & the square black mass of the Barn. The view is rather bleak & uninteresting. —

It is quite extraordinary, the scrupulous degree to which the coast must formerly have been guarded. There are alarm houses, alarm guns & alarm stations on every peak.— I was much struck with the number of forts & picket houses on the line leading down to Prosperous Bay; one would suppose this at least must be an easy descent. I found it, however, a mere goat path, & in one spot the use of ropes which are fixed into rings in the cliff, were almost indispensable. — At the present day two artillery men are kept there, for what use it is not easy to conjecture. Prosperous Bay, although with so flourishing a name, has nothing more attractive than a wild sea beach & black utterly barren rocks. In|746| some other situations, which were formerly no doubt important, a couple of invalids were stationed; really the places are sufficient to kill the poor men with ennui & melancholy.— The only inconvenience I suffered in my walks was from the impetuous winds. One day I noticed a curious fact; standing on the edge of a plain terminated by a great cliff of about a thousand feet elevation, I saw at the distance of a few yards, right to windward, some Tern struggling against a very strong breeze, whilst where I stood the air was quite calm. Approaching close to the brink I stretched out my arm, which immediately felt the full force of the wind. An invisible barrier of two yards wide, separated a strongly agitated from a perfectly calm air. — The current meeting the bold face of the cliff must have been reflected upwards at a certain angle, beyond which there would be an eddy, or a calm. —

[1] For Syms Covington's account of Napoleon's tomb, see an extract from his diary and a drawing that he made of the tomb, in *Beagle Record* pp. 359–62.

[2] As pointed out by Sulloway (1983), this is the first occasion in the Diary on which CD definitely abandoned the spelling 'broard'.

[3] The passage from 'These numerous species . . .' to '. . . may be in progress' was written on a separate unnumbered page, and marked to be inserted here.

[4] A note in the margin runs: 'Covington. Two pages marked XX(a) and XX(b).' These pages have been lost, but in *Journal of Researches* pp. 581-3 a passage is interpolated dealing with the changes in the fauna and flora of the island since the introduction of goats and hogs in 1502, and the presence of a species of land shell, *Bulimus*, in the higher parts.

14th   I so much enjoyed my rambles amongst the rocks & mountains, that I almost felt sorry on the morning of the 14th, to descend to the town. — Before noon I was on board, & the Beagle made sail for Ascension. —

19th   Reached the anchorage in the afternoon, & received some letters. This alone with such a surrounding scene, was capable of producing pleasant sensations. Those who have beheld a volcanic Island, situated within an arid climate,|747| will be able at once to picture to themselves the aspect of Ascension. They will imagine smooth conical hills of a bright red colour, with their summits generally truncated, rising distinct out of a level surface of black horrid lava. — A principal mound in the centre of the Island seems the father of the lesser cones. It is called Green Hill, its name is taken from the faintest tinge of that colour, which at this time was barely perceptible from the anchorage. To complete this desolate scene, the black rocks on the coast are lashed by a wild turbulent sea. The settlement is near the beach, it consists of several houses & barracks, placed irregularly but well built of white freestone. The only inhabitants are Marines & some negroes liberated from slave ships, who are paid & victualled by government: there is not a private person on the island. Many of the Marines appeared well contented with their situation: they think it better to serve their one & twenty years on shore, let it be what it may, than in a Ship. — With which choice, if I was a Marine, I should most heartily agree.

20th   The next morning I ascended Green Hill, 2840 ft high, & walked from thence across the Is^d to the windward point. — A good cart road leads from the coast settlement to the houses, gardens & fields placed near the summit of the central mountain. On the road side are milestones & cisterns, where each|748| thirsty passer by can drink some good water. Similar care is displayed in each part of the establishment, & especially in the management of the Springs, so that a single drop of water shall not be lost. Indeed the whole Isl^d may be compared to [a] huge Ship kept in first rate order. I could not help, when admiring the active industry which has created such effects out of such means, at the same time regretting that it was wasted on so poor & trifling an end. — M. Lesson has remarked with justice that the English nation alone would ever have thought of making the Is^d of Ascension a productive spot; any other people would have held it, without any further views, as a mere fortress in the ocean.

Near the coast, nothing grows, a little inland, an occasional green Castor oil plant & a few grasshoppers, true friends of the desert, may be met with. On the central elevated parts, some grass is scattered over the surface, much resembling the worse parts of the Welsh mountains. But scanty as it appears, about six hundred sheep, many goats, a few cows & horses, all thrive well. Of native animals, rats, mice, land-crabs are abundant:—of Birds the guinea-fowl imported from the C. Verd's, swarm in great numbers.— The Is^d is entirely destitute of trees, in which & in every other respect it is|749| very far inferior to St Helena. Mr Dring tells me that the witty people of the latter place say "We know we live on a rock, but the poor people at Ascension live on a cinder": the distinction is in truth very just.

21st & 22nd   On the two succeeding days I took long walks & examined some rather curious points in the mineralogical composition of some of the Volcanic rocks, to which I was guided by the kindness of Lieut. Evans. One day I walked to the SW extremity of the Isl^d: the day was clear & hot, & I saw the Island not smiling with beauty, but staring with naked hideousness.— The lava streams are covered with hummocks, & are rugged to a degree which geologically speaking is not of easy explanation. The intervening spaces are concealed with layers of pumice, ashes, & volcanic sandstone. In some parts, rounded volcanic bombs, which must have assumed this form when projected red hot from the crater, lie strewed on the surface. When passing this end of the Isl^d at sea, I could not imagine the cause of the white patches, with which the whole plain was mottled: I now found out it was owing to the number of seafowl, which sleep in such full confidence, as even in midday to allow a man to walk up to & seize hold of them. These birds were the only living creatures I this day saw. On the beach a great sea, although the breeze was light, was tumbling over the|750| broken lava rocks.— The ocean is a raging monster, insult him a thousand miles distant, & his great carcase is stirred with anger through half an hemisphere.—

23rd   In the afternoon put to sea.— When in the offing, the Ships head was directed in W.S.W. course—a sore discomfiture & surprise to those on board who were most anxious to reach England. I did not think again to see the coast of S. America; but I am glad our fate has directed us to Bahia in Brazil.—

August 1ˢᵗ Anchored in Bahia de todos los Santos. The first aspect of the
city & its outskirts, with the beauties of which we were formerly so
much delighted, had lost part of its charms. The novelty & surprise
were gone, & perhaps our memories had, in the long interval, exaggerated the colours of the scenery. There existed, however, as we
afterwards discovered, a more true reason, in the loss of some of the
finest Mango trees, which during the late disturbances of the negroes
had been cut down. We staid here four days, in which time I took
several long walks. I was glad to find my enjoyment of tropical scenery,
from the loss of novelty, had not decreased even in the slightest
degree.[1]— The elements of the scenery are so simple, that they are
worth mentioning as a proof on what|751| trifling circumstances exquis-
ite natural beauty depends. The country may be described as a quite
level plain of about three hundred feet elevation which has been in
every part worn into flat-bottomed valleys. This structure is remarkable
in a granitic land, but it is nearly universal in all those softer formations,
of which plains usually are composed. The whole surface is covered by
various kinds of stately trees, interspersed with patches of cultivated
ground, amidst which stand houses, convents & Chapels. — It must be
remembered that within the tropics, the wild luxuriance of nature is not
lost, even in the vicinity of large cities; the natural vegetation of the
hedges & hill sides overpowers in picturesque effect, the artificial labor
of man. Hence in but few parts, the bright red soil affords a strong
contrast to the universal clothing of green. From the edges of the plain
there are distant glimpses either of the ocean or of the great bay,
bordered by low wooded shores, & on the surface of which numerous
boats & canoes show their white sails. Excepting from these points, the
range of vision is very limited; following the level pathways, on each
hand alternate peeps into the wooded valleys below can alone be
obtained. Lastly I must add, that the houses & especially the sacred
edifices are built in a peculiar & rather fantastick style of architecture.
They are all white-washed, so that when illuminated by the brilliant
sun of midday & as seen against the pale blue|752| sky of the horizon,
they stand out more like shadows than substantial buildings. Such are
the elements, but to paint their effects is an hopeless endeavour. —
Learned naturalists describe these scenes of the Tropics by naming a
multitude of objects & mentioning some characteristic feature of each.
To a learned traveller, this possibly may communicate some definite
ideas; but who else from seeing a plant in an herbarium can imagine its
appearance when growing in its native soil? Who, from seeing choice

plants in a hot house, can multiply some into the dimensions of forest trees, or crowd others into an entangled mass? Who, when examining in a cabinet the gay butterflies, or singular Cicadas, will associate with these objects the ceaseless harsh music of the latter, or the lazy flight of the former—the sure accompaniments of the still glowing noon day of the Tropics— It is at these times, when the sun has attained its greatest height, that such views should be beheld. Then the dense splendid foliage of the Mango hides the ground with its darkest shade, whilst its upper branches are rendered the more brilliant by the profusion of light. In the temperate zones, as it appears to me, the case is different, the colours there are not so dark, or rich, & hence the declining sun, which casts forth red, purple or yellow rays, is best adapted to add beauties to the scenery of those climes. — |753|

When quietly walking along the shady pathways & admiring each successive view, one wishes to find language to express ones ideas: epithet after epithet is found too weak to convey to those who have not had an opportunity of experiencing these sensations, a true picture of the mind. — I have said the plants in a hot-house fail to communicate a just idea of the vegetation, Yet I must recur to it: the land is one great wild, untidy, luxuriant hot house, which nature made for her menagerie, but man has taken possession of it, & has studded it with gay houses & formal gardens. — How great would be the desire in every admirer of nature to behold, if such was possible, another planet; yet at the distance of a few degrees from his native country, it may be truly said, the glories of another world are open to him. — In the last walk I took, I stopped again and again to gaze on such beauties, & tried to fix for ever in my mind, an impression which at the time I knew must sooner or later fade away. The forms of the Orange tree, the Cocoa nut, the Palms, the Mango, the Banana, will remain clear & separate, but the thousand beauties which unite them all into one perfect scene, must perish: yet they will leave, like a tale heard in childhood, a picture full of indistinct, but most beautiful figures. |754|

[1]Followed by a deleted sentence: 'I can truly say that I have never in my life relished a keener pleasure, than whilst gazing on some of these charming views.'

6th In the afternoon weighed anchor & stood out to sea. —

12th The weather having been unfavourable, we altered course & ran for Pernambuco. We anchored outside; but in a short time a pilot came on board & took us into the inner harbor, where we lay close to the town.[1] Pernambuco is built on some narrow, low, sand banks, which are separated from each other by shoal channels of salt water. The three parts of the town are connected together by two long bridges, built on wooden piles. The town is in all parts disgusting, the streets narrow, ill-paved, filthy, the houses very tall & gloomy. The number of white people, which during the morning may be met with in the streets, appears to be about in the proportion of foreigners in any other nation; all the rest are black or of a dusky colour. The latter as well as the Brazilians are far from prepossessing in their appearance: the poor negroes, wherever they may be, are cheerful, talkative & boisterous. There was nothing in the sight, smell or sounds within this large town, which conveyed to me any pleasing impressions. The season of heavy rains scarcely had come to an end & hence the surrounding country, which is scarcely elevated about the level of the sea, was flooded with water.|755| I failed in all my attempts to take any long walks. — I was however enabled to observe that many of the country houses in the outskirts were like those of Bahia, of a gay appearance which harmonized well with the luxuriant character of the tropical vegetation.

The flat swampy land is surrounded at the distance of a few miles by a semicircle of low hills, or rather by the edge of a country elevated perhaps two hundred feet above the sea. The old city of Olinda stands on one extremity of this range. One day I took a canoe & proceeded up one of the channels to visit it; I found the old town from its situation both sweeter & cleaner than that of Pernambuco. — I must commemorate, as being the first time during the four & a half years we have been wandering about, that I met with a want of politeness amongst any class of people; I was refused in a sullen manner at two different houses, & obtained with difficulty from a third permission to pass through their gardens to an uncultivated hill for the purpose of taking a view of the country. I feel quite glad this happened in the land of the "Brava Gente"; for I bear them no good will. — A Spaniard would have been ashamed at the very thought of refusing such a request, or of behaving to any one with rudeness. — The channel by which we came to & returned from Olinda is bordered on each side by Mangroves|756| which spring like a miniature forest out of the greasy mud banks. the bright green color of these bushes always reminds me of the rank grass in a Church-yard: both are nourished by putrid exhalations; the one speaks of death past, the other too often of death to come. —

The most curious thing which I saw in the neighbourhood of Pernambuco, is the reef that forms the harbor. It runs for a length of several miles in a perfectly straight line, parallel to & not far distant from the shore; it varies in width from thirty to sixty yards; it is quite dry at low water, has a level smooth surface, & is composed of obscurely stratified hard sandstone: hence at the first sight it is difficult to credit that it is the work of nature & not of art. Its utility is great; close within the inner water, there is a good depth of water, & ships lie moored to old guns, which are fixed in holes on the summit. — A light-house stands on one extremity, & around it the sea breaks heavily. In entering the harbor, a ship passes within thirty yards round this point, & amidst the foam of the breakers; close by, on the other hand, are other breakers, which thus form a narrow gateway: it is almost fearful to behold a ship running, as it appears, headlong into such dangers.

With respect to the origin of the reef, I believe,|757| a bar composed of sand & pebbles formerly existed beneath the water, when the low land on which the town now stands was occupied by a large bay; & that this bar was first consolidated, & then elevated. These two distinct processes are of so common occurrence in S. America, that I now feel none of that surprise, with which such facts would formerly have startled me.[2] There is another & slightly different explanation, which possesses equal probability, namely that a long spit of sand like some that now exist on the neighbouring coast, had its central part consolidated, & then by a slight change in the set of currents the loose matter was removed, the hard nucleus alone remaining. Although the swell of the open ocean breaks heavily on the outer side of the narrow & insignificant line of reef, yet there is no record of its decay. This durability is the most curious circumstance connected with its existence: it appears to be owing to a layer of calcareous matter, formed by the successive growth of several kinds of organic bodies, chiefly serpulae, balani, corallinae, but no true corals. It is a process strictly analogous to the formation of peat, & like that substance, its effects are to preserve from degradations the matter on which it rests. — In true coral reefs, when the upper extremities of the living mass are killed by the rays of the sun, they become enveloped & protected by a nearly similar process. It is probable that if a Breakwater such as that of Plymouth, was built in these tropical seas, it would be imperishable, that is, as imperishable as any part of the solid land, which all, some day, must suffer decay & renovation.

[1] Followed by a deleted sentence: 'On the ensuing day I took up my residence on shore in a Brazilian inn. —'
[2] The sentence which follows was written on a separate unnumbered page, and marked to be inserted here.

17th I was delighted on the 17th to get on board the ship & in the afternoon to leave the shores of Brazil. We lie close hauled to the wind, & therefore there is a considerable pitching motion; I suffer very much from |758| sea-sickness. — But it is on the road to England; in truth some such comfort is necessary to support the tedious misery of loss of time, health & comfort. —

21st We crossed the Equator. —

31st–September 4th After a most excellent passage, we came to an anchor early in the morning at Porto Praya. We found lying there, as commonly is the case, some Slaving vessels. The weather, during our short stay of four days was very fine, but as this was the beginning of the unhealthy season, I confined my walks to short distances.

I have nothing to say about the place; as some rain had fallen, a most faint tinge of green was just distinguishable. Our old friend the great Baobab tree was clothed with a thick green foilage, which much altered its appearance. As might be expected, I was not so much delighted with St Jago, as during our former visit; but even this time I found much in its Natural History very interesting. It would indeed be strange if the first view of desert volcanic plains, (a kind of country so utterly different from anything in England) and the first sensations on entering an ardent climate, did not excite the most vivid impressions in the mind of every one, who takes pleasure in beholding the face of nature.

4th We were all very glad in the evening of the 4th to wish farewell to the irregular mountains of St Jago, as they disappeared in the|759| evening shades. I confess, I feel some good will to the Island; I should be ungrateful if it was otherwise; for I shall never forget the delight of first standing in a certain lava cavern & looking at the swell of the Atlantic lashing the rugged shores. —

9th Crossed the Tropic of Cancer.

20th In the morning we were off the East end of the Island of Terceira, and a little after noon reached the town of Angra. The island is moderately lofty & has a rounded outline with detached conical hills evidently of volcanic origin. The land is well cultivated, & is divided into a multitude of rectangular fields by stone walls, extending from the water's edge to high up on the central hills. There are few or no trees, & the yellow stubble land at this time of year gives a burnt up and unpleasant character to the scenery. Small hamlets & single white-washed houses are scattered in all parts. In the evening a party went on shore;—[1] We found the city a very clean & tidy little place, containing about 10,000 inhabitants, which includes nearly the fourth part of the total number on the island. There are no good shops, & little signs of activity, excepting the intolerable creaking of an occasional bullock waggon. The churches are very respectable, & there were formerly a good many convents: but Dom Pedro destroyed several;|760| he levelled three nunneries to the ground, & gave permission to the nuns to marry, which, excepting by some of the very old ones, was gladly received. — Angra was formerly the capital of the whole archipelago, but it has now only one division of the islands under its government, and its glory has departed. The city is defended by a strong castle & line of batteries which encircle the base of Mount Brazil, an extinct volcano with sloping sides, which overlooks the town. — Terceira was the first place that received Dom Pedro, & from this beginning he conquered the other islands & finally Portugal. A loan was scraped together in this one island of no less than 400,000 dollars, of which sum not one farthing has ever been paid to these first supporters of the present right royal & honourable family.

[1]Followed by deleted words: 'to the town or rather city of Angra, the capital of the neighbouring islands'.

21st The next day the Consul kindly lent me his horse & furnished me with guides to proceed to a spot, in the centre of the island, which was described as an active crater. — Ascending in deep lanes, bordered on each side by high stone walls, for the three first miles, we passed many houses and gardens. We then entered on a very irregular plain country, consisting of more recent streams of hummocky basaltic lava. The rocks are covered in some parts by a thick brushwood about three feet high, and|761| in others by heath, fern, & short pasture: a few broken down old stone walls completed the resemblance with the mountains of Wales. I saw, moreover, some old English friends amongst the insects, and of birds, the starling, water wagtail, chaffinch and blackbird. There are no houses in this elevated and central part, and the ground is only used for the pasture of cattle and goats. On every side, besides the ridges of more ancient lavas, there were cones of various dimensions, which yet partly retained their crater-formed summits, and where broken down showed a pile of cinders such as those from an iron foundry. — When we reached the so called crater, I found it a slight depression, or rather a short valley abutting against a higher range, and without any exit. The bottom was traversed by several large fissures, out of which, in nearly a dozen places, small jets of steam issued, as from the cracks in the boiler of a steam engine.

The steam close to the irregular orifices, is far too hot for the hand to endure it;—it has but little smell, yet from everything made of iron being blackened, and from a peculiar rough sensation communicated to the skin, the vapour cannot be pure, and I imagine it contains some muriatic acid gas.— The effect on the surrounding trachytic lavas is singular, the solid stone being entirely converted either into |762|pure, snow white, porcelain clay, or into a kind of bright red or the two colours marbled together: the steam issued through the moist and hot clay. This phenomenon has thus gone on for many years; it is said that flames once issued from the cracks. During rain, the water from each bank, must flow into these cracks; & it is probable that this same water, trickling down to the neighbourhood of some heated subterranean lava, causes this phenomenon.— Throughout the island, the powers below have been unusually active during the last year; several small earthquakes have been caused, and during a few days a jet of steam issued from a bold precipice overhanging the sea, not far from the town of Angra.

I enjoyed my day's ride, though I did not see much worth seeing: it was pleasant to meet such a number of fine peasantry; I do not recollect ever having beheld a set of handsomer young men, with more good humoured pleasant expressions.[1] The men and boys are all dressed in a plain jacket & trowsers, without shoes or stockings; their heads are barely covered by a little blue cloth cap with two ears and a border of red; this|763| they lift in the most courteous manner to each passing stranger. Their clothes although very ragged, appeared singularly clean, as well as their persons; I am told, that in almost every cottage, a visitor will sleep in snow white sheets & will dine off a clean napkin. Each man carries in his hand a walking staff about six feet high; by fixing a large knife at each extremity, they can make this into a formidable weapon.— Their ruddy complexions, bright eyes & erect gait, made them a picture of a fine peasantry: how different from the Portugeese of Brazil!— The greater number, which we this day met, were employed in the mountains gathering sticks for fire-wood.— A whole family, from the father to the least boy, might be seen, each carrying his bundle on his head to sell in the town. Their burthens were very heavy; this hard labour & the ragged state of their clothes too plainly bespoke poverty, yet I am told, it is not the want of food, but of all luxuries, a case parallel to that of Chiloe.— Hence, although the whole land is not cultivated, at the present time numbers emigrate to Brazil, where the contract to which they are bound, differs but little from slavery. It seems

a great pity that so fine a population should be compelled [to] leave a
land of plenty, where every article of food, meat, vegetables & fruit, — is
exceedingly cheap & most abundant,[2]|764| but the labourer finds his
labour of proportionally little value. —

[1] Followed by deleted sentence: 'A surprising number of the boys had white or lightly
coloured hair, which from its strangeness to our eyes made it the more pleasing.'
[2] The following page is numbered both 764 and 765.

22nd   I staid the greater part of the day on board.[1]

[1] The entry for this day has been marked for deletion.

23rd   Another day I set out early in the morning to visit the town of Praya
seated on the NE end of the island. — The distance is about fifteen miles;
the road ran the greater part of the way not far from the coast. The
country is all cultivated & scattered with houses & small villages. I
noticed in several places, from the long traffic of the bullock waggons,
that the solid lava, which formed in parts the road, was worn into ruts
of the depth of twelve inches. This circumstance has been noticed with
surprise, in the ancient pavement of Pompeii, as not occurring in any of
the present towns of Italy. At this place the wheels have a tire sur-
mounted by singularly large iron knobs, perhaps the old Roman wheels
were thus furnished. The country during our morning's ride, was not
interesting, excepting always the pleasant sight of a happy peasantry.
The harvest was lately over, & near to the houses the fine yellow heads
of Indian corn, were bound, for the sake of drying, in large bundles to
the stems of the poplar trees. These seen from a distance, appeared
weighed down by some beautiful fruit, — the very emblem of fertility. —
One part of the road crossed a broad stream of lava, which from its|766|
rocky & black surface, showed itself to be of comparatively recent
origin; indeed the crater whence it had flowed could be distinguished.
The industrious inhabitants, have turned this space into vineyards, but
for this purpose it was necessary to clear away the loose fragments &
pile them into a multitude of walls, which enclosed little patches of
ground a few yards square; thus covering the country with a network
of black lines. —

   The town of Praya is a quiet forlorn little place; Many years since a
large city was here overwhelmed by an earthquake. It is asserted the
land subsided, and a wall of a convent now bathed by the sea is shown
as a proof: the fact is probable, but the proof not convincing. I returned
home by another road, which first leads along the Northern shore, &
then crosses the central part of the Island. — This North Eastern

extremity is particularly well cultivated, & produces a large quantity of fine wheat. The square, open fields, & small villages with white washed churches, gave to the view as seen from the heights, an aspect resembling the less picturesque parts of central England. — We soon reached the region of clouds, which during our whole visit have hung very low & concealed the tops of the mountains. For a couple of hours we crossed the elevated central part, which is not inhabited & bears a desolate|767| appearance. When we descended from the clouds to the city, I heard the good news that observations had been obtained, & that we should go to sea the same evening.

The anchorage is exposed to the whole swell of the Southern ocean, & hence during the present boisterous time of year is very disagreeable & far from safe. —

24th In the morning, we were off the Western end of St Michaels; to the capital of which we were bound in quest of letters. A contrary wind detained us the whole day,

25th but by the following morning, we were off the city, & a boat was sent on shore. — The Isl^d of St Michaels is considerably larger & three times more populous & enjoys a more extensive trade than Terceira. — The chief export is the fruit, for which a fleet of vessels annually arrives. Although several hundred vessels are loaded with oranges, these trees on neither island appear in any great numbers. No one would guess that this was the great market for the numberless oranges imported into England. St Michaels has much the same open, semi-green, cultivated patchwork appearance as Terceira. The town is more scattered; the houses & churches there & throughout the country are white washed & look from a distance neat and pretty. The land behind the|768| town is less elevated than at Terceira, but yet rises considerably; it is thickly studded or rather made up of small mammiformed hills, each of which has sometime been an active Volcano. — In an hours time the boat returned without any letters, and then getting a good offing from the land, we steered, thanks to God, a direct course for England. —

Our voyage having come to an end, I will take a short retrospect of the advantages and disadvantages the pain & pleasure of our five years' wandering. If a person should ask my advice before undertaking a long voyage, my answer would depend upon his possessing a decided taste for some branch of knowledge, which could by such means be acquired. No doubt it is a high satisfaction to behold various countries, and the many races of Mankind, but the pleasures gained at the time do not counterbalance the evils. It is necessary to look forward to a harvest, however distant it may be, when some fruit will be reaped, some good effected. Many of the losses which must be experienced are obvious, such as that of the society of all old friends, and of the sight of those places with which every dearest remembrance is so intimately connected. These losses however, are at the time|769| partly relieved by the exhaustless delight of anticipating the long wished for day of return. If, as poets say, life is a dream, I am sure in a long voyage these are the visions which best pass away the long night. Other losses, although not at first felt, after a period tell heavily, those are the want of room, of seclusion, of rest—the jading feeling of constant hurry—the privation of small luxuries, the comforts of civilization, domestic society, and lastly even of music & the other pleasures of imagination. When such trifles are mentioned, it is evident that the real grievances (excepting from accidents) of a sea life are at an end. The short space of sixty years has made a most astonishing difference in the facility of distant navigation. Even in the time of Cook, a man who left his comfortable fire side for such expeditions, did undergo privations: a yatch with every luxury of life might now circumnavigate the globe. Besides the vast improvements in ships & naval resources, the whole Western shores of America are thrown open; and Australia is become a metropolis of a rising continent. How different are the circumstances to a man shipwrecked at the present day in the Pacific, to what they would have been in the time of Cook: since his voyage|770| a hemisphere has been added to the civilized world.

If a person suffers much from sea sickness, let him weigh it heavily in the balance: I speak from experience, it is no trifling evil cured in a week.[1] If he takes pleasure in naval tactics, it will afford him full scope for his taste; but even the greater number of sailors, as it appears to me, have little real liking for the sea itself;[2] It must be borne in mind how large a proportion of the time during a long voyage is spent on the water, as compared to the days in harbour.[3] And what are the boasted glories of the illimitable ocean? A tedious waste, a desert of water as the Arabian calls it. No doubt there are some delightful scenes; a moonlight night, with the clear heavens, the dark glittering sea, the white sails filled by the soft air of a gently blowing trade wind, a dead calm, the heaving surface polished like a mirror, and all quite still excepting the occasional flapping of|771| the sails. It is well once to behold a squall, with its rising arch, and coming fury, or the heavy gale and mountainous waves. I confess however my imagination had painted something more grand, more terrific in the full grown storm. It is a finer sight on the canvass of Vandervelde, and infinitely finer when beheld on shore, when the waving trees, the wild flight of the birds, the dark shadows & bright lights, the rushing torrents all proclaim the strife of the unloosed elements. At sea, the albatross and petrel fly as if the storm was their proper sphere, the water rises and sinks as if performing its usual task, the ship alone and its inhabitants seem the object of wrath. On a forlorn & weather-beaten coast the scene is indeed different, but the feelings partake more of horror than of wild delight.

Let us now look at the brighter side of the past time. The pleasure derived from beholding the scenery and general aspect of the various countries we have visited, has decidedly been the most constant and highest source of enjoyment. It is probable that the picturesque beauty of many parts of Europe far exceeds anything we have beheld. But there is a growing pleasure in comparing the character of scenery in|772| different countries, which to a certain degree is distinct from merely admiring their beauty. It more depends on an acquaintance with the individual parts of each view: I am strongly induced to believe that as in Music, the person who understands every note will, if he also has true taste, more thoroughily enjoy the whole; so he who examines each part of [a] fine view may also thoroughily comprehend the full and combined effect. Hence a traveller should be a botanist, for in all views plants form the chief embellishment. Group masses of naked rocks, even in the wildest forms; for a time they may afford a sublime spectacle, but they will soon grow monotomous; paint them with bright and varied colours, they will become fantastick; clothe them with vegetation, they must form, at least a decent, if not a most beautiful picture.

When I said that the scenery of Europe was probably superior to anything which we have beheld, I must except, as a class by itself, that of the intertropical regions. The two can not be compared together; but I have already too often enlarged on the grandeur of these latter climates. As the force of impression frequently depends on preconceived ideas, I may add that all mine were taken from the vivid descriptions in the|773| Personal Narrative[4] which far exceed in merit anything I have ever read on the subject. Yet with these high wrought ideas, my feelings were very remote from partaking of a tinge of disappointment on first landing on the coast of Brazil.

Among the scenes which are deeply impressed on my mind, none
exceed in sublimity the primeval forests, undefaced by the hand of
man, whether those of Brazil, where the powers of life are predomi-
nant, or those of Tierra del Fuego, where death & decay prevail. Both
are temples filled with the varied productions of the God of Nature: —
No one can stand unmoved in these solitudes, without feeling that
there is more in man than the mere breath of his body. — In calling up
images of the past, I find the plains of Patagonia most frequently cross
before my eyes. Yet these plains are pronounced by all most wretched
& useless. They are only characterized by negative possessions; — without habitations, without water, without trees, without mountains, they
support merely a few dwarf plants. Why then, and the case is not
peculiar to myself, do these arid wastes take so firm possession of the
memory? Why have not the still more level, greener & fertile Pampas,
which are serviceable to mankind, produced an equal impression?|774|
I can scarcely analyse these feelings. — But it must be partly owing to
the free scope given to the imagination. They are boundless, for they
are scarcely practicable & hence unknown: they bear the stamp of
having thus lasted for ages, & there appears no limit to their duration
through future time. If, as the ancients supposed, the flat earth was
surrounded by an impassable breadth of water, or by deserts heated to
an intolerable excess, who would not look at these last boundaries to
man's knowledge with deep, but ill defined sensations. — Lastly of
natural scenery, the views from lofty mountains, though certainly in
one sense not beautiful, are very memorable. I remember looking down
from the crest of the highest Cordillera; the mind, undisturbed by
minute details, was filled by the stupendous dimensions of the surrounding masses. —

Of individual objects, perhaps no one is more sure to create astonishment, than the first sight, in his native haunt, of a real barbarian, — of
man in his lowest and most savage state. One's mind hurries back over
past centuries, & then asks could our progenitors be such as these?
Men, — whose very signs & expressions are less intelligible to us than
those of the domesticated animals; who do not possess the instinct of
those animals, nor yet appear|775| to boast of human reason, or at least
of arts consequent on that reason. I do not believe it is possible to
describe or paint the difference of savage and civilized man. It is the
difference between a wild and tame animal: and part of the interest in
beholding a savage is the same which would lead every one to desire to
see the lion in his desert, the tiger tearing his prey in the jungle, the

rhinoceros on the wide plain, or the hippopotamus wallowing in the mud of some African river. —

Amongst the other most remarkable spectacles, which we have beheld, may be ranked, — the stars of the Southern hemisphere, the water-spout — the glacier leading its blue stream of ice in a bold precipice overhanging the sea — a lagoon island, raised by the coral forming animalcule — an active volcano — the overwhelming effects of a violent earthquake. — These latter phenomena perhaps possess for me a higher interest, from their intimate connection with the geological structure of the world. The earthquake must however be to everyone a most impressive event; the solid earth, considered from our earliest childhood as the very type of solidity, has oscillated like a thin crust beneath our feet; and in seeing the most beautiful|776| and laboured works of man in a moment overthrown, we feel the insignificance of his boasted power. —

It has been said that the love of the chace is an inherent delight in man, — a relic of an instinctive passion. — if so, I am sure the pleasure of living in the open air, with the sky for a roof, and the ground for a table, is part of the same feeling. It is the savage returning to his wild and native habits. I always look back to our boat cruizes & my land journeys, when through unfrequented countries, with a kind of extreme delight, which no scenes of civilization could create. — I do not doubt every traveller must remember the glowing sense of happiness, from the simple consciousness of breathing in a foreign clime, where the civilized man has seldom or never trod.

There are several other sources of enjoyment in a long voyage, which are perhaps of a more reasonable nature. The map of the world ceases to be a blank; it becomes a picture full of the most varied and animated figures. Each part assumes its true dimensions: large continents are not looked at in the light of islands, or islands considered as mere specks, which in truth are larger than many kingdoms of Europe. — Africa, or North & South|777| America, are well-sounding names and easily pronounced, but it is not till having sailed for some weeks along small portions of their coasts, that one is thoroughily convinced, how large a piece of our immense world, these names imply. —

From seeing the present state, it is impossible not to look forward with high expectation to the future progress of nearly an entire hemisphere. The march of improvement, consequent on the introduction of Christianity through the South Sea, probably stands by itself on the records of the world. It is the more striking when we remember that but sixty years since, Cook, whose most excellent judgment none will dispute, could foresee no prospect of such change. Yet these changes have now been effected by the philanthropic spirit of the English nation.

In the same quarter of the globe Australia is rising, or indeed may be said to have risen, into a grand centre of civilization, which, at some not very remote period, will rule the empress of the Southern hemisphere. It is impossible for an Englishman to behold these distant colonies, without a high pride and satisfaction. To hoist the British flag seems to draw as a certain consequence wealth, prosperity and civilization. —|778|

In conclusion, —it appears to me that nothing can be more improving to a young naturalist, than a journey in distant countries. It both sharpens and partly also allays that want and craving, which as Sir J. Herschel remarks, (*Note in margin:* Discourse on the Study of Natural Philosophy, p. 3.), a man experiences, although every corporeal sense is fully satisfied. The excitement from the novelty of objects, and the chance of success stimulates him on to activity. Moreover as a number of isolated facts soon become uninteresting, the habit of comparison leads to generalization; on the other hand, as the traveller stays but a short space of time in each place, his description must generally consist of mere sketches instead of detailed observation. Hence arises, as I have found to my cost, a constant tendency to fill up the wide gaps of knowledge by inaccurate & superficial hypotheses.

But I have too deeply enjoyed the voyage not to recommend to any naturalist to take all chances, and to start on travels by land if possible, if otherwise on a long voyage. He may feel assured he will meet with no difficulties or dangers (excepting in rare cases) nearly so bad as he before hand imagined. — In a moral point of view, the effect ought to be, to teach him good humoured patience, unselfishness, the habit of acting for himself, and of making the best of everything, or contentment:|779| in short, he should partake of the characteristic qualities of the greater number of sailors. — Travelling ought also to teach him to distrust others; but at the same time he will discover how many truly goodnatured people there are, with whom he never before had, nor ever again will have any further communication, yet who are ready to offer him the most disinterested assistance.

---

[1] Followed by deleted words: 'as most people suppose. I speak from experience, as well I may, suffering now more than I did three years ago.'
[2] Followed by deleted words: 'if not compelled to it by necessity, visions of glory when very young & the force of habit when old, are the sole bonds of attraction.'
[3] Followed by deleted sentence: 'In our five years the excess of days, during the whole

of which the anchor has been down, over the remainder, has scarcely equalled fifty.'
[4]By Alexander von Humboldt. See note 2 for entry of 16 January 1831, p. 24.

October 2nd   After a tolerably short passage, but with some very heavy weather, we came to an anchor at Falmouth. — To my surprise and shame I confess the first sight of the shores of England inspired me with no warmer feelings, than if it had been a miserable Portugeese settlement.[1] The same night (and a dreadfully stormy one it was) I started by the Mail for Shrewsbury. —

[1]A note in the margin reads: 'Mem: Freycinet remarks after his troubles'. See Louis Claude Desaulses de Freycinet, *Voyage autour du monde, entrepris par ordre du Roi.* 4 vols. Paris, 1824–6.

4th   The Beagle proceeded ʦo Plymouth; where she lay till the 17th. —[1]

[1]On the following day, CD wrote from Shrewsbury to his uncle Josiah Wedgwood II: 'The Beagle arrived at Falmouth on Sunday evening, & I reached home late last night. My head is quite confused with so much delight, but I cannot allow my sisters to tell you first, how happy I am to see all my dear friends again. I am obliged to return in three or four days to London, where the Beagle will be paid off, & then I shall pay Shrewsbury a longer visit. I am most anxious once again to see Maer, & all its inhabitants, so that in the course of two or three weeks, I hope in person to thank you, as being my first Lord of the Admiralty.' See *Correspondence* 1: 504.

On 6 October he wrote to FitzRoy: 'I arrived here yesterday morning at Breakfast time, & thank God, found all my dear good sisters & father quite well. — . . . I wish with all my heart, I was writing to you, amongst your friends instead of at that horrid Plymouth. But the day will soon come and you will be as happy as I am now — I do assure you I am a very great man at home — the five years voyage has certainly raised me a hundred per cent. I fear such greatness must experience a fall. — . . . I thought when I began this letter I would convince you what a steady & sober frame of mind I was in. But I find I am writing most precious nonsense. Two or three of our labourers yesterday immediately set to work, and got most excessively drunk in honour of the arrival of Master Charles. — Who then shall gainsay if Master Charles himself chooses to make himself a fool. Good bye — God bless you — I hope you are as happy, but much wiser than your most sincere but unworthy Philos. Chas. Darwin.'

18th   Sailed for the Thames, calling on her way at Portsmouth & Deal, & got up the river to Greenwich on the 28th. —

November 7th   She moved down to Woolwich, where on the 17th she was paid off. —

The Beagle was put into commission on the 4th of July 1831; thus having completed the unusually long period of five years and one hundred and thirty six days.

On the following pages Crew Lists of the 'Ships Company of H.M.S. Beagle, Octr 1836', and 'In Schooner Constitution', are added in another hand.

Biographical register

Adanson, Michel (1727–1806). French naturalist and traveller.

Aldunate Toro, José Santiago (1796–1864). Chilean General and Governor of Chiloe.

Anson, George (1697–1762). British naval officer. Circumnavigated the globe, 1740–4. First Lord of the Admiralty, 1751–6 and 1757–62. Admiral of the Fleet, 1761.

Archer, Mr. Joint manager of farm at Walerawang near Sydney.

Aston, Mr. British Minister at Rio de Janeiro in 1832.

Baker, Mr. Missionary at Pahia, Bay of Islands, New Zealand.

Baker, Thomas (d. 1845) Naval officer. Captain RN, 1797; Rear-Admiral, 1821. Admiral commanding South American station, 1829–33. KCB 1831.

Banks, Joseph (1743–1820). Naturalist and patron of science. Accompanied Cook on the voyage of HMS *Endeavour* to the South Seas in 1768–71. President of the Royal Society, 1778–1820. Bt 1781. FRS 1766.

Basket, Fuegia (1821–1883). Fuegian girl brought to England in 1830 by FitzRoy, and returned to Tierra del Fuego in 1833. Name in Alikhoolip language was Yorkicushlu. Married York Minster.

Beaufort, Francis (1774–1857). Naval officer and hydrographer. Post-Captain RN, 1810; Rear-Admiral, 1846. Hydrographer of the Navy, 1829–55. Originator of the Beaufort scale of wind strengths. KCB 1848. FRS 1814.

Beechey, Frederick William (1796–1856). Naval officer and geographer. Captain RN, 1827. President of the Royal Geographical Society, 1855–6.

Bell, Colonel. CD's host at Cape of Good Hope. Husband of Lady Caroline Bell.

Bingley, Mr. English merchant and manager of estate at Potrero Seco near Copiapó.

Boat Memory (*c.* 1810–30). Fuegian brought to England by FitzRoy in 1830. Died of smallpox in Royal Hospital, Plymouth, November 1830.

Bolivar, Simon (1783–1830). General and patriot leader. Instrumental in independence movements in Venezuela, Colombia, Ecuador, Peru and Bolivia.

Bonplan, Aimé (1773–1858). French botanist and traveller.

Bougainville, Louis Antoine de (French sea captain and navigator. Circumnavigated the globe in *La Boudeuse*, 1767–8.

Bridges, Esteban Lucas (1874–1949). Second son of Thomas Bridges.

Bridges, Thomas (1842–98). English missionary who first established the Church of England Mission to the Fuegians in 1867.

Brisbane, Matthew (d. 1833). First official British Resident in Falkland Islands. Murdered at Port Louis by gauchos and Indians on 26 August 1833.

Busby, James (1801–71). First official British Resident in New Zealand.

Button, Jemmy (b. 1816). Fuegian boy brought to England by FitzRoy in 1830, and returned to Tierra del Fuego in 1833. Name in Yahgan language was Orundellico. Was still alive in 1863.

Bynoe, Benjamin (1804–65). Naval surgeon. Assistant and then Acting Surgeon on the *Beagle*, 1832–7; Surgeon, 1837–43. FRCS 1844.

Cairnes, Mr. British merchant in Rio de Janeiro.

Caldcleugh, Alexander (d. 1858). British trader and plant collector living in Santiago. Owner of copper mines at Panuncillo. Author of *Travels in South America during the years 1819 . . . 21*, London 1825. FRS 1831.

Cavendish, Sir Thomas (1555–92). English gentleman adventurer and mariner. Discovered Port Desire on 17 December 1586, and made the third circumnavigation of the globe 1586–8.

Chaffers, Edward Main. Master on the *Beagle*, 1831–6.

Chesterfield, Earl of (1694–1773). Politician, wit and letter writer.

Chetwode, Capt. Commandant of troops stationed at Bathurst, New South Wales.

Clarke, Mr. Missionary at Waimate, Bay of Islands, New Zealand.

Clavac, le Compte de. French artist and traveller.

Cochrane, Lord, 10th Earl of Dundonald (1775–1860). Naval officer. Captain RN, 1801; Rear-Admiral, 1854. Commanded Chilean navy, 1818–22.

Cole, Sir Galbraith Lowry (1772–1842). While Governor of the Cape of Good Hope, 1828–33, he built many new roads.

Cook, James (1728–79). Circumnavigator. Captain RN, 1775. Commanded voyages to the South Seas and around the world in 1768–71, 1772–5 and 1776–9. FRS 1776.

Corfield, Richard Henry (1804–97). English merchant living in Valparaiso. Attended Shrewsbury School 1816–19, overlapping CD in 1818–19. CD stayed at his house in the Almendral on several occasions, including his period of illness in September and October 1834.

Covington, Syms (c. 1816–61). Appointed as CD's personal servant 22 May 1833, and remained in his service until 25 February 1839. Emigrated to Australia and settled as postmaster at Pambula, Twofold Bay, New South Wales.

Cuvier, Georges (1769–1832). French systematist, comparative anatomist, palaeontologist and administrator.

Daniell, John Frederic (1790–1845). Meteorologist, chemist and physicist. Professor of Chemistry at King's College, London, 1831–45. FRS 1813.

Darwin, Caroline Sarah (1800–88). CD's sister. Married Josiah Wedgwood III in 1837.

Darwin, Emily Catherine (1810–66). CD's sister, known in the family as Catherine or Catty. Married Charles Langton as his second wife in 1863.

Darwin, Erasmus Alvey (1804–81). CD's elder brother. Studied science and medicine at Cambridge and Edinburgh, but never practised.

Darwin, Robert Waring (1766–1848). CD's father. Third son of Erasmus Darwin by his second wife. Physician with a large practice in Shrewsbury. Resided at The Mount, Shrewsbury, which he built *c*. 1796–8. Married Susannah, daughter of Josiah Wedgwood I, in 1796. FRS 1788.

Darwin, Susan Elizabeth (1803–66). CD's sister, called by him Granny. Did not marry.

Davies, Richard. Missionary at Waimate, Bay of Islands, New Zealand, since 1819. Not in holy orders, but ran the farm at which the natives were taught agriculture.

Davis, John (1550–1605). English sea captain and navigator. Discoverer in the *Desire* of the Falkland Islands on 14 August 1592.

Derbishire, Alexander. Mate on the *Beagle*, 1831–2. Returned to England, April 1832.

Dickson, Mr. (d. 1833). Englishman (or Irishman according to FitzRoy) resident as storekeeper at Port Louis in the Falkland Islands, and left in charge of British flag in January 1833. Murdered by gauchos and Indians, 26 August 1833.

Dixon, Sir Manley (d. 1837). Naval officer. Commander-in-Chief at Plymouth, 1830–3.

Douglas, Charles D. Surveyor and pilot, long resident in Chiloe.

Doverton , Sir W. Resident in St Helena.

Drake, Sir Francis (1540–96). English adventurer and admiral. Circumnavigated the globe in the *Golden Hind*, 1577–80.

Dumas, Sr. Chief of Police in Montevideo.

Earle, Augustus (1793–1838). Official artist on the *Beagle*, 1831–2.

Edwards, Joaquin. Major-domo of copper mines at Panuncillo.

Edwards, José Maria. Anglo-Chilean, son of the owner of silver mines at Arqueros.

Ellis, William (1794–1872). Missionary in the South Seas and author.

Evans, Lieut. Officer of Marines posted on Ascension Island.

Falkner, Thomas (1707–84). Jesuit missionary in Patagonia, 1740–68.

Figuireda, Manoel Joaquem da. Estate owner at Socêgo near Campos in Brazil.

FitzRoy, Robert (1805–65). Naval officer, hydrographer and meteorologist. Commander RN, 1828; Captain, 1835; Rear-Admiral, 1857; Vice-Admiral, 1863. Commanded HMS *Beagle* from December 1828 to October 1830, and from June 1831 to November 1836. MP for Durham, 1841–3. Governor-General of New Zealand, 1843–5. Meteorological Statist to the Board of Trade, 1854–65. FRS 1851.

Forsyth, Charles. Midshipman on the *Beagle*, 1832–6.

Foster, Henry (1796–1831). Naval officer and navigator. Commanded HMS *Chanticleer* from 1828 till he was drowned in Panama in 1831. FRS 1824.

Fox, Henry Stephen (1791–1846). British diplomat. Minister Plenipotentiary, Buenos Aires, 1831–2; Rio de Janeiro, 1833–6; Washington, DC, 1836–44.

Fox, William Darwin (1805–80). CD's second cousin and close friend. BA, Christ's College, Cambridge, 1829. Rector of Delamere, Cheshire, 1838–73.

Frankland, Mr. Surveyor General in Tasmania.

Freire Serrano, Ramón (1787–1851). Chilean General. Very active in Chilean struggle for independence, and twice (1823–6; 1827) Supreme Director of Chile.

Frutez, Sr. Politician in Montevideo.

Fuentes, Juan. Estate owner near Maldonado.

Gay, Claude (1800–73). French naturalist and traveller. Professor of physics and chemistry in Santiago, 1828–42.

Gill, Mr. Civil engineer in Lima.

Gonzales, Francisco. Guide employed by CD at Maldonado.

Gonzales, Mariano. Guide employed by CD in Chile.

Gore, Philip Yorke. British Chargé d'Affaires in Buenos Aires, 1832–4.

Gould, John (1804–81). Ornithologist and artist. Taxidermist to the Zoological Society of London, 1826–81. FRS 1843.

Grenville, Mr. English estate owner near Montevideo. Fought in Chilean and Brazilian navies.

Grey, Charles, 2nd Earl (1764–1845). Statesman and Prime Minister of England.

Hall, Basil (1788–1844). Naval officer and anthropologist. Travelled in South America and Mexico, 1820–2. FRS 1816.

Hamond, Robert Nicholas (1809–83). Mate on HMS *Druid*, loaned to the *Beagle* in November 1832, and returned to England in May 1833.

Hardy, Mr. Owner of copper mines at Guasco.

Harris, James. British trader and sea captain at the Rio Negro in Patagonia. Acted as pilot for the *Beagle*, and hired *La Paz* and *La Liebre* to FitzRoy for surveys of the shallow coastal waters.

Harris, William Snow (1791–1867). Physicist and inventor of the lightning conductor and other electrical devices. Kt 1847. FRS 1831.

Head, Francis Bond (1793–1875). Colonial governor and author. Travelled in South America as manager of the Rio Plata Mining Association, 1825–6. Bt 1836.

Hellyer, Edward H. (d. 1833). Clerk on the *Beagle*, 1831–3. Drowned in the Falkland Islands, 4 March 1833.

Henslow, John Stevens (1796–1861). Professor of Mineralogy at Cambridge University, 1822–7; Professor of Botany, 1827–61. Curate of Little St Mary's Church, Cambridge, 1824–32. Vicar of Cholsey–cum–Moulsford, Berkshire, 1832–7. Rector of Hitcham, Suffolk, 1837–61. FRS 1818.

Herschel, John Frederick William (1792–1871). Astronomer, mathematician, chemist and philosopher. Made astronomical observations at the Cape of Good Hope, 1834–8. Master of the Mint, 1850–5. Bt 1838. FRS 1813.

Hoare, Rev. Old Cambridge friend of CD's, preaching at Plymouth.

Hood, Thomas Samuel. British Consul General at Montevideo.

Humboldt, Friedrich Wilhelm Heinrich Alexander von (1769–1859). German naturalist and traveller. Explored South America, 1799–1804. Corresponding Member, Royal Society, 1815.

Johnson, Charles Richardson. Mate on the *Beagle*, 1832–6.

Kent, William. Assistant Surgeon on the *Beagle*, 1833–6.

King, Philip Gidley (1817–1904). Eldest son of P. P. King. Midshipman on the *Beagle*, 1831–6. Left the ship in 1836 to stay with his father in Australia. Drew diagrams of the layout of the *Beagle* for John Murray in 1890.

King, Philip Parker (1793–1856). Naval officer and hydrographer. Commander of the *Adventure* and *Beagle* for the first surveying voyage to South America, 1826–30. Settled in Australia, his father having been first Governor of New South Wales. Rear-Admiral RN, 1855. FRS 1824.

Kirby, William (1759–1850). Clergyman and entomologist. One of the first Fellows of the Linnean Society, 1788. FRS 1818.

Kotzebue, Otto von (1787–1846). Russian explorer and navigator.

Lavalleja, Juan Antonio (1784–1853). Uruguayan soldier and patriot leader. Military Governor of Montevideo.

Lawrie, Mr. Scottish merchant in Rio de Janeiro.

Lawson, Nicholas E. Englishman serving the Republic of the Equator (Ecuador) as Governor of the Galapagos Islands.

Le Dilly, M. French sea captain whose whaler *Le Magellan* was wrecked in the Falkland Islands on 12 January 1833, and whose crew members were rescued by the *Beagle*.

Lennon, Patrick. Irish merchant in Rio de Janeiro.

Lesson, René Primevère (1794–1849). French naturalist on the voyage of *La Coquille* round the world, 1822–5.

Liesk, Mr. Settler in Cocos Keeling Islands.

Lloyd, John Augustus (1800–54). Civil engineer and surveyor. Surveyor-General, Mauritius, 1831–49. FRS 1830.

Lopez, Sr. Governor of Santa Fé province.

Low, William. Scottish trader and sea captain for many years in the waters around Patagonia, Falkland Islands and Tierra del Fuego. Sold his schooner *Unicorn*, then renamed *Adventure*, to FitzRoy in March 1833.

Lumb, Edward. English merchant in Buenos Aires.

Lyell, Charles (1797–1875). Uniformitarian geologist. Professor of Geology, King's College, London, 1831–3. President of the Geological Society, 1834–6 and 1849–50. Kt 1848. 1st Bt 1864. FRS 1826.

Macarthur, James (1798–1867). Large land owner and politician in New South Wales. Son of the introducer of the Merino sheep.

McCormick, Robert (1800–90). Naval surgeon, explorer and naturalist. Surgeon on the *Beagle*, 1831–2.

Maclear, Thomas (1794–1879). Astronomer. Astronomer Royal at the Cape of Good Hope, 1834–70. Kt 1860. FRS 1831.

Magellan, Ferdinand (1480–1521). Portuguese sea captain and the first circumnavigator. Appointed Captain General of an Armada by King Charles I of Spain with the aim of discovering a route to the South Seas by sailing westwards round the world. Although Magellan himself was killed in the Philippines, one of his ships, the *Vittoria*, completed the circumnavigation.

Martens, Conrad (1801–78). Landscape painter. Official artist on the *Beagle*, 1833–4. Settled in Australia at Sydney, 1835.

Martin, John (1789–1854). Historical and landscape painter.

Matthews, Richard (1811–93). Missionary from the Church Missionary Society to the Fuegians. Landed at Woollya 23 January 1834 and taken off again 6 February. Sailed in the *Beagle* to New Zealand, where he remained as a missionary with his brother at Waimate.

May, Jonathan. Carpenter on the *Beagle*, 1831–6.

Mellersh, Arthur. Midshipman on the *Adventure*, 1828–30; on the *Beagle*, 1833–4. Mate on the *Beagle*, 1832–6. Vice-Admiral, 1878.

Miranda, Commandante. Subordinate of General Rosas.

Musters, Charles (d. 1832). Volunteer 1st Class on the *Beagle*, 1831–2. Fourth son of John Masters of Colwick Hall, Notts. Died of fever in Rio de Janeiro.

Narbrough, John (1640–88). British naval commander. Commissioner of the Navy, 1680–7. Sailed through the Straits of Magellan in command of HMS *Sweepstakes* in November 1670. Kt 1673.

Nott, Mr. Missionary in Tahiti since 1790s.

O'Higgins, Don Ambrosio (*c.* 1720–1801). Soldier and statesman. Viceroy of Chile, 1789–96; of Peru, 1796–1801. Father of

O'Higgins, Bernardo (1780–1846). Liberator of Chile and President of the Congress.

Owen, Richard (1804–92). Comparative anatomist. Assistant conservator of the Hunterian Museum, Royal College of Surgeons, 1827; Hunterian Professor, 1836–56. Superintendent of the Natural History Departments, British Museum, 1856–84.

Paget, Charles Henry (b. 1806) Naval officer. Captain RN, 1829. Commanded HMS *Samarang*, 1831–5.

Parry, Mr. British merchant at Montevideo.

Peacock, George (1791–1858). Tutor in mathematics at Trinity College, Cambridge, 1823–39. Lowndean Professor of Geometry and Astronomy at Cambridge University, 1837–58. Dean of Ely, 1839–58. FRS 1818.

Pedro I, Dom. (1798–1834). Succeeded his father as Regent of Brazil, 1821; became Emperor, 1822; abdicated in favour of Dom Pedro II, 1831.

Pedro, Don. Chilean Governor of Chiloe, 1835.

Pimiento, Sebastian. Estate owner in Maldonado district.

Pomare IV, Aimata. Queen of Tahiti, 1827–77.

Ponsonby, John, Lord (*c.* 1770–1855). Diplomatist. Envoy-extraordinary and Minister Plenipotentiary in Buenos Aires, 1826–8.

Price, Mr. British merchant in Valparaiso whom CD met in Rio.

Pritchard, George (1796–1883). Missionary at Papiete in Tahiti. British Consul in Tahiti, 1837–44; in Samoa, 1844–57.

Rennie, John (1794–1874). Civil engineer and architect. Engineer to the Admiralty. Completed the breakwater at Plymouth begun by his father. Kt 1831.

Rocha, Antonio da. Owner of house at Chacera o Macâco near Rio de Janeiro.

Rolor, General. Rebel commander in Argentina.

Rosas, Juan Manuel de (1793–1877). Argentinian cattle rancher, soldier and dictator. Ruled Argentina as Governor of Buenos Aires, 1829–32 and 1835–52. Led campaign against the Indians, 1833–5. Overthrown 1852 and retired to Swaythling in Hampshire, where CD once met him.

Ross, Capt. Naval Commissioner at Plymouth, 1831.

Ross, Capt. Trader living in Cocos Keeling Islands.

Rous, Mr. British Consul in Concepcion, 1835.

Roussin, Albin Reine, Baron (1781–1854). French naval officer and diplomat. Surveyed the coast of Brazil, 1819. Admiral, 1840.

Rowlett, George (1797–1834). Purser on the *Beagle*, 1831–4. Died at sea, 27 June 1834.

Rugendas, Johann Moritz (1802–58). German artist who travelled extensively in South America.

Sarmiento, Don Pedro de Sarmiento y Gamboa (*c.* 1532–1592). Spanish Admiral and historian. Commanded fleet sent to fortify the Straits of Magellan in 1581.

Scott, Mr. British Attaché in Rio de Janeiro, 1832.

Sedgwick, Adam (1785–1873). Woodwardian Professor of Geology at Cambridge University, 1818–73. Canon of Norwich, 1834–73. FRS 1821.

Shongi. Maori chief at Bay of Islands, New Zealand.

Smith, Lieut. Naval officer left in charge of the Falkland Islands by the Captain of HMS *Challenger*, March 1834.

Smith, Andrew (1797–1872). Military surgeon, naturalist and explorer. Served in South Africa, 1821–37. Director General of the Army Medical Department, 1851. KCB 1859. FRS 1857.

Smith, Charles Hamilton (1776–1859). Soldier, naturalist, writer and artist.

Solander, Daniel Carl (1733–82). Swedish botanist. Accompanied Cook and Banks in HMS *Endeavour*, 1768–71. Assistant librarian, British Museum, 1763; Keeper of the Natural History Department, 1773.

Spencer, Richard (1779–1839). Naval officer and colonist. Government Resident at King George's Sound, Western Australia. Kt 1833.

Stewart, Peter Benson. Mate on the *Beagle*, 1831–6.

Stokes, John Lort (1812–85). Mate and Assistant Surveyor on the *Beagle*, 1831–6. Midshipman RN, 1826; Lieutenant, 1831; Commander, 1841; Rear-Admiral, 1877.

Stokes, Pringle (d. 1828). Naval officer. Commander of the *Beagle* from 1826 to August 1828, when he committed suicide on board at Port Famine.

Sulivan, Bartholomew James (1810–90). Naval officer and hydrographer. Lieutenant on the *Beagle*, 1831–6. Surveyed the Falkland Islands, 1838–46. Rear-Admiral 1877. KCB 1869.

Talbot, Charles (b. 1801). Naval officer. Captain RN, 1830. Commanded HMS *Warspite*, 1830–42. Rescued Brazilian Royal Family from an insurrection on 6 April 1831.

Thompson, John Vaughan (1779–1847). Zoologist and inventor of plankton net.

Usborne, Alexander Burns. Master's Assistant on the *Beagle*, 1831–5. Surveyed the coast of Peru in the *Constitucion*, 1835–6, and returned to England via Cape Horn.

Vernon, Colonel Harcourt. British traveller whom CD met at Buenos Aires.

Vidal, Alexander Thomas Emeric. Naval officer and hydrographer. Captain RN, 1825.

Walford, Capt. Englishman living at Lirquen, near Concepcion.

Weddell, James (1787–1834). Navigator and explorer. Named Wigwam Cove in Tierra del Fuego during a voyage to Lat. 74° 15' in 1822–4.

Wedgwood, Emma (1808–96). Youngest daughter of Bessy and Josiah Wedgwood II. Married CD, her cousin, on 29 January 1839.

Wedgwood, Frances Mackintosh (Fanny) (1800–89). Daughter of James and Catherine Mackintosh. Married Hensleigh Wedgwood in 1832.

Wedgwood, Hensleigh (1803–91). CD's cousin, son of Josiah Wedgwood II. BA, Christ's College, Cambridge, 1824; Fellow, 1829–30. Philologist and barrister. Metropolitan police magistrate at Lambeth, 1832–7. Registrar of Metropolitan Carriages, 1838–49. Married Frances Mackintosh in 1832.

Wedgwood, Josiah II (1769–1843). Master-Potter of Etruria, Stoke-on-Trent. Owner of Maer Hall, Staffordshire. MP for Stoke-on-Trent, 1832–4. CD's uncle and father of his future wife, Emma.

Wickham, John Clements (1798–1864). First lieutenant on the *Beagle*, 1831–6. Commander of the *Beagle* while surveying the coast of Australia, 1837–41. Police magistrate in New South Wales, 1843–57. Government Resident at Moreton Bay, 1857–60.

Williams, Rev. William. Missionary at Bay of Islands in New Zealand. Younger brother of Henry Williams (1792–1867), leader of the group of missionaries, who was away at the time of CD's visit.

Wilson, Belford Hinton (1804–58). British Consul in Lima, 1832–7; Chargé d'Affaires, 1837–41. Consul General in Venezuela, 1842–52.

Wilson, Mr. Missionary at Matavai in Tahiti since 1797.

Wood, Alexander Charles (b. 1810). Colonial land and emigration commissioner. Robert FitzRoy's cousin.

York Minster (b. 1804). Fuegian brought to England by FitzRoy in 1830, and returned to Tierra del Fuego in 1833. Name in Alikhoolip language was Elleparu. Killed in a quarrel some time before 1863.

Index

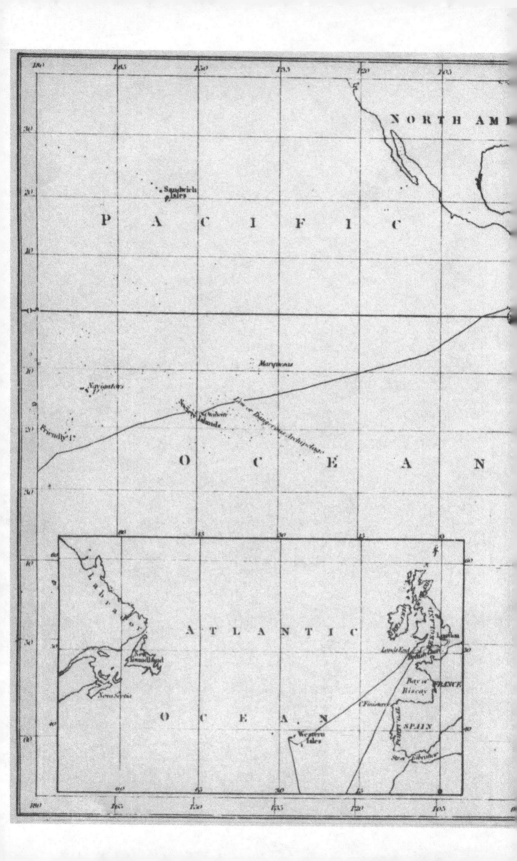

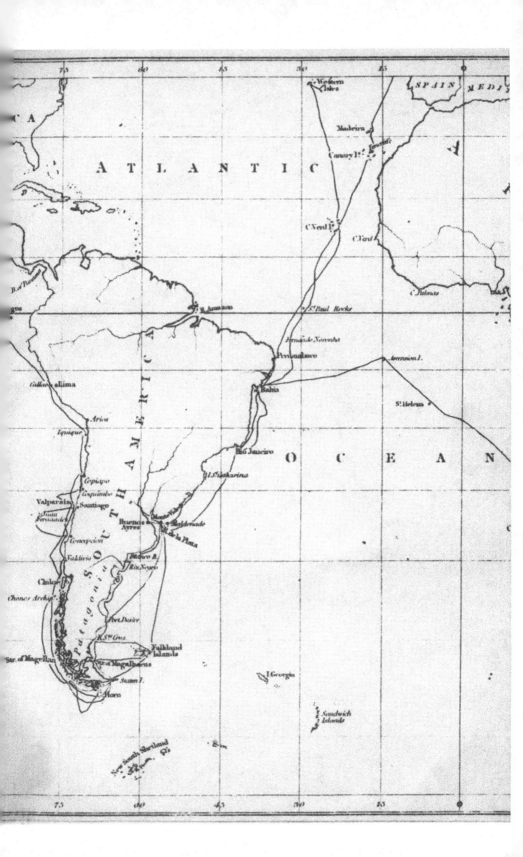

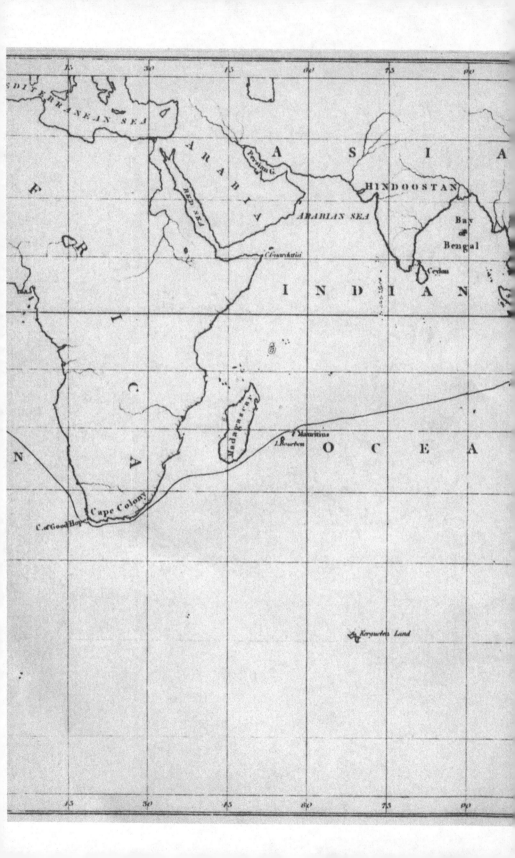

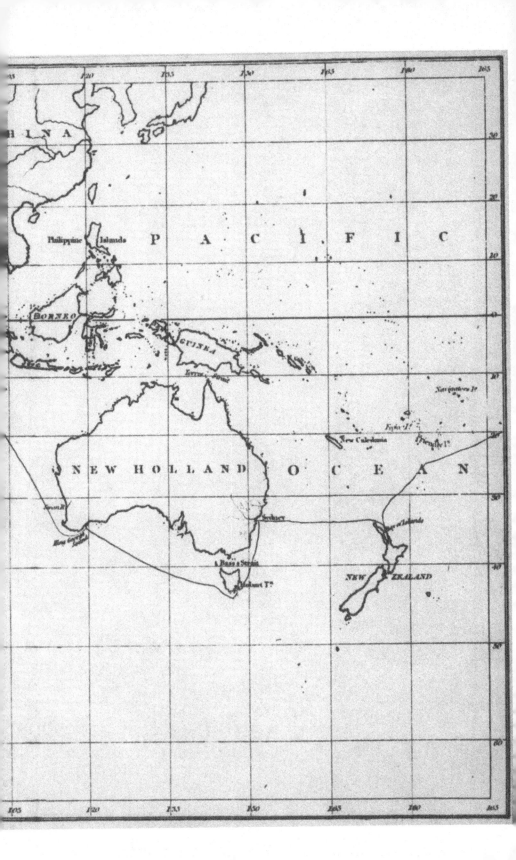